MW00760644

Silicon Photonics Bloom

Silicon Photonics Bloom

Editors

Ozdal Boyraz
Qiancheng Zhao

MDPI • Basel • Beijing • Wuhan • Barcelona • Belgrade • Manchester • Tokyo • Cluj • Tianjin

Editors
Ozdal Boyraz
University of California
USA

Qiancheng Zhao
University of California
USA

Editorial Office
MDPI
St. Alban-Anlage 66
4052 Basel, Switzerland

This is a reprint of articles from the Special Issue published online in the open access journal *Micromachines* (ISSN 2072-666X) (available at: https://www.mdpi.com/journal/micromachines/special_issues/Silicon_Photonics_Bloom).

For citation purposes, cite each article independently as indicated on the article page online and as indicated below:

LastName, A.A.; LastName, B.B.; LastName, C.C. Article Title. *Journal Name* **Year**, *Article Number*, Page Range.

ISBN 978-3-03936-908-9 (Hbk)
ISBN 978-3-03936-909-6 (PDF)

Contents

About the Editors

Ozdal Boyraz received his M.S. and Ph.D. degrees from the University of Michigan, Ann Arbor in 1997 and 2001, respectively. After two years of industry experience, he joined the University of California, Los Angeles as a postdoctoral research fellow in 2003. In 2005, he joined UCI Electrical Engineering Department as a tenure-track faculty, and he continues his research in the same department since then. His research areas include Optoelectronic Devices, Integrated Optics, Optical Communications Systems, Free Space Communications and Cube-Sats, Remote Sensing, and Optical Signal Processing. He has over 170 journal and conference publications and 5 issued and 3 pending patents. His awards and recognitions include 2010 DARPA Young Faculty Award, UCLA Best Postdoctoral Researcher Award and IEICE Best Paper Award.

Qiancheng Zhao is a postdoctoral researcher in the Department of Electrical and Computer Engineering at University of California, Santa Barbara since 2019. Before joining UCSB, he worked as a signal integrity engineer in Apple Inc. He received his Ph.D. degree in the Department of Electrical Engineering and Computer Science from University of California, Irvine in 2017. His research focuses on silicon nitride planar waveguides, ultra-low-loss integrated optical waveguides, integrated optical reference cavity, and frequency stabilization. He has published more than 30 papers and serves as a reviewer for 11 journals. He is a member of OSA.

Editorial

Editorial for the Special Issue on Silicon Photonics Bloom

Qiancheng Zhao [1,*] and Ozdal Boyraz [2,*]

[1] Department of Electrical and Computer Engineering, University of California, Santa Barbara, Santa Barbara, CA 93106, USA

[2] Department of Electrical Engineering and Computer Science, University of California, Irvine, Irvine, CA 92697, USA

* Correspondence: qczhao@ucsb.edu (Q.Z.); oboyraz@uci.edu (O.B.)

Received: 18 June 2020; Accepted: 25 June 2020; Published: 10 July 2020

Silicon (Si) photonics debuted in the mid-1980s through the pioneering work done by Soref et al. While early work mainly focused on waveguides, switches, and modulators, significant momentum surged around the mid-2000s when great breakthroughs were achieved in GHz Si modulators, Raman Si lasers, and germanium (Ge)-on-Si epitaxial integration. Today, 30 years later, silicon photonics has experienced tremendous growth and become the backbone of integrated photonics, evolving from single passive components to hybrid functionalized architectures. Its scope is far beyond the traditional group IV elements, extending to compounds like silicon nitride, silicon oxynitride, and silicon carbon, in addition to heterogeneous integrations with III-V/II-VI elements, chalcogenide, graphene, crystals, polymers, etc. The popularity of Si photonics is partially attributed to its compatibility with the mature complementary metal–oxide–semiconductor (CMOS) technology that allows for low-cost and large-scale manufacturing. With the recent injection of government and private funding, more and more foundries, equipped with well-established and market-proven product development kits, will spring up, promoting a bloom in Si photonics in the new era.

This Special Issue of *Micromachines*, entitled "Silicon Photonics Bloom", has 10 research papers and 2 review articles, covering the scale from material preparation [1,2], to single device design [3–7], to photonic integration [8–11], to system architecture [12]. The demonstrated devices and components include source generation [1,5,6,11], modulators [7], switches [4,8], gratings [3,10], and couplers [9,12] and are applied to applications such as dispersion control [3], photonic memory [4], optic communication [8,10], polarization management [9], and photonic computing [12]. The spectrum of the contributed research spans a wide range, from visible [1,2,6], to telecom wavelength [3,4,7–10], to mid-IR [11], to terahertz frequencies [5].

Revolutionary technology usually starts from fundamental breakthroughs, especially in materials, in which new properties prompt unprecedented discoveries and innovations. Studies on the material properties are the cornerstones of silicon photonics, not only because new materials enable novel functionalities, but also because the accuracy of the material properties directly impacts the design of the photonic devices. Song et al. [1] studied SiC_xO_y material, particularly on the effect of nitrogen doping on the photoluminescence of the amorphous SiC_xO_y films. Nitrogen doping creates defect centers in the SiC_xO_y bandgap. By varying the doping concentration, the defect center energy level could be adjusted, yielding photoluminescence from red to orange, as well as blue photoluminescence. Similar to the SiC_xO_y, the luminescent properties of the SiN_xO_y film are also studied in this Special Issue. In the review article [2] by Shi et al., the luminescence properties and fabrication methods of the SiN_xO_y films are summarized, and their applications as barrier materials in non-volatile semiconducting memory, optical devices, and anti-scratch coating are enumerated with abundant state-of-the-art examples. The review has an in-depth elaboration of the preparation of the SiN_xO_y film, serving as a solid reference for fabrication.

The merits of using Si in integrated photonics come not only from the fabrication compatibility to CMOS technology, but also from its versatility in tuning its optical parameters, which renders itself suitable for active devices. The refractive index of Si can be tuned thermally, as utilized in [3]; in this study, Klitis et al. demonstrate active group delay control in a Si Bragg grating by creating a thermal gradient along the grating length through the metal heaters. By varying the distance between the metal heaters and the waveguides, the thermal gradient profile can be adjusted, which effectively changes the Si refractive index along the grating. Both blue and red chirps can be realized using a single design, and specific dispersion compensation can be achieved by a nanometer bandwidth filter. In addition to the thermo-optic effect, the optical constants of Si can also be varied by using the carrier injection method. Inoue et al. [7] developed a novel phase modulator based on the carrier plasma effect. The fin-type electrodes are placed at self-imaging positions of a silicon multimode interference waveguide to reduce scattering losses and relax the fabrication tolerance. The measured propagation losses and spectral bandwidth were 0.7 dB and 33 nm on a 987-μm-long phase shifter. The π-shift current of the modulator was 1.5 mA.

The active Si components serve as the building blocks for complex photonic integrations. By integrating phase shifter and multimode interferometers (MMI), a silicon-on-insulator (SOI)-based polarization controller is proposed and experimentally demonstrated in [9]. Geometrical analysis based on phasors and a Poincare sphere shows that the component can be configured as either a polarization compensator or a polarization controller. Active MZIs and micro-ring filters are also used in the work [10]. Huang et al. experimentally presented a 100 Gb/s silicon photonic WDM transmitter which consists of a passive bidirectional grating coupler and active Si components. The bidirectional grating coupler works as a beam splitter, and the split light is connected to two arms of an MZI for modulation. The modulated light is coupled to the bus waveguide through a micro-ring resonator. Four channels around 1550 nm with a channel spacing of 2.4 nm are demonstrated at 25Gb/s each channel. An even larger-scale integration of Si phase shifters can build photonic processors. In [12], researchers from Shanghai Jiao Tong University proposed a two-dimensional self-coupled optical waveguide (SCOW) mesh photonic processor to work as a rectangular unitary multiport interferometer. This photonic processor can accomplish arbitrary optical unitary transformation that has wide applications in quantum signal processing and photonic machine learning.

Although taking the spotlight of making modulators, Si is not an appropriate material as a light source due to its indirect bandgap. However, efforts to make Si luminous have never stopped. Yamada et al. [6] reported a Si-based quantum dot light-emitting diode with a peak wavelength at 620 nm. The Si-based quantum dots have the potential to replace cadmium-based quantum dots which are toxic. The quantum dots are sandwiched in multilayer structures, emitting pale-orange color with 0.03% external quantum efficiency. Moving to lower frequencies, such as the terahertz regime, an integrated THz impulse source can be realized by coupling a Si optical waveguide to a germanium-based photoconductive antenna [5]. The phosphorus-doped Ge thin film has a faster transient response speed due to the carrier lifetime reduction and antenna gap narrowing. The device has the advantage of low-cost fabrication and compact integration with on-chip excitation at the laser wavelength of 1550 nm.

In the above approach, Si itself acts as a power delivery medium rather than an active material. The functionality of generating THz wave is achieved by heterogeneous integration with Ge. Heterogeneous integration with other materials equips Si photonics with more capabilities which cannot be realized by Si intrinsically. Another example of heterogeneous integration is a photonic memristive switch [4], implemented by the integration of Si waveguides with a phase-change material $Ge_2Sb_2Te_5$ (GST) segment. The GST material exhibits distinct optical refractive indices and extinction coefficients in amorphous and crystalline phases. The "self-holding" capability renders the material suitable for low-energy applications, as it does not require continuous power to keep the phases.

Beyond heterogeneous integration, Si-based compounds such as silicon nitride (Si_3N_4) also play an important role in integrated photonics. Compared to Si, Si_3N_4 does not suffer from two-photon absorption and carrier absorption, rendering itself suitable for nonlinear applications. Si_3N_4 used in frequency comb generation is mentioned in the review article [11]. The review of the frequency comb source mainly focuses on using the semiconductor mode-lock laser as the heart of the system, pumping the nonlinear integrated waveguides for supercontinuum generation. Carrier–envelope offset detection, stabilization, requirements for lasers, and material nonlinearity are covered in this review, followed by a future outlook on heterogeneous integration of semiconductor mode-lock lasers to achieve fully on-chip stabilized frequency combs in the near-IR region. Besides nonlinear optics, using Si_3N_4 as the waveguide material also gives one a greater degree of freedom in a Si platform. As shown in the work done by Sharma et al. [8], Si is used as a MEMS platform for highly efficient planar optical switching, whereas light is conducted in silicon nitride waveguides. Inverted tapers were introduced to increase the butt-coupling efficiency of the Si_3N_4 waveguides. Different MEMS designs were simulated and optimized. The optimum design was fabricated by commercial services and tested.

While it is impossible to cover all the research areas of silicon photonics, this Special Issue provides a humble selection of related topics with state-of-the-art results, hoping to demonstrate the recent achievements in multiple aspects. We would like to take this opportunity to thank all the contributing authors for their excellent work presented in this Special Issue. Our appreciation also goes to all the reviewing experts who dedicated their time to provide valuable comments and helped improve the quality of the submitted papers. The unconditional and generous support from the editorial staff of *Micromachines* is also highly appreciated.

Conflicts of Interest: The authors declare no conflict of interest.

References

1. Song, J.; Huang, R.; Zhang, Y.; Lin, Z.; Zhang, W.; Li, H.; Song, C.; Guo, Y.; Lin, Z. Effect of Nitrogen Doping on the Photoluminescence of Amorphous Silicon Oxycarbide Films. *Micromachines* **2019**, *10*, 649. [CrossRef] [PubMed]
2. Shi, Y.; He, L.; Guang, F.; Li, L.; Xin, Z.; Liu, R. A Review: Preparation, Performance, and Applications of Silicon Oxynitride Film. *Micromachines* **2019**, *10*, 552. [CrossRef] [PubMed]
3. Klitis, C.; Sorel, M.; Strain, M.J. Active On-Chip Dispersion Control Using a Tunable Silicon Bragg Grating. *Micromachines* **2019**, *10*, 569. [CrossRef] [PubMed]
4. Wang, N.; Zhang, H.; Zhou, L.; Lu, L.; Chen, J.; Rahman, B.M.A. Design of Ultra-Compact Optical Memristive Switches with GST as the Active Material. *Micromachines* **2019**, *10*, 453. [CrossRef] [PubMed]
5. Chen, P.; Hosseini, M.; Babakhani, A. An Integrated Germanium-Based THz Impulse Radiator with an Optical Waveguide Coupled Photoconductive Switch in Silicon. *Micromachines* **2019**, *10*, 367. [CrossRef] [PubMed]
6. Yamada, H.; Shirahata, N. Silicon Quantum Dot Light Emitting Diode at 620 nm. *Micromachines* **2019**, *10*, 318. [CrossRef] [PubMed]
7. Inoue, D.; Ichikawa, T.; Kawasaki, A.; Yamashita, T. Silicon Optical Modulator Using a Low-Loss Phase Shifter Based on a Multimode Interference Waveguide. *Micromachines* **2019**, *10*, 482. [CrossRef] [PubMed]
8. Sharma, S.; Kohli, N.; Brière, J.; Ménard, M.; Nabki, F. Translational MEMS Platform for Planar Optical Switching Fabrics. *Micromachines* **2019**, *10*, 435. [CrossRef] [PubMed]
9. Preite, M.V.; Sorianello, V.; de Angelis, G.; Romagnoli, M.; Velha, P. Geometrical Representation of a Polarisation Management Component on a SOI Platform. *Micromachines* **2019**, *10*, 364. [CrossRef] [PubMed]
10. Huang, B.; Zhang, Z.; Zhang, Z.; Cheng, C.; Zhang, H.; Zhang, H.; Chen, H. 100 Gb/s Silicon Photonic WDM Transmitter with Misalignment-Tolerant Surface-Normal Optical Interfaces. *Micromachines* **2019**, *10*, 336. [CrossRef] [PubMed]

11. Malinowski, M.; Bustos-Ramirez, R.; Tremblay, J.-E.; Camacho-Gonzalez, G.F.; Wu, M.C.; Delfyett, P.J.; Fathpour, S. Towards On-Chip Self-Referenced Frequency-Comb Sources Based on Semiconductor Mode-Locked Lasers. *Micromachines* **2019**, *10*, 391. [CrossRef] [PubMed]
12. Lu, L.; Zhou, L.; Chen, J. Programmable SCOW Mesh Silicon Photonic Processor for Linear Unitary Operator. *Micromachines* **2019**, *10*, 646. [CrossRef] [PubMed]

Review

A Review: Preparation, Performance, and Applications of Silicon Oxynitride Film

Yue Shi [1,†], Liang He [2,†], Fangcao Guang [1], Luhai Li [1], Zhiqing Xin [1] and Ruping Liu [1,*]

[1] School of Printing and Packaging Engineering, Beijing Institute of Graphic Communication, Beijing 102600, China

[2] State Key Laboratory of Advanced Technology for Materials Synthesis and Processing, Wuhan University of Technology, Wuhan 430070, China

* Correspondence: liuruping@bigc.edu.cn; Tel.: +86-10-6026-1602

† Yue Shi and Liang He contributed equally to this work.

Received: 15 July 2019; Accepted: 14 August 2019; Published: 20 August 2019

Abstract: Silicon oxynitride (SiN_xO_y) is a highly promising functional material for its luminescence performance and tunable refractive index, which has wide applications in optical devices, non-volatile memory, barrier layer, and scratch-resistant coatings. This review presents recent developments, and discusses the preparation methods, performance, and applications of SiN_xO_y film. In particular, the preparation of SiN_xO_y film by chemical vapor deposition, physical vapor deposition, and oxynitridation is elaborated in details.

Keywords: silicon oxynitride; thin film; photoluminescence; chemical vapor deposition; physical vapor deposition

1. Introduction

Silicon oxynitride (SiN_xO_y) is an important inorganic material widely studied for its outstanding electronic and mechanical performance. SiN_xO_y is the intermediate phase between silicon dioxide (SiO_2) and silicon nitride (Si_3N_4) [1,2], which possesses high durability at high temperature, high resistance to thermal-shock and oxidation, high density, excellent mechanical performance, and a low dielectric constant [3–8]. Due to these unique performance, SiN_xO_y has great potential for high-temperature related applications, for example, it is typically used in non-linear optics [9] and as mechanical component owing to its high strength, thermal insulation, and electronic and chemical resistance [10,11]. When Si_3N_4 thin film is utilized as the core material of waveguide devices, Si_3N_4 waveguides have better tolerance to sidewall roughness and geometric variations than silicon waveguides, and its relative refractive index difference can increase to 0.62, allowing for much tighter bending radius. Tighter bending radius will not only reduce the bending radius and improve the device integration, but also narrow the waveguide and reduce the input loss [12–16]. However, Si_3N_4 has low fracture toughness and poor electrical insulation, limiting its wide applications. SiO_2 film has a wide range of applications, from microelectronics [17,18] to optical waveguides [19], due to its low dielectric constant, low defect density, and low residual stress. However, SiO_2 does not perform as an encapsulation layer since oxygen, sodium, and boron can diffuse within it [20,21], which is different from Si_3N_4. For these mentioned applications, SiN_xO_y is a more promising candidate, in addition, it has tunable optical and electrical performance. By varying the chemical composition of SiN_xO_y, during the fabrication process, its refractive index and dielectric constant will be tuned [22–24]. Except the tunable refractive index, SiN_xO_y also has adjustable thin film stress [25] and exhibits photoluminescence (PL) in the visible light range at room temperature [26]. Thus, SiN_xO_y is highly attractive in integrated circuits (IC) [27], barrier layers [28,29], non-volatile memory [30], optical waveguides [31], organic light emitting diode (OLED) [32], and anti-scratch coatings [33,34]. This review presents a discussion and summary of the preparation methods, performance, and applications of SiN_xO_y film.

2. Performance of SiN_xO_y Film

2.1. Luminescent Performance

With the development of semiconductor technology, silicon-based micro/nano devices with applications in optoelectronics and IC are in the rapid development. Optoelectronic integration technology urgently requires high-efficiency and high-intensity luminous materials, and the currently-existing silicon integration technology is utilized to develop high-performance optoelectronic devices/systems. Previously, the research on PL of porous silicon and nano-scale silicon at room temperature has aroused widespread attention in this field [34–36]. However, porous silicon exhibits several disadvantages such as degradation and poor stability, and it is the most important that it isvery difficult to use it on standard CMOS circuits, thin film sensors, or flexible substrates [37,38]. In addition, nano-scale silicon has problems such as insufficient density, difficulty in controlling size and distribution, unbalanced carrier injection efficiency, complicated luminescence mechanism, and non-radiative recombination [39–43]. As known to all, SiN_xO_y film is an important protective and barrier film with its luminescence characteristics, high mechanical performance, and high reliability [44,45]. Therefore, it is of great significance to study the luminescence characteristics of SiN_xO_y film. Some reported results showed that the luminescence mechanism of SiN_xO_y film is generally divided into three types: defect-state radiation composite luminescence [46], band-tail (BT) radiation composite luminescence [47] and quantum dot radiation composite luminescence [48].

The light-emitting performance of SiN_xO_y film is effected by its composition, because its composition has significant influences on the formation of Si-O, Si-N, Si-H, and N-H bonds, resulting in the changes of absorption peak position of SiN_xO_y film [49,50]. As a result, many researchers studied the atom ratio of N and O in SiN_xO_y film [51,52]. During the preparation of SiN_xO_y film, it is found that the luminescence performance of the SiN_xO_y film can be adjusted by changing the flow rate of nitrogen (N_2) and concentration of oxygen. In order to investigate the effect of flow rate of N_2 on the evaporated SiN_xO_y film, Lee et al. [53] prepared a SiN_xO_y film on a poly(ethylene naphthalate) (PEN) substrate by ion-beam assisted electron beam evaporation at room temperature. It is found that when the flow rate of N_2 is 40 sccm, the refractive index of SiN_xO_y film increased to 1.535, and the SiN_xO_y film density increased to ~2.5 g cm^{-3}, whereas the surface roughness and optical transmittance of SiN_xO_y film decreased. In preparation of the SiN_xO_y film, adjusting the flow rate of N_2 has a significant improvement on the luminescent performance of the doped SiN_xO_y film. Among them, Labbé et al. [54] prepared a nitrogen-rich SiN_xO_y film doped with Tb^{3+} by reactive magnetron co-sputtering under N_2 with different flow rates and annealing conditions. The influences of flow rate of N_2 on the atomic composition of SiN_xO_y film, the N excess (N_{ex}) in the sedimentary layer and the deposition rate are shown in Figure 1a. For the synthesis, the reverse flow of N_2 during deposition is studied. Through the characterization results, the researchers carefully identified different vibration modes of Si-N and Si-O bonds, especially the 'non-phase' tensile vibration mode of Si-O bonds. The highest PL intensity of Tb^{3+} is obtained by optimizing the nitrogen incorporation and annealing condition. The aggregation effect in Si_3N_4 matrix is significantly reduced, thus allowing the higher concentration of optically active Tb^{3+}, which promoted its luminescence applications. In related researches, Ehré and co-workers [55] prepared a cerium (Ce)-doped SiN_xO_y film by magnetron sputtering under N_2 atmosphere. Their results showed that a broad and strong PL peak is red-shifted at the N_2 flow rate of 2 sccm. The peak of PL band shifts to 450 nm, and the results showed that the PL strength is 90 times higher than the BT strength of the sample deposited at a low flow rate of N_2. The effects of flow rate of N_2 on PL of Ce-doped SiN_xO_y film, PL spectrum (solid line), and PL excitation spectrum (dotted line) at flow rate of N_2 (2 sccm) are shown in Figure 1b,c. As described above, it is possible to adjust not only the luminescence characteristics of the SiN_xO_y film by changing the flow rate of N_2 but also the oxygen content, such as Huang et al. [56] demonstrated the strong PL of SiN_xO_y film by adjusting the oxygen content. With the oxygen content in the SiN_xO_y film increasing from 8% to 61%, the PL changed from red light to orange light and white light, and they indicated that the change in PL performance of SiN_xO_y film is due to the change in the center of the defect luminescence and the change in the main phase structure from Si_3N_4 to SiN_xO_y and SiO_2. Similarly, the researchers studied the effect of oxygen concentration on PL of SiN_xO_y film doped with other components. Steveler et al. [57] prepared an Er^{3+}-doped amorphous SiN_xO_y film by reactive evaporation. They found that the PL of Er^{3+} is observed only in the samples with oxygen concentration equal to or less than 25%, and they indicated that oxygen will make Er^{3+} ions optically active and can be indirectly excited in the presence of excess Si. It is further confirmed that, when both the amounts of oxygen and nitrogen are equal to or about 25%, Er^{3+} related PL increases with the increase of annealing temperature.

Processing conditions of the preparation of SiN_xO_y film also have influences on the PL performance of SiN_xO_y film, such as the annealing process. For instance, the SiN_xO_y film prepared by reactive sputtering on a silicon substrate, and then vacuum-annealed at 900 °C for 1 hour, is amorphous, composed of mixed Si-N and Si-O bonds, and blue and green emission are observed in the PL spectra of this prepared SiN_xO_y film, as shown in Figure 1d [46], so the SiN_xO_y film integrated with a top electrode is used for a electroluminescent device.

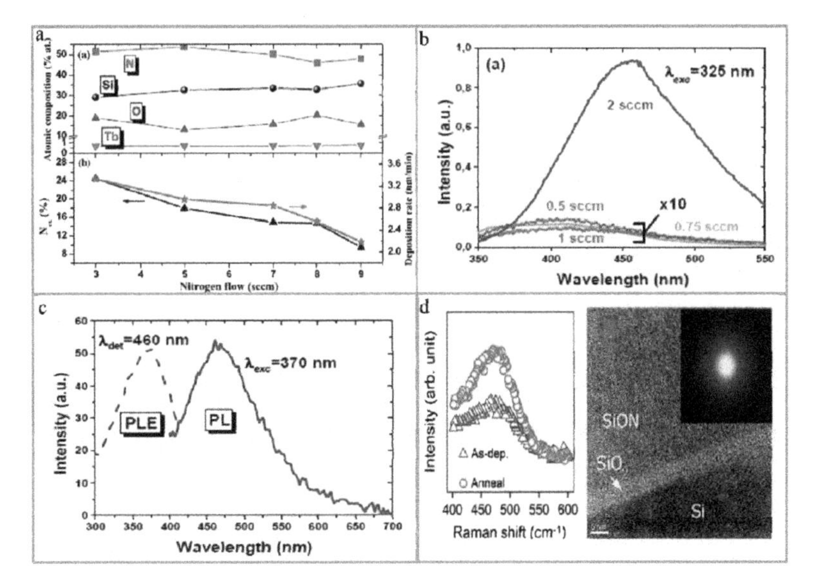

Figure 1. The luminescent performance of SiN_xO_y film. (**a**) Above part shows the effect of flow rate of N_2 on the atomic composition of the Tb-doped nitrogen-rich SiN_xO_y film, and below part shows the effect of the flow rate of N_2 on the nitrogen excess parameter (Nex) and deposition rate of the deposited layer [53]. Copyright 2017, *Nanotechnology*. (**b**) Effect of flow rate of N_2 on the PL of the Ce-doped SiN_xO_y film. (**c**) PL spectrum (solid line) and photoluminescence excitation spectrum (PLE) (dashed line) of a Ce-doped SiN_xO_y film at flow rate of N_2 of 2 sccm [55]. Copyright 2018, *Nanoscale*. (**d**) Effect of annealing temperature on SiN_xO_y film [46]. Copyright 2013, *J. Lumin.*

2.2. Adjustable Refractive Index

Adjustable refractive index means that the refractive index can be continuously tuned along the normal direction of the surface of the SiN_xO_y film by changing the proportion of the reaction gas [58]. In order to prepare graded-index SiN_xO_y film for applications in optical waveguide materials, gradient-index films, and anti-reflection films, researchers conducted extensive research on the refractive index-adjustable performance of SiN_xO_y film, such as the research on controlling the flow rate and ratio of reactive gas, reaction process, etc. [59–61]. Hänninen et al. [62] found that the refractive index and extinction coefficient of the SiN_xO_y film decrease with the increase of oxygen and nitrogen contents in the SiN_xO_y film, which could be adjusted by controlling the ratio of reactive gas. Therefore, in order to control the ratio, they applied reactive high-power pulsed magnetron sputtering to synthesize a SiN_xO_y film using N_2O as a single source, providing oxygen and nitrogen for SiN_xO_y film's growth. The characterization results showed that the synthesized SiN_xO_y film has the characteristics of silicon-rich, amorphous, and randomly-chemical-bonding structure. Furthermore, Himmler et al. [32] used reactive magnetron sputtering to deposit SiN_xO_y film and found that the refractive index of SiN_xO_y film depends on its oxygen and nitrogen contents, so they adjusted the refractive index by controlling the content ratio of oxygen/nitrogen. It is found that the reaction gases are differently incorporated into the layer due to different plasma conditions in the coating region, so there is a higher nitrogen incorporation and a higher refractive index in plasma regions with a high plasma density, while plasma regions with lower plasma density will result in a higher oxygen bonding and a lower refractive index. In addition to controlling the proportion of gas, the preparation method also has a certain influence on the refractive index of the SiN_xO_y film. Farhaoui et al. [63] used the reactive gas pulse in sputtering process (RGPP) to adjust the composition of SiN_xO_y film from oxide to nitride by controlling the average flow rate of O_2. Compared with the conventional reaction process (CP), not only did the deposition rate increase, but also a wide range of SiN_xO_y films' refractive indexes varying within the same range could be obtained through this pulse process. Moreover, extinction coefficient of the SiN_xO_y film is low, and this SiN_xO_y film can be used for multi-layer anti-reflection coating (ARC). Nakanishi et al. [64] also introduced argon (Ar) in the preparation of SiN_xO_y film by pulsed direct current (DC) reactive magnetron sputtering. They found that the higher the Ar concentration is, the more stable the SiN_xO_y film's formation and the higher the deposition rate is. The researchers claimed that a large amount of sputtered silicon atoms reach the substrate at high Ar concentration, causing the oxidation probability of the SiN_xO_y film to decrease and the refractive index of the SiN_xO_y film to gradually change with the percentage of oxygen in the reactive gas. The tunable refractive index of SiN_xO_y film makes it superior to Si_3N_4 and SiO_2 in optical device applications. Furthermore, it also provides a new strategy for development of optical devices.

3. Preparation of SiN_xO_y Film

At present, the preparation methods of SiN_xO_y film are mainly classified into chemical vapor deposition (CVD), physical vapor deposition (PVD), high-temperature nitridation, and ion implantation [65]. The descriptions and comparisons of these preparation processes are as follows.

3.1. *CVD Method*

CVD is a vapor phase growth method for preparing materials by introducing one or more compounds containing a constituent film element. During this growth process, the reactive gas is purged into a reaction chamber in which a substrate is placed, and a depositing process on the gas–phase or gas–solid interface is executed to generate solid sediments [66]. CVD based methods are mainly divided into plasma enhanced CVD (PECVD), low-pressure CVD (LPCVD), photochemical vapor deposition [67], thermal CVD [68], etc. Among them, PECVD and LPCVD are the most commonly employed methods. Additionally, PECVD can be extended to radio frequency PECVD (RF-PECVD) [69], electron cyclotron resonance PECVD (ECR-PECVD) [70,71], and inductively coupled PECVD (IC-PECVD) [72].

3.1.1. PECVD

PECVD is a method for preparing a semiconductor thin film which is subjected to chemical reaction deposition on a substrate using a glow discharge in a deposition chamber [73]. The preparation process of SiN_xO_y film via PECVD is generally described as follows: at low temperature (<400 °C), ammonia (NH_3), pure silane (usually SiH_4), N_2, and nitrous oxide (N_2O) are generally employed in a

PECVD chamber with a certain power. There are generally some differences in the composition of the precursor gases reported in studies. In general, the flow rates of NH_3, pure silane, and N_2 remain the same, and the total flow rate is controlled by adjusting the flow rate of N_2O [74]. Generally, SiN_xO_y film is deposited on silicon substrate or quartz substrate, wherein the substrate's temperature is kept at room temperature, but some composite films are deposited on other composite layers by PECVD, such as the preparation of SiN_xO_y and Si_3N_4 by Park et al. [75]. For composite film, they deposited a SiN_xO_y film directly on the deposited Si_3N_4 layer.

For SiN_xO_y film by PECVD, NH_3 is often used as the reaction gas of the nitrogen source, and SiH_4 is used as the reaction gas of the silicon source. Although NH_3 reacts with SiH_4 easily, the SiN_xO_y film produced by NH_3 at a lower temperature has a higher hydrogen content, which causes the decreased electrical performance of SiN_xO_y film, so some studies used RF-PECVD, with N_2 and SiH_4 as precursor gases to prepare SiN_xO_y film with lower hydrogen content, and some studies used RF-PECVD, with N_2, SiH_4 and NH_3 as the front gases [76]. For example, Kijaszek et al. [77] used RF-PECVD and maintained the RF of 13.56 MHz, pressure, power, and substrate temperature (350 °C), and controlled the composition of SiN_xO_y film by the flow ratio of different gaseous precursors: NH_3, 2% SiH_4/98% N_2 and N_2O, wherein the flow rate of 2% SiH_4/98% N_2 and N_2O remained the same, and the flow rate of NH_3 is adjusted to control the SiN_xO_y film's performance. When the flow rate of NH_3 is low, the SiN_xO_y film's hydrogen content is also lowered, and the electrical performance of the SiN_xO_y film is improved.

ECR-PECVD is also included in the PECVD method. Okazaki et al. [78] deposited a SiN_xO_y film under hydrogen-free conditions by ECR-PECVD at a low temperature of ~200 °C with O_2 and N_2 as reaction gases, and the obtained deposition rate is ~0.1 μm min^{-1}. The deposited SiN_xO_y film has good optical performance. Furthermore, Wood [79] used an ECR-PECVD system to deposit SiN_xO_y dielectric film at low substrate temperature. The electrical performance of these films is found to be comparable with those deposited in systems using ion-assisted PVD and sputtering systems. Furthermore, thin film electroluminescence devices containing ECR SiN_xO_y dielectrics exhibit high brightness and excellent breakdown characteristics.

3.1.2. LPCVD

For LPCVD, a gas source under low pressure is decomposed to deposit SiN_xO_y film directly on a substrate. Since the mean free path of the reactive gas molecules increases at a low pressure, the diffusion coefficient increases. Thereby, the transmission speeds of gaseous reactants and by-products are increased, the aggregation of impurities on the substrate is reduced to some extent, and the film is more uniform. It has the advantages of structural integrity, few pinhole defects, and high deposition rate, therefore it is suitable for large-area production [80]. Kaghouche et al. [81] deposited a SiN_xO_y film on a single crystal silicon wafer using LPCVD at a high temperature of 850 °C with N_2O, NH_3 and dichlorosilane (SiH_2Cl_2) as precursor gases. In the synthesis, the control variable experiment is carried out by adjusting the flow ratio of NH_3/N_2O via keeping the flow rate of SiH_2Cl_2 as constant. Additionally, the deposition duration remains constant to maintain similar annealing conditions during the deposition process. Finally, the thickness of the obtained SiN_xO_y film is generally in the range of 300–400 nm.

However, the LPCVD has its disadvantages of low heating rate, long reaction time, and high deposition temperature (generally >550 °C), which limit its applications to some extent. A comparison of different CVD methods is summarized in Table 1 [76,82–85].

Table 1. Comparison of materials, ratios, and deposition conditions in different chemical vapor deposition (CVD) methods.

Deposition Method	Precursor Gases	Ratio of Precursor Gases	Deposition Condition	Reference
PECVD	SiH_4, N_2O	SiH_4/N_2O = 0.05–0.125	200 °C, 97.09 Pa	[82]
RF-PECVD	SiH_4, N_2O, NH_3	$(NH_3 + SiH_4)/N_2O$ = 0.64–3.22	120 Pa	[76]
ECR-PECVD	N_2, O_2, SiH_4	O_2/N_2 = 0.03–0.1	-	[83]
IC-PECVD	N_2, Ar, SiH_4	N_2/Ar = 0.0625–0.5	90–250 °C, 1–6 Pa	[84]
LPCVD	N_2O, NH_3 SiH_2Cl_2	N_2O/NH_3 = 4.8-	860 °C, 53.2 Pa	[85]

3.1.3. High Temperature Thermochemical Vapor Deposition (HTCVD)

The HTCVD method uses a direct heating method to decompose or chemically react to obtain a solid film on the surface of a substrate [86]. Since the HTCVD method uses direct heating to provide activation energy for the gas, it does not require complicated equipment, and the operation is relatively facile. In addition, the reaction rate is high, and the formed film has less impurity of hydrogen and a dense structure [80]. However, the heating temperature of the HTCVD method is generally high (≥700 °C), which easily causes deformation and internal structural change of the substrate, reducing the mechanical performance of the substrate and the bonding force between film and substrate.

For optimization of this method, a rapid thermal CVD (RTCVD) is investigated [68]. RTCVD refers to the formation of a single-layer SiN_xO_y film by a rapid annealing method to seal the substrate, thereby avoiding the substrate being affected by high temperature [76]. The RTCVD method can deposit advanced dielectric films on III-V substrates under high temperature. This method not only has the advantages of HTCVD, but also prevents the V group elements from sublimating due to high temperature. Lebland et al. [87] deposited SiN_xO_y film on III-V substrates by RFCVD. They controlled the deposition rate and stoichiometry of the SiN_xO_y film by adjusting the partial pressure of N_2O and temperature. It is found that a deposition rate of up to 10 nm s^{-1} is obtained at 750 °C, and the InP substrate does not degrade, solving the contradiction between the high deposition temperature using the direct CVD method and the degradation of the V group element.

3.1.4. Photochemical Vapor Deposition (Photo-CVD)

Photo-CVD is a novel low-temperature deposition method using ultraviolet (UV) light or laser to photodecompose a reaction gas to obtain a solid film [88]. It has the following advantages: greatly reducing the substrate temperature (≤250 °C), avoiding the damage caused by high-energy particle radiation on the surface of the film, making the surface of the film smooth, and reducing the by-product [89]. However, the main disadvantages of the photo-CVD are its high cost, and that the formed film is not stable.

The CVD method has certain advantages and many extension methods, Table 2 shows the pros and cons of four different CVD approaches [80,89–92], among them PECVD and LPCVD are the most commonly used deposition methods. Even so, it is difficult to avoid the existence of hydrogen in preparing SiN_xO_y film by PECVD or LPCVD, and the hydrogen deposited in the raw material is difficult to remove due to the low deposition temperature of the CVD method, further affecting the performance of SiN_xO_y film [90]. To this end, researchers have proposed various methods to reduce the hydrogen content of SiN_xO_y film. Among them, the thermal oxidation treatment by Hallam et al. [91] is highly promising, they deposited a SiN_xO_y film by PECVD and demonstrated that the SiN_xO_y film with Si-H peak wave number of >2200 cm^{-1} has an open circuit voltage of up to 80 mV during thermal annealing. They pointed out that the result is due to an increase in oxygen content. In addition, increasing the flow of raw materials and the annealing temperature will also reduce the hydrogen content to various degrees.

Table 2. Advantages and disadvantages of SiN_xO_y film prepared by PECVD, LPCVD, HTCVD and Photo-CVD methods.

Method	Advantages	Disadvantages
PECVD	Flexible operation method, High process repeatability, High step coverage, Low deposition temperature (<400 °C) [92]	High cost, High H content in film
LPCVD	Uniform film, Complete structure, Less pinhole defects, High deposition speed, Large-area preparation [80]	Low heating rate, Long reaction time, High deposition temperature (generally >550 °C)

Table 2. *Cont.*

Method	Advantages	Disadvantages
HTCVD	Simple operation and operation, High reaction rate, Low H content in film and dense structure [80]	High deformation, Impaired interface performance
Photo-CVD	Low reaction temperature (≤250 °C), Smooth film surface, Less by-products [89]	High cost, Low film stability

3.2. PVD

PVD is a physical method of vaporizing a material source into gaseous atoms, molecules, or ionized ions under vacuum conditions, and depositing a film on the substrate by sputtering or plasma technology [93]. The main processes of PVD are vacuum evaporation, sputter coating, ion plating, arc plasma coating, etc. [94–96]. Among them, sputter coating is a widely used and mature method, which means that under vacuum conditions, the surface of the target material is bombarded with the particles having the function, then the surface atoms of the target are obtained with sufficient energy to escape, and the sputtered target is deposited on the substrate to form a film [97]. The large-scale magnetron sputtered coating developed on the basis of sputter coating has high deposition rate, good process repeatability, easy automation, etc. [90].

Magnetron sputtering coating method is also divided into several categories such as reactive pulsed magnetron sputtering, pulsed DC magnetron sputtering, rotatable dual magnetron pulsed DC reactive magnetron sputtering, intermediate frequency (MF) magnetron sputtering, etc. [98]. Tang et al. [99] deposited SiN_xO_y film by reactive pulsed magnetron sputtering, and the effects of nitrogen ratio on the optical, structural, and mechanical performance of SiN_xO_y film are investigated. They found that with the increase of nitrogen ratio, the refractive index of the SiN_xO_y film increased from 1.487 to 1.956, its surface roughness decreased from 1.33 to 0.97 nm, its hardness increased from 13.51 to 19.74 GPa, and its Young's modulus increased from 110.41 to 140.49 GPa. When the SiN_xO_y film is applied to the anti-reflection coating, the hardness of the coating is greatly improved. Additionally, Simurka et al. [100] deposited a SiN_xO_y film on a glass substrate by pulsed DC magnetron sputtering with a constant flow of Ar (38–94 sccm) as the working gas and a constant gas flow ratio of oxygen (2–5 sccm) and nitrogen (20 sccm) is employed. Through experimental characterizations, it is found that as the sputtering power increases, the density, refractive index, hardness, and Young's modulus of the film increase slightly. Therefore, in the preparation process, it is very important to select the appropriate sputtering power according to the requirements. Furthermore, in the preparing process of SiN_xO_y film, not only a reasonable sputtering power should be set, but also the voltage is important. Himmler et al. [32] deposited a single layer of SiO_x, SiN_y, and SiN_xO_y using a rotatable dual magnetron pulsed DC reactive magnetron sputtering, as shown in Figure 2a, showing a layout of deposition zone within the roll-to-roll coating tool. Experimental results demonstrated that the ratio of oxygen/nitrogen in the layer is determined not only by the reactive gas, but also by the voltage. Furthermore, Li et al. [101] deposited a multilayer film of hydrogenated silicon nitride (SiN_x:H)/silicon nitride (SiN_x)/SiN_xO_y for silicon crystal solar cells using intermediate frequency (MF) magnetron sputtering, and the process design and application of passivation film on anti-reflection surface of silicon solar cell in laboratory are shown in Figure 2b. The multilayer film prepared by this process has excellent cell efficiency when applied to solar cells.

Although the SiN_xO_y film prepared by the PVD method has a lower hydrogen content than that of the SiN_xO_y film by CVD, the PVD is superior to the CVD for preparing SiN_xO_y film. Moreover, the sputtering method has a low deposition temperature and is easy to control. However, it is difficult to perform rapid film deposition on a large-scale substrate by sputtering method [102] and the sputtering method is prone to "target poisoning" [103], that is the added reaction gas generates a composite material on the target, thereby changing the sputtering yield and the deposition rate. Finally, the film stoichiometry is changed.

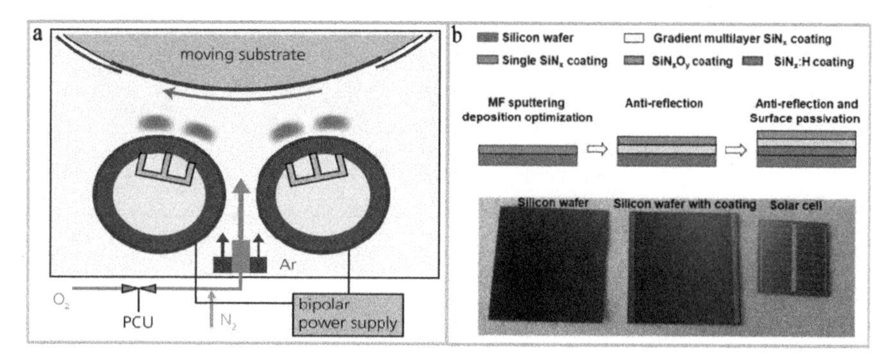

Figure 2. Physical vapor deposition (PVD). (**a**) Schematic of the deposition zone inside the roll-to-roll tool [32]. Copyright 2018, *Surf. Coat. Technol.* (**b**) The procedure design and application for laboratory silicon solar cell of the anti-reflection and surface passivation film [101]. Copyright 2017, *Mater. Sci. Semicond. Process.*

3.3. Oxynitridation

Oxynitridation is generally a method of reacting gas such as N_2O, NO, NH_3, N_2, and O_2 with Si, SiO_2, or Si_3N_4 under suitable conditions to obtain a SiN_xO_y film.

The high-temperature nitridation is a more commonly used method, which refers to the preparation of SiN_xO_y film by introducing N into SiO_2 film by means of thermal nitriding [104], rapid annealing [105], and plasma nitriding [106] under certain conditions with reactive gas containing nitrogen (e.g., N_2O, NO, NH_3, N_2) [107]. Among them, a SiO_2 film which can be obtained by a thermal oxidation method [108] and a sol–gel method [109]. Different performance of membranes prepared under different oxidation/nitridation conditions in the nitrogen oxidation process are obtained. When the temperature is 800–1200 °C, NH_3 is introduced as a nitrogen source for high-temperature nitridation of SiO_2 by thermal nitridation, although a simple and smooth SiN_xO_y film can be obtained, H is inevitably introduced. H doping in the SiN_xO_y film affects its electrical performance, such as the formation of a large number of electron traps in the SiN_xO_y film. In order to reduce the effect of H on membrane performance, reoxidation is an effective method [110]. Furthermore, a nitrogen-containing gas without H can be used as a nitrogen source, such as N_2O, to form a Si-O-N film in situ generation without introducing a H atom. However, the high temperature conditions of the high temperature nitridation process have a serious influence on the substrate and are prone to defects. In addition, the SiN_xO_y film obtained by this method has poor uniformity and poor compactness.

Three methods for preparing SiN_xO_y film are introduced above, and the advantages and disadvantages of different preparation methods are summarized in Table 3 [110–113].

Table 3. Preparation methods of SiN_xO_y film.

Method	Advantages	Disadvantages
CVD	High deposition rate Low deposition temperature Uniform film	Hydrogen content has an effect on electrical conductivity [111]
PVD	Low hydrogen content	Low deposition rate [112]
Oxynitridation	Relatively simple operation, large-scale preparation	Target poisoning is common Film thickness is difficult to control Toxicity of raw gas [110] Low N_2 nitriding degree [113]

The deposition process of the SiN_xO_y film is reviewed. The properties of the film microstructure for different growth conditions and post-growth thermal annealing will be illustrated by Figure 3. As shown in Figure 3a–c, when the ratio of Ar/SiH_4 and $O_2/(O_2 + N_2)$ is 4.5 and 0.1, respectively, the porosity of the SiN_xO_y film gradually decreases. However, at this state, the SiN_xO_y film is still porous and pore-oriented. The effect of the rate reduction is small. As shown in Figure 3d, by changing Ar/SiH_4 from 4.5 to 41 and RF to 0.9 KW, a very significant decrease in porosity is observed, showing a very dense SiN_xO_y film, and it is shown by Figure 3e when Ar/SiH_4 changes from 4.5 to 41, and when the RF is 0.9 KW, its water vapor transmission rate (WVTR) is the lowest, and it has good protection function [72]. Figure 3f,g shows HRTEM images of different types of Ce^{3+} and Tb^{3+} co-doped SiN_xO_y film and structure (III) multilayer designs. It is worth noting that no ripples and layer mixing are observed. Furthermore, the fact that the morphological structure remained unchanged even after annealing at 1180 °C for 1 h confirmed the manufacturing process and showed the robustness of the sample (Figure 3g) [114]. As shown in Figure 3h, (I) and (II) are infrared transmission photographs of the bond pairs and the annealed pairs, respectively. The results show that the crack length of the composite is obviously shortened after annealing at 120 °C, and the bond strength is obviously improved. After the sample is annealed at 300 °C for 2 h, the bond strength is sufficient to withstand peeling, and the infrared transmission photograph is shown in Figure 3 h(III). As shown in the AFM chart (Figure 3i), the SiN_xO_y film has a smooth and uniform surface with a root mean square (RMS) roughness of 0.162 nm, which is sufficient for direct soldering without complicated grinding and polishing processes. Figure 3j shows the corresponding depth profiles of N, O, and Si. Obviously, you can find two layers. From the surface to a depth of about 40 nm, the amounts of O and N are about 40% and 20%, respectively. Starting from 40 nm, the O concentration gradually decreases compared with the first layer, but the concentration of N increases. After annealing at 1100 °C, the higher structural disorder confirmed by XRD analysis in the SiN_xO_{yy} film can be attributed to the presence of O. This helps to inhibit polycrystallization of the buried insulator during post-annealing in conventional CMOS processes [115]. By analyzing the effects of different growth conditions and the properties of the thermally annealed film microstructure by post-growth on the performance of SiN_xO_y film, it is helpful to effectively control the parameters as needed in later applications.

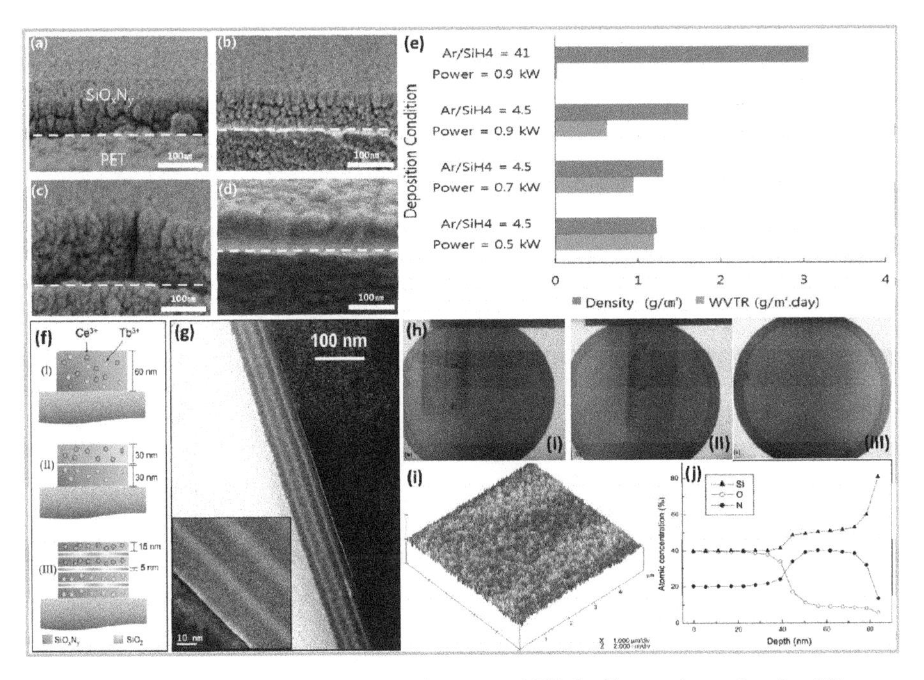

Figure 3. The surface morphology and performance of SiN_xO_y film are changed under different conditions. **(a–d)** FESEM cross-section images of SiN_xO_y film deposited at Ar/SiH_4, $O_2/(O_2+N_2)$ of 4.5

and 0.1, respectively with RF power of (a) 0.5 kW, (b) 0.7 kW, (c) 0.9 kW, and (d) at Ar/SiH_4, $O_2/(O_2 + N_2)$ of 41 and 0.1, respectively with RF power of 0.9 kW. (**e**) Effect of Ar/SiH_4 ratio and RF power variation on the density and WVTR of 100 nm-thick SiN_xO_y film deposited at $O_2/(O_2 + N_2)$ ratio of 0.1 [72]. Copyright 2019, *Thin Solid Films.* (**f**) Sample description, being (I) a Ce^{3+} and Tb^{3+} co-doped SiN_xO_y film, 60 nm thick with 42 at.% of N; (II) a bilayer composed of a Ce^{3+} doped SiN_xO_y single layer and a Tb^{3+}- doped SiN_xO_y single layer, each 30 nm thick; and (III) a multilayer made of two layers of Tb^{3+}- doped SiN_xO_y and two layers of Ce^{3+}-doped SiN_xO_y (each sub-layer with a thickness of 15 nm), separated by SiO_2 spacers of 5 nm. All SiN_xO_y layers have a nitrogen content of 42 at. %. (**g**) HRTEM image of the multilayer design (structure (III), as-deposited sample). The inset shows a magnified region of the multilayer close to the Si substrate, for a sample annealed at 1180 °C for 1 h [114]. Copyright 2016, *J. Appl. Phys.* (**h**) Infrared transmission image of a bonded pair after the crack-opening test: (I) as-bonded, (II) after annealing at 120 °C, (III) after annealing at 300 °C. (**i**) AFM image of SiN_xO_y film synthesized by plasma immersion ion implantation (PIII). (**j**) Depth profiles of Si, O, and N in SiN_xO_y film acquired by sputtering XPS [115]. Copyright 2005, *Appl. Surf. Sci.*

4. Applications of SiN_xO_y Film

As an intermediate of SiN_x and SiO_2, SiN_xO_y film has an adjustable dielectric constant, refractive index, and extinction coefficient by controlling the ratio of nitrogen/oxygen in the chemical composition [116]. The gas ratio and process parameters of the various reactants involved in the preparation of SiN_xO_y are different, so that the stoichiometry of each element in SiN_xO_y is different, and the limit forms may be a-Si, SiO_2, and Si_3N_4.As shown in the Figure 4, a series of possible forms of the SiN_xO_y film is given, as well as changes in the properties of different film forms as the O content and H content increase, are mainly reflected in changes in transparency, band gap width, refractive index, and insulation [117]. SiN_xO_y film has performance between SiO_2 and Si_3N_4. Due to its excellent photoelectric performance, it has been widely used in optical devices, dielectric gate dielectric materials, and optical waveguide materials [118]. The SiN_xO_y film also has high chemical stability, high resistance to impurity diffusion, and water vapor permeability, which is highly promising for applications in barrier devices such as gas barriers [119]. In addition, the SiN_xO_y film has a small defect density and is advantageous in applications as a storage medium. The applications of SiN_xO_y film in microelectronic devices, optical devices, barrier materials, and non-volatile memory will be introduced in this section.

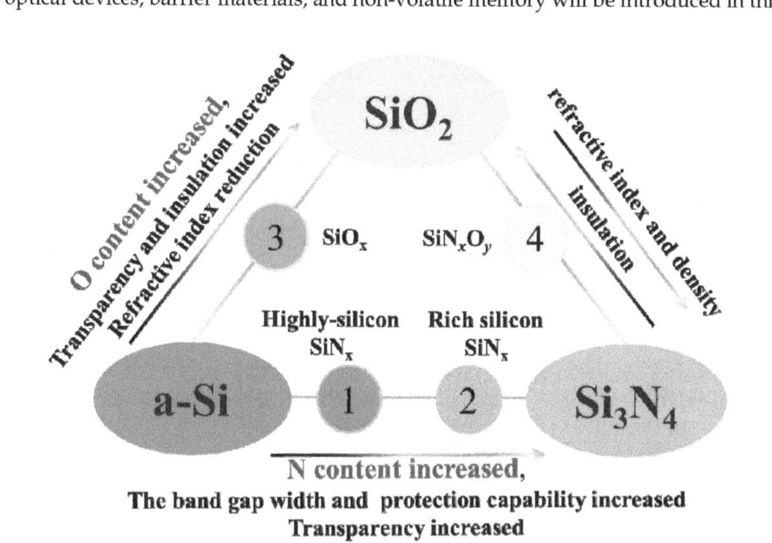

Figure 4. A series of possible forms of the SiN_xO_y film and their performance.

4.1. Application of Barrier Material

Most organic conductive polymers and chemically reactive electrodes degrade when exposed to water or oxygen, causing failure of electronic devices [120]. Therefore, in order to enhance the service life of device, it is necessary to use a barrier material for encapsulation layer of the device with a low water vapor transmission rate and oxygen permeability. Currently, glass or metal is often used for device packaging, however, these rigid materials with high spring constant will greatly limit the widespread applications of devices, especially in flexible devices. On this basis, researchers studied thin-film encapsulation (TFE) technology [121]. In recent years, the SiN_xO_y film has also been widely used as encapsulating layers. However, SiN_xO_y film has some problems served as an encapsulation layer. The basic degradation mechanism of water vapor or oxygen on a single layer of SiN_xO_y film is studied, and it is found that water vapor diffuses into the SiN_xO_y film through the percolation channel and nano-defects to react with SiON:H, making Si-N gradually changes to Si-O bond under repeated erosion of water vapor, eventually becoming a fully oxidized film. Therefore, the degradation of SiO_2:H becomes a way for water vapor to diffuse, and because of the lower density of SiO_2:H, the diffusion rate of water vapor in the SiN_xO_y film is further enhanced [116]. In addition, the difference in moisture resistance of the SiN_xO_y film is explained by film oxidation and surface defect density. Oku and co-workers [122] indicated that the surface of SiN_xO_y film easily changes to a low-quality film containing an excessive amount of H_2O molecule and O-H bond, which reduces the waterproofness, after pressure cooker test (PCT) test. Therefore, it is generally formed by depositing a multilayer film to reduce film defects and the probability of film pores communicating with the atmosphere, thereby forming a perfect barrier layer. On the other hand, SiN_xO_y film deposited at a higher temperature has higher performance, and its defects are less, however, the high-temperature treatment should be avoided in some devices, and the deposited SiN_xO_y film has poor barrier performance under low temperature, which is not conducive to the protection of the device. Therefore, Satoh et al. [123] developed a AlO_x/SiON composite bilayer structure deposited at a relatively low temperature (<160 °C), in which the AlO_x layer is used as a barrier layer, and the SiN_xO_y layer ensured the chemical and thermal stability of the film to protect the device from damage during processing. The test results indicated that the device did not show any degradation at 85 °C, and 85% relative humidity (RH) is obtained.

Additionally, some researchers have developed other methods, such as changing the preparation process and controlling the preparation parameters, to optimize the barrier performance of SiN_xO_y film to air and water vapor. Shim et al. [124] deposited a thin SiN_xO_y layer on the surface of poly(ether sulfone) (PES) membrane by PECVD using a mixture of hexamethyldisiloxane (HMDSO) and ammonia. SiN_xO_y is used as gas barrier layer, and a silicon-based coating layer of an organic/inorganic hybrid structure is added between the surface of PES and SiN_xO_y film as a buffer layer for both. The results showed that the undercoat layer is indispensable in the composite film. Under the action of the undercoat layer, the dense inorganic SiN_xO_y layer has an excellent oxygen barrier performance of 0.2 cm^3 m^{-2} day^{-1} at a critical coating thickness of ~20 nm. Moreover, the presence of the undercoat layer not only prevents the film from being cracked when the composite film is highly curved, but also maintains the initial gas barrier performance of the curved film. Iwamori et al. [125] prepared a SiN_xO_y-based transparent gas barrier material by depositing a SiN_xO_y film on a polyethylene terephthalate (PET) substrate by reactive sputtering in a nitrogen plasma. The results showed that the SiN_xO_y film has a lower oxygen transmission rate than that of the SiO_2 film, because the use of nitrogen plasma instead of oxygen suppresses the formation of defects induced by peroxide, and the increase in nitrogen content greatly enhances the density of SiN_xO_y film, forming a composite film with excellent gas barrier performance. Liu et al. [126] deposited a SiN_xO_y film on PES by RF magnetron sputtering in Ar/N_2 atmosphere as a barrier to prevent water vapor permeation. The experimental results showed that the RF power is 250 W. At the fixed film pressure of 1.6 Pa, with the deposition time of 30 min, the 100% N_2 content, and the absence of substrate bias, the water vapor transmission rate of the composite film is two orders of magnitude smaller than that of the uncoated PES.

These studies provided some new research ideas for the development of high-quality SiN_xO_y barrier material, and have important research significance for the applications of SiN_xO_y film in packaging materials, protective layer materials, and gas barriers.

4.2. Application of Non-Volatile Semiconducting Memory

Non-volatile semiconducting memory (NVSM) means that when the current is turned off, the stored data does not disappear, which can be applied to many fields [127–129]. Wrazien et al. [130] conducted a lot of work on non-volatile memory. They studied NVSM with silicon-oxide-nitride-oxide-silicon (SONOS) structure, which theoretically proves its storage capacity under low pressure and 150 °C can be as long as 10 years and the erase/write cycle is 10^5 times. Moreover, the SONOS based device has an oxynitride charge storage layer. In recent years, resistive memory is considered as a non-volatile memory with broad application prospects because of its excellent durability, high data transmission speed, and low power consumption [131]. Resistive random access memory (RRAM) is also known as a memristor-an electronic component with memory function [132]. Currently, excellent resistance switching characteristics of silicon-based dielectrics are often applied in RRAM devices. Recently, with the development of SiN_xO_y film, it has been found that the defect density of SiN_xO_y film is low, and the high operating voltage is more advantageous when used as a storage medium. Chen et al. [133] proposed that the SiN_xO_y film has non-uniformities such as operating voltage and current, and improved uniformity of the RRAM by doping. They prepared SiN_xO_y film with different oxygen concentrations by reactive magnetron sputtering. The resistance conversion performance and conduction mechanisms of $Cu/SiN_xO_y/ITO$ device are investigated. The SiN_xO_y is deposited under O_2 with a flow rate of 0.8 sccm in their investigation. The fabricated $Cu/SiN_xO_y/ITO$ device showed reliable resistance conversion behavior, including high durability and retention. Zhang et al. [134] prepared a SiN_xO_y film with ultra-low power a-$SiN_{0.69}O_{0.53}$:H by PECVD and pass N_2O into a vacuum chamber. An a-$SiN_{0.69}O_{0.53}$:H film is prepared under high resistance state and low power conversion. The RRAM cells based on the a-$SiN_{0.69}O_{0.53}$:H film shows an ultra-low current compared with the pure a-$SiN_{0.62}$:H-based RRAM, effectively reducing the operating current.

Recently, Wang et al. [135] developed a diffusion type memristor composed of two electrodes and a SiN_xO_y film embedded with nano silver particles, wherein the film is placed between the two electrodes. The SiN_xO_y film is an insulator, but after electrification, under the action of heat and electric power, the positions of the silver particles arranged neatly on the film begin to scatter, gradually diffuse, and penetrate the film to finally form a cluster of conductive filaments. Current is passed from one electrode to the other. After turning off the power, the temperature drops and the nano silver particles are rearranged. The work of the memristor developed by the researchers is very similar to that of calcium ions in biological synapses, so the device can simulate the short-term plasticity of neurons, and the schematic diagram and circuit diagram of RRAM simulating synaptic memory are shown in Figure 5. In addition, the diffusion memristor can also be used as a selector with large transient nonlinearity, which has great research significance and application prospects.

4.3. Application of Optical Devices

By controlling the ratio of oxygen and nitrogen in the SiN_xO_y film, the refractive index and extinction coefficient of the SiN_xO_y film can be controlled, and it can be used as an optical waveguide material and an anti-reflection film.

Optical waveguide materials are generally required to have low transmission loss, single mode transmission, and the ability to fabricate a variety of active and passive devices on the same platform. The outstanding feature of planar optical waveguides based on SiN_xO_y and SiN is their ultra-low loss. Planar waveguides with attenuations less than 0.1 dB m^{-1} have been implemented, such as Baudzus et al. [136] studied phase shifters based on electro-optic (EO) polymers and SiN and SiN_xO_y waveguide materials systems, and found that SiN_xO_y and SiN can be combined with EO polymers and a fast adaptive phase shifter with very low attenuation can be made. The phase shifter has an attenuation of 0.8 dB cm^{-1} at 1550 nm and an EO efficiency factor of 27%, which can achieve lower loss and has important research significance.

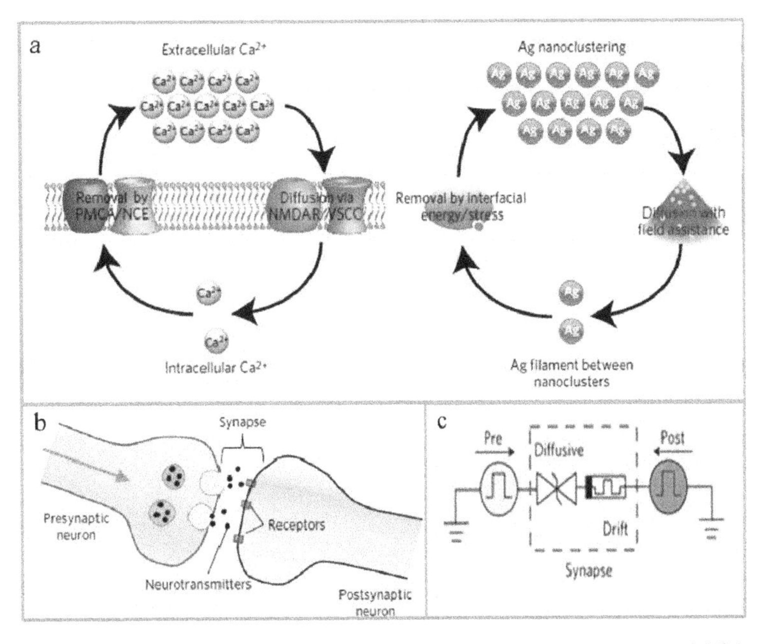

Figure 5. Resistive random-access memory (RRAM) that mimics synaptic memory. (**a**) Schematic illustration of the analogy between Ca^{2+} and Ag dynamics. (**b**) Illustration of biological synaptic connections between presynaptic and postsynaptic neurons and circuit diagrams of electrical synapses [135]. Copyright 2016, *Nat. Mater.*

Optical waveguide materials, such as SiN_xO_y and SiN, also have applications in resonators. Compared with the straight waveguide structure, the ring resonator (RR) has the advantages of small footprint, high wavelength selectivity, and accumulated light intensity in the cavity, so that a large absorption is obtained in the case of a short absorption medium. Therefore, the microring resonator structure can achieve high responsiveness in the photodetector, achieving large extinction in the modulator. In recent years, the excellent performance of silicon waveguide photodetectors with integrated graphene has been experimentally verified. In a recent study of SiN microring resonators, Wang et al. [137] used a graphene-photonic integrated circuits (PICs) structure to fabricate microring resonators on the Si_3N_4 platform. Compared with the graphene-Si_3N_4 straight waveguide, the absorption of the ring resonator is increased, and the length of the required graphene is greatly shortened, and the quality factor of the prepared resonator is 0.282×10^5–0.038×10^5. In 2018, Jia et al. [138] developed a SiN_xO_y-based optical waveguide ring resonator (OWRR), which employed a liquid source CVD (LSCVD) method to deposit a SiN_xO_y film with a good refractive index. It is found that the SiN_xO_y-based resonator achieves a measured waveguide loss of 4.07 dB cm^{-1} and a quality factor of 0.93×10^5 in the transverse electric (TE) mode, which provides a new idea for the preparation of other SiN_xO_y-based optical devices. The use of SiN and SiN_xO_y described above in ring resonators exhibits different ranges of quality factors, and in specific applications, is effectively selected based on the required bandwidth. In order to improve the controllability of the deposition process and the precise control of the SiN_xO_y film performance, PECVD and LPCVD methods are both used in the preparation process. When the refractive index is less than 1.7, it is suitable for deposition by PECVD. When the refractive index is greater than 1.7, the dependence of the refractive index of the SiN_xO_y film on the gas flow rate in PECVD deposition is large, which is not conducive to precise control. It must be deposited by LPCVD. The product rate is 1–10 nm/min, suitable for growing thin films with a thickness of 10–500 nm. The two deposition methods described above have thickness and refractive index uniformity and process repeatability of the resulting SiN_xO_y film deposited in the respective refractive index ranges, which can meet the requirements of high performance waveguides [139]. In addition to these, Trenti et al. [140] not only confirmed the low loss of SiN_xO_y, but also assumed that SiN_xO_y has thermal and nonlinear optical properties, and that SiN_xO_y is an excellent platform for nonlinear and quantum optical integrated photonic circuit design.

As an anti-reflective film, SiN_xO_y film can greatly reduce light reflection and loss, and increase efficiency when applied to photovoltaic devices such as solar cells and OLEDs. The excellent optical properties of SiN_xO_y film can increase the transmittance of light and reduce the reflectivity. Sapphire substrate is widely used in various photoelectric applications on glass substrate. However, sapphire has a low transmittance. In order to increase the transmittance of sapphire substrate and reduce the refractive index, Loka et al. [141] used RF magnetron sputtering to deposit a layer of SiN_xO_y film on the surface of sapphire, followed by annealing at a high temperature of 1099.85°C.The visible light transmittance is comparable with that of sodium-calcium glass. In this study, it is worth noting that after high temperature annealing, the nitrogen in the SiN_xO_y film disappears and the film is converted into SiO_n. It is believed that the formation of SiO_n and the high temperature annealing by increasing the oxygen content help to lower the refractive index and reflectivity. Moreover, SiN_xO_y film is used as an anti-reflection film on solar cells and OLEDs, which not only improves efficiency, but also protects the device due to its high resistance to corrosion and oxidation. However, in most preparation processes, composite films of SiN_xO_y and other components are usually double or even multilayered in order to obtain higher performance. In these studies, Nguyen et al. [142] developed a dual stack of SiN_xO_y/Al_2O_3 for n-type c-Si solar cells. They deposited a SiN_xO_y layer on the surface of Al_2O_3 as a coating by PECVD. It is found that the Al_2O_3/a-SiN_x:H phase is excellent due to its excellent anti-reflection and front surface passivation of the Al_2O_3/SiON stack. The refractive index and high positive fixed charge of the SiN_xO_y cap layer in the Al_2O_3/SiN_xO_y stack are reduced, but the lifetime is significantly improved, and the optical properties are excellent, such as the average transmission in the entire wavelength range of 300–1100 nm. The rate is 93.8%, the average absorption rate is 0.33%, the energy conversion efficiency is increased from 17.55% to 18.34%, the short-circuit current density and open circuit voltage are also improved. Sahouane et al. [143] further investigated the application of multilayer anti-reflective coatings to reduce optical losses in solar cells. They deposited multilayer silicon nitride film and SiN_xO_y film by PECVD and found that the best reflectivity of six layers of film is 1.05%, and that of four layers is 3.26%. It is proved that multilayer deposition can greatly reduce reflectivity, reduce loss, and improve efficiency. Parashar et al. [144] found that the hybrid plasma structure formed by adding self-assembled silver-aluminum alloy nanoparticles to the SiN_xO_y film can also significantly reduce the reflectivity of silicon. The results showed that SiN_xO_y/Ag_2Al nanoparticles reduce the average reflectance of silicon from 22.7% to 9.2%, and the SiN_xO_y cover layer reduces the reflectivity of silicon from 9.2% to 3.6% in the wavelength range of 300–1150 nm. By further enhancing photon management, the reflectivity decreased from 22.7% to ~3.6% in a 35 nm SiN_xO_y/Ag_2Al nanoparticles /25 nm SiN_xO_y hybrid plasma structure, and when used as a battery, photocurrent and cell efficiency are improved to some extent.

4.4. Application of Anti-Scratch Coating

With the development of flexible optoelectronic devices, more and more transparent polymers are developed as flexible substrate in devices, and the polymer is widely used because of its advantages of light weight, low cost, high transparency, and easy to process and design [145,146]. However, for long-term use, the polymer is limited in many applications due to its poor scratch resistance, fragility, or influence on device. Researchers have further protected the device by depositing a coating on the surface of the transparent polymer to enhance its scratch resistance. In recent years, silica-based coatings (SiN_xO_y film) have been considered as promising transparent coatings for protection of transparent polymers. Lin et al. [147,148] deposited organo-silicon oxynitride ($SiO_xC_yN_z$) film on a flexible polycarbonate (PC) substrate by low-temperature tetramethylsilane (TMS)-O_2-N_2 plasma polymerization at room temperature. They also studied the effect of N_2 flow rate on scratch resistance of PC substrates, and found that the optimal flow rate of N_2 is 3 sccm, and the scraping rate is 0% after scraping 200 times with steel wool under the pressure of 300 g. The $SiO_xC_yN_z$ film covered PC substrate has very good scratch resistance and provides a hard and smooth surface for flexible PC substrates. In addition, they also deposited $SiO_xC_yN_z$ film on reinforced carbon fiber reinforced polymer composites (FCFRPCs) by atmospheric pressure plasma low temperature polymerization to enhance the scratch resistance of the composites. There is a large amount of scratching (100%) on the original FCFRPC. By depositing $SiO_xC_yN_z$ film, the scratch resistance of FCFRPCs is significantly improved. After scratch testing, there are no scratches (0%), which is of great significance for the development of FCFRPCs. Furthermore, Zhang et al. [149] prepared a SiN_xO_y coating on glass and PET films by thermally annealing the inorganic polymer perhydropolysilazane (PHPS) between 60 and 200 °C. The results showed that the glass and PET film covered by SiN_xO_y film have higher hardness, stronger hydrophobicity and excellent adhesion. Additionally, the coated PET film exhibited high transparency and excellent scratch resistance in the visible wavelength range, and is advantageous as a hydrophobic scratch-resistant coating in optical devices, as shown in Figure 6.

The high hardness and high scratch resistance of the SiN_xO_y film make it widely used as a coating on transparent polymer and glass, which enhances the reliability of devices and has important research significance.

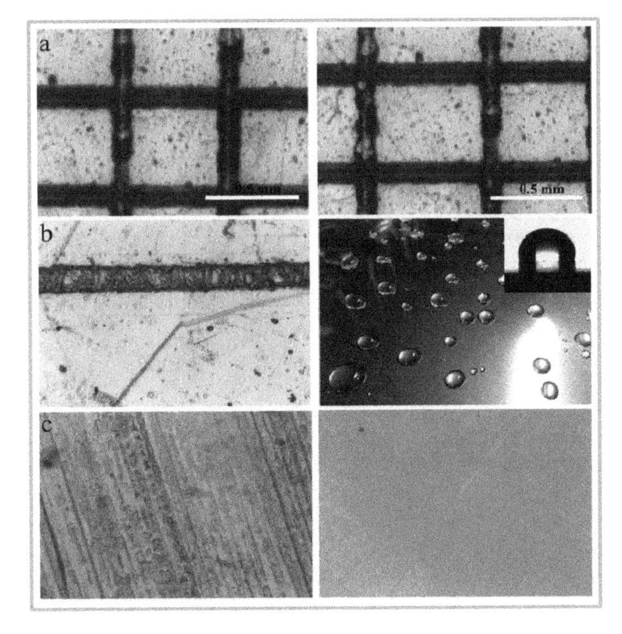

Figure 6. Scratch-resistant coating properties of the coated polyethylene terephthalate (PET) film. (**a**) Optical pictures of the coated films after cut-cross tape test (the left side shows a magnification of 1 time, and the right side shows a magnification of 10 times). (**b**) The left side shows the sample under the load of 1000 g left for the pencil scratch marks over 9 hours, and the water droplets on the coated PET film are shown in the right image. (**c**) Optical images of naked PET film. The PET film after coating is observed on the left and the right shows the coated PET film after wear test [149]. Copyright 2015, *Polym. Int.*

5. Conclusions

The SiN_xO_y film has important applications in optical devices, non-volatile memory, barrier materials, and scratch-resistant materials due to its good optoelectronic performance, mechanical strength, chemical stability, and barrier performance. The review focuses on the optical performance of SiN_xO_y, including luminescent performance and adjustable refractive index. This is a characteristic of SiN_xO_y film that is superior to conventional silicon-containing films such as Si_3N_4 and SiO_2. Moreover, several methods for preparing SiN_xO_y film, including chemical deposition methods, sputtering methods, and nitrogen oxidation methods, which are mainly used at present, are reviewed and compared, and the advantages and disadvantages of different methods will help us design and select the preparation methods correctly.

As a new type of thin film, SiN_xO_y film has a wide range of research space in many aspects. We believe that with the maturity and development of various preparation methods, the application prospect of SiN_xO_y film will be brighter.

Author Contributions: Y.S. and L.H. summarized and wrote the related research progress. R.L. revised the review article. All the authors participated in the discussion, writing, and revision of this review.

Funding: This work was supported by the Beijing Municipal Science and Technology Commission (Z181100004418004), the National Natural Science Foundation of China (61501039), the Beijing Natural Science Foundation (2162017), the Research and Development Program of Beijing Institute of Graphic Communication (Ec201808, Ea201803), the Programme of Key Technologies of Thin Film Printing Electronics (04190118002/042), and the Programme of Flexible Printed Electronic Materials and Technology Innovation (04190118002/092).

Conflicts of Interest: The authors declare no conflict of interest.

References

1. Iwase, Y.; Horie, Y.; Daiko, Y.; Honda, S.; Iwamoto, Y. Synthesis of a Novel Polyethoxysilsesquiazane and Thermal Conversion into Ternary Silicon Oxynitride Ceramics with Enhanced Thermal Stability. *Materials* **2017**, *10*, 1391. [CrossRef] [PubMed]
2. Meziani, S.; Moussi, A.; Mahiou, L.; Outemzabet, R. Compositional analysis of silicon oxide/silicon nitride thin films. *Mater. Sci.* **2016**, *34*, 315–321. [CrossRef]
3. Larker, R. Reaction Sintering and Properties of Silicon Oxynitride Densified by Hot Isostatic Pressing. *J. Am. Ceram. Soc.* **1992**, *75*, 62–66. [CrossRef]
4. Ohashi, M.; Tabata, H.; Kanzaki, S. High-temperature flexural strength of hot-pressed silicon oxynitride ceramics. *J. Mater. Sci. Lett.* **1988**, *7*, 339–340. [CrossRef]
5. Li, S.Q.; Pei, Y.C.; Yu, C.Q.; Li, J.L. Mechanical and Dielectric Properties of Porous Si_2N_2O-Si_3N_4 in Situ Composites. *Ceram. Int.* **2009**, *35*, 1851–1854.
6. Ohashi, M.; Kanzaki, S.; Tabata, H. Processing, Mechanical Properties, and Oxidation Behavior of Silicon Oxynitride Ceramics. *J. Am. Ceram. Soc.* **1991**, *74*, 109–114. [CrossRef]
7. Ohashi, M.; Kanzaki, S.; Tabata, H. Effect of additives on some properties of silicon oxynitride ceramics. *J. Mater. Sci.* **1991**, *26*, 2608–2614. [CrossRef]
8. Rocabois, P.; Chatillon, C.; Bernard, C. Thermodynamics of the Si-O-N System: II, Stability of Si_2N_2O(s) by High-Temperature Mass Spectrometric Vaporization. *J. Am. Ceram. Soc.* **1996**, *79*, 1361–1365. [CrossRef]
9. Ikeda, K.; Saperstein, R.E.; Alic, N.; Fainman, Y. Thermal and Kerr nonlinear properties of plasma-deposited silicon nitride/silicon dioxide waveguides. *Opt. Express* **2008**, *16*, 12987–12994. [CrossRef]
10. Feng, Y.; Gong, H.; Zhang, Y.; Wang, X.; Che, S.; Zhao, Y.; Guo, X. Effect of BN content on the mechanical and dielectric properties of porous BNp/Si_3N_4 ceramics. *Ceram. Int.* **2016**, *42*, 661–665. [CrossRef]
11. Lee, S.J.; Baek, S. Effect of SiO_2 content on the microstructure, mechanical and dielectric properties of Si_3N_4 ceramics. *Ceram. Int.* **2016**, *42*, 9921–9925. [CrossRef]
12. Philipp, H.; Andersen, K.; Svendsen, W.E.; Ou, H. Amorphous silicon rich silicon nitride optical waveguides for high density integrated optics. *Electron. Lett.* **2004**, *40*, 419–421. [CrossRef]
13. Wang, J.; Zhang, L.; Chen, Y.; Geng, Y.; Hong, X.; Li, X.; Cheng, Z.; Zhenzhou, C. Saturable absorption in graphene-on-waveguide devices. *Appl. Phys. Express* **2019**, *12*, 032003. [CrossRef]

14. Wang, J.Q.; Cheng, Z.; Chen, Z.; Xu, J.-B.; Tsang, H.K.; Shu, C. Graphene photodetector integrated on silicon nitride waveguide. *J. Appl. Phys.* **2015**, *117*, 144504. [CrossRef]
15. Levy, J.S.; Saha, K.; Okawachi, Y.; Foster, M.A.; Gaeta, A.L.; Lipson, M. High-Performance Silicon-Nitride-Based Multiple-Wavelength Source. *IEEE Photonics Technol. Lett.* **2012**, *24*, 1375–1377. [CrossRef]
16. Gruhler, N.; Benz, C.; Jang, H.; Ahn, J.-H.; Danneau, R.; Pernice, W.H.P. High-quality Si_3N_4 circuits as a platform for graphene-based nanophotonic devices. *Opt. Express* **2013**, *21*, 31678–31689. [CrossRef] [PubMed]
17. Bhatt, V.; Chandra, S. Silicon dioxide films by RF sputtering for microelectronic and MEMS applications. *J. Micromech. Microeng.* **2007**, *17*, 1066–1077. [CrossRef]
18. Ho, S.-S.; Rajgopal, S.; Mehregany, M. Thick PECVD silicon dioxide films for MEMS devices. *Sens. Actuators A Phys.* **2016**, *240*, 1–9. [CrossRef]
19. Chen, J.; Zheng, Y.; Xue, C.; Zhang, C.; Chen, Y. Filtering effect of SiO_2 optical waveguide ring resonator applied to optoelectronic oscillator. *Opt. Express* **2018**, *26*, 12638–12647. [CrossRef]
20. Brown, D.M.; Gray, P.V.; Heumann, F.K.; Philipp, H.R.; Taft, E.A. Properties of $Si_xO_yN_z$ Films on Si. *J. Electrochem. Soc.* **1968**, *115*, 311–317. [CrossRef]
21. Gunning, W.J.; Hall, R.L.; Woodberry, F.J.; Southwell, W.H.; Gluck, N.S. Codeposition of continuous composition rugate filters. *Appl. Opt.* **1989**, *28*, 2945–2948. [CrossRef] [PubMed]
22. Lipiński, M.; Kluska, S.; Czternastek, H.; Zięba, P. Graded SiO_xN_y layers as antireflection coatings for solar cells application. In Proceedings of the 8th International Conference on Intermolecular and Magnetic Interaction in Matter, Nałeczów, Poland, 8–10 September 2005; pp. 1009–1016.
23. Hayafuji, Y.; Kajiwara, K. Nitridation of Silicon and Oxidized-Silicon. *J. Electrochem. Soc.* **1982**, *129*, 2102–2108. [CrossRef]
24. Sombrio, G.; Franzen, P.L.; Maltez, R.; Matos, L.G.; Pereira, M.B.; Boudinov, H. Photoluminescence from SiN_xO_y films deposited by reactive sputtering. *J. Phys. D Appl. Phys.* **2013**, *46*, 235106. [CrossRef]
25. Temple-Boyer, P.; Hajji, B.; Alay, J.; Morante, J.; Martinez, A. Properties of SiO_xN_y films deposited by LPCVD from $SiH_4/N_2O/NH_3$ gaseous mixture. *Sens. Actuators A Phys.* **1999**, *74*, 52–55. [CrossRef]
26. Augustine, B.H.; Hu, Y.Z.; Irene, E.A.; McNeil, L.E. An annealing study of luminescent amorphous silicon-rich silicon oxynitride thin films. *Appl. Phys. Lett.* **1995**, *67*, 3694–3696. [CrossRef]
27. Dong, J.; Du, P.; Zhang, X. Characterization of the Young's modulus and residual stresses for a sputtered silicon oxynitride film using micro-structures. *Thin Solid Films* **2013**, *545*, 414–418. [CrossRef]
28. Lim, J.W.; Jin, C.K.; Lim, K.Y.; Lee, Y.J.; Kim, S.-R.; Choi, B.-I.; Kim, T.W.; Kim, D.H.; Hwang, D.K.; Choi, W.K. Transparent high-performance SiO_xN_y/SiO_x barrier films for organic photovoltaic cells with high durability. *Nano Energy* **2017**, *33*, 12–20. [CrossRef]
29. Trinh, T.T.; Jang, K.; Nguyen, V.D.; Dao, V.A.; Yi, J. Role of SiO_xN_y surface passivation layer on stability improvement and kink effect reduction of ELA poly silicon thin film transistors. *Microelectron. Eng.* **2016**, *164*, 14–19. [CrossRef]
30. Park, J.-H.; Shin, M.-H.; Yi, J.-S. The Characteristics of Transparent Non-Volatile Memory Devices Employing Si-Rich SiOX as a Charge Trapping Layer and Indium-Tin-Zinc-Oxide. *Nanomaterials* **2019**, *9*, 784. [CrossRef]
31. Soman, A.; Antony, A. Broad range refractive index engineering of Si_xN_y and SiO_xN_y thin films and exploring their potential applications in crystalline silicon solar cells. *Mater. Chem. Phys.* **2017**, *197*, 181–191. [CrossRef]
32. Himmler, A.; Fahland, M.; Linß, V. Roll-to-roll deposition of silicon oxynitride layers on polymer films using a rotatable dual magnetron system. *Surf. Coat. Technol.* **2018**, *336*, 123–127. [CrossRef]
33. Rats, D.; Martinu, L.; Von Stebut, J. Mechanical properties of plasma-deposited SiO_xN_y coatings on polymer substrates using low load carrying capacity techniques. *Surf. Coat. Technol.* **2000**, *123*, 36–43. [CrossRef]
34. Wang, J.; Bouchard, J.P.; Hart, G.A.; Oudard, J.F.; Paulson, C.A.; Sachenik, P.A.; Price, J.J. Silicon oxynitride based scratch resistant anti-reflective coatings. In Proceedings of the Advanced Optics for Defense Applications: UV through LWIR III (2018), Orlando, FL, USA, 15–16 April 2018; Volume 10627.
35. Vivaldo, I.; Moreno, M.; Torres, A.; Ambrosio, R.; Rosales, P.; Carlos, N.; Calleja, W.; Monfil, K.; Benítez, A. A comparative study of amorphous silicon carbide and silicon rich oxide for light emission applications. *J. Lumin.* **2017**, *190*, 215–220. [CrossRef]
36. Shcherban, N.D. Review on synthesis, structure, physical and chemical properties and functional characteristics of porous silicon carbide. *J. Ind. Eng. Chem.* **2017**, *50*, 15–28. [CrossRef]

37. Chang, I.M.; Pan, S.C.; Chen, Y.F. Light-induced degradation on porous silicon. *Phys. Rev. B* **1993**, *48*, 8747–8750. [CrossRef] [PubMed]
38. Dudel, F.P.; Gole, J.L. Stabilization of the photoluminescence from porous silicon: The competition between photoluminescence and dissolution. *J. Appl. Phys.* **1997**, *82*, 402–406. [CrossRef]
39. Qin, G.; Jia, Y. Mechanism of the visible luminescence in porous silicon. *Solid State Commun.* **1993**, *86*, 559–563. [CrossRef]
40. Qin, G.G.; Li, Y.J. Photoluminescence mechanism model for oxidized porous silicon and nanoscale-silicon-particle-embedded silicon oxide. *Phys. Rev. B* **2003**, *68*, 085309. [CrossRef]
41. Torchynska, T.; Cano, A.D.; Rodríguez, M.M.; Khomenkova, L.; Khomenkova, L. Hot carriers and excitation of Si/SiOx interface defect photoluminescence in Si nanocrystallites. *Phys. B Condens. Matter* **2003**, *340*, 1113–1118. [CrossRef]
42. Wolkin, M.V.; Jorne, J.; Fauchet, P.M.; Allan, G.; Delerue, C. Electronic States and Luminescence in Porous Silicon Quantum Dots: The Role of Oxygen. *Phys. Rev. Lett.* **1999**, *82*, 197–200. [CrossRef]
43. Prokes, S.M.; Glembocki, O.J. Role of interfacial oxide-related defects in the red-light emission in porous silicon. *Phys. Rev. B* **1994**, *49*, 2238–2241. [CrossRef] [PubMed]
44. Godinho, V.; De Haro, M.J.; García-López, J.; Goossens, V.; Terryn, H.; Delplancke-Ogletree, M.; Fernández, A.; De Haro, M.C.J. SiO_xN_y thin films with variable refraction index: Microstructural, chemical and mechanical properties. *Appl. Surf. Sci.* **2010**, *256*, 4548–4553. [CrossRef]
45. Godinho, V.; Rojas, T.C.; Fernández, A. Magnetron sputtered a-SiO_xN_y thin films: A closed porous nanostructure with controlled optical and mechanical properties. *Microporous Mesoporous Mater.* **2012**, *149*, 142–146. [CrossRef]
46. Jou, S.; Liaw, I.-C.; Cheng, Y.-C.; Li, C.-H. Light emission of silicon oxynitride films prepared by reactive sputtering of silicon. *J. Lumin.* **2013**, *134*, 853–857. [CrossRef]
47. Jan, V.; Michael, G.; Sebastian, G.; Daniel, H.; Margit, Z. Photoluminescence performance limits of Si nanocrystals in silicon oxynitride matrices. *J. Appl. Phys.* **2017**, *122*, 144303.
48. Goncharova, L.V.; Nguyen, P.H.; Karner, V.L.; D'Ortenzio, R.; Chaudhary, S.; Mokry, C.R.; Simpson, P.J. Si quantum dots in silicon nitride: Quantum confinement and defects. *J. Appl. Phys.* **2015**, *118*, 224302. [CrossRef]
49. Augustine, B.H.; Irene, E.A. Visible light emission from thin films containing Si, O, N, and H. *J. Appl. Phys.* **1995**, *78*, 4020–4030. [CrossRef]
50. Claassen, W.A.P.; Pol, H.A.J.T.; Goemans, A.H.; Kuiper, A.E.T. Characterization of Silicon-Oxynitride Films Deposited by Plasma-Enhanced CVD. *J. Electrochem. Soc.* **1986**, *133*, 1458–1464. [CrossRef]
51. Alayo, M.; Pereyra, I.; Scopel, W.; Fantini, M.; Scopel, W.; Fantini, M. On the nitrogen and oxygen incorporation in plasma-enhanced chemical vapor deposition (PECVD) SiO_xN_y films. *Thin Solid Films* **2002**, *402*, 154–161. [CrossRef]
52. Zhang, P.; Zhang, L.; Wu, Y.; Wang, S.; Ge, X. High photoluminescence quantum yields generated from N-Si-O bonding states in amorphous silicon oxynitride films. *Opt. Express* **2018**, *26*, 31617–31625. [CrossRef]
53. Lee, D.K.; Shin, H.J.; Sohn, S.H. Characteristics of Silicon Oxynitride Barrier Films Grown on Poly(ethylene naphthalate) by Ion-Beam-Assisted Deposition. *Jpn. J. Appl. Phys.* **2010**, *49*, 05EA14. [CrossRef]
54. Labbé, C.; An, Y.T.; Zatryb, G.; Portier, X.; Podhorodecki, A.; Marie, P.; Frilay, C.; Cardin, J.; Gourbilleau, F. Structural and emission properties of Tb^{3+}-doped nitrogen-rich silicon oxynitride films. *Nanotechnology* **2017**, *28*, 115710. [CrossRef] [PubMed]
55. Ehré, F.; Labbé, C.; Dufour, C.; Jadwisienczak, W.M.; Weimmerskirch-Aubatin, J.; Portier, X.; Doualan, J.L.; Cardin, J.; Richard, A.L.; Ingram, D.C.; et al. The Nitrogen concentration effect on Ce doped SiO_xN_y emission: Towards optimized Ce3+ for DEL applications. *Nanoscale* **2018**, *10*, 3823–3837. [CrossRef] [PubMed]
56. Huang, R.; Lin, Z.; Guo, Y.; Song, C.; Wang, X.; Lin, H.; Xu, L.; Song, J.; Li, H. Bright red, orange-yellow and white switching photoluminescence from silicon oxynitride films with fast decay dynamics. *Opt. Mater. Express* **2014**, *4*, 205–212. [CrossRef]
57. Steveler, E.; Rinnert, H.; Vergnat, M. Photoluminescence of erbium in SiO_xN_y alloys annealed at high temperature. *J. Alloy. Compd.* **2014**, *593*, 56–60. [CrossRef]
58. Kim, D.S.; Yoon, S.G.; Jang, G.E.; Suh, S.J.; Kim, H.; Yoon, D.H. Refractive index properties of SiN thin films and fabrication of SiN optical waveguide. *J. Electroceram.* **2006**, *17*, 315–318. [CrossRef]

59. Zhu, Y.; Gu, P.F.; Shen, W.D.; Zou, T. Study of Silicon Oxynitride Film Deposited by RF Magnetron Sputtering. *Acta Opt. Sin.* **2005**, *25*, 567.
60. Hu, J.L.; Hang, L.X.; Zhou, S. Preparation of Low Refractive Index Optical Thin Film by PECVD Technology. *Surf. Technol.* **2013**, *42*, 95–97.
61. Xue, J.; Hang, L.X.; Lin, H.X. Study on controlled refractive index of optical thin films by PECVD. *Opt. Tech.* **2014**, *40*, 353–356.
62. Hänninen, T.; Schmidt, S.; Jensen, J.; Hultman, L.; Högberg, H. Silicon oxynitride films deposited by reactive high power impulse magnetron sputtering using nitrous oxide as a single-source precursor. *J. Vac. Sci. Technol. A* **2015**, *33*, 05E121. [CrossRef]
63. Farhaoui, A.; Bousquet, A.; Smaali, R.; Moreau, A.; Centeno, E.; Cellier, J.; Bernard, C.; Rapegno, R.; Réveret, F.; Tomasella, E. Reactive gas pulsing sputtering process, a promising technique to elaborate silicon oxynitride multilayer nanometric antireflective coatings. *J. Phys. D Appl. Phys.* **2017**, *50*, 015306. [CrossRef]
64. Nakanishi, Y.; Kato, K.; Omoto, H.; Tomioka, T.; Takamatsu, A. Stable deposition of silicon oxynitride thin films with intermediate refractive indices by reactive sputtering. *Thin Solid Films* **2012**, *520*, 3862–3864. [CrossRef]
65. Yadav, A.; Polji, R.H.; Singh, V.; Dubey, S.; Rao, T.G. Synthesis of buried silicon oxynitride layers by ion implantation for silicon-on-insulator (SOI) structures. *Nucl. Instrum. Methods Phys. Res. Sect. B Beam Interact. Mater. Atoms* **2006**, *245*, 475–479. [CrossRef]
66. Jasinski, J.M.; Meyerson, B.S.; Scott, B.A. Mechanistic Studies of Chemical Vapor Deposition. *Annu. Rev. Phys. Chem.* **1987**, *38*, 109–140. [CrossRef]
67. Ouvry, S. Random Aharonov–Bohm vortices and some exactly solvable families of integrals. *J. Stat. Mech. Theory Exp.* **2005**, *2005*, 09004. [CrossRef]
68. Pandey, R.; Patil, L.; Bange, J.; Patil, D.; Mahajan, A.; Patil, D.; Gautam, D.; Mahajan, A. Growth and characterization of SiON thin films by using thermal-CVD machine. *Opt. Mater.* **2004**, *25*, 1–7. [CrossRef]
69. Balaji, N.; Nguyen, H.T.T.; Park, C.; Ju, M.; Raja, J.; Chatterjee, S.; Jeyakumar, R.; Yi, J. Electrical and optical characterization of SiO_xN_y and SiO_2 dielectric layers and rear surface passivation by using SiO_2/SiO_xN_y stack layers with screen printed local Al-BSF for c-Si solar cells. *Curr. Appl. Phys.* **2018**, *18*, 107–113. [CrossRef]
70. Piqueras, J.; Hernandez, M.J.; Garrido, J.; Martinez, J. Compositional and electrical properties of ECR-CVD silicon oxynitrides. *Semicond. Sci. Technol.* **1997**, *12*, 927–932.
71. Ramírez, J.M.; Wójcik, J.; Berencén, Y.; Ruiz-Caridad, A.; Estrade, S.; Peirò, F.; Mascher, P.; Garrido, B. Amorphous sub-nanometre Tb-doped SiO_xN_y /SiO_2 superlattices for optoelectronics. *Nanotechnology* **2015**, *26*, 85203. [CrossRef]
72. Bang, S.-H.; Suk, J.-H.; Kim, K.-S.; Park, J.-H.; Hwang, N.-M. Effects of radio frequency power and gas ratio on barrier properties of SiO_xN_y films deposited by inductively coupled plasma chemical vapor deposition. *Thin Solid Films* **2019**, *669*, 108–113. [CrossRef]
73. Chen, W.; Lee, A.; Deng, W.; Liu, K. The implementation of neural network for semiconductor PECVD process. *Expert Syst. Appl.* **2007**, *32*, 1148–1153. [CrossRef]
74. Varanasi, V.G.; Ilyas, A.; Velten, M.F.; Shah, A.; Lanford, W.A.; Aswath, P.B. Role of Hydrogen and Nitrogen on the Surface Chemical Structure of Bioactive Amorphous Silicon Oxynitride Films. *J. Phys. Chem. B* **2017**, *121*, 8991–9005. [CrossRef] [PubMed]
75. Park, S.; Park, H.; Kim, D.; Nam, J.; Yang, J.; Lee, D.; Min, B.K.; Kim, K.N.; Park, S.J.; Kim, S.; et al. Continuously deposited anti-reflection double layer of silicon nitride and silicon oxynitride for selective emitter solar cells by PECVD. *Curr. Appl. Phys.* **2017**, *17*, 517–521. [CrossRef]
76. Kalisz, M.; Szymanska, M.; Dębowska, A.K.; Michalak, B.; Brzozowska, E.; Górska, S.; Smietana, M. Influence of biofunctionalization process on properties of silicon oxynitride substrate layer. *Surf. Interface Anal.* **2014**, *46*, 1086–1089. [CrossRef]
77. Kijaszek, W.; Oleszkiewicz, W.; Zakrzewski, A.; Patela, S.; Tłaczała, M. Investigation of optical properties of silicon oxynitride films deposited by RF PECVD method. *Mater. Sci. Pol.* **2016**, *34*, 868–871. [CrossRef]
78. Okazaki, K.; Nishi, H.; Tsuchizawa, T.; Hiraki, T.; Ishikawa, Y.; Wada, K.; Yamamoto, T.; Yamada, K. Optical coupling between SiO_xN_y waveguide and Ge mesa structures for bulk-Si photonics platform. In Proceedings of the 12th International Conference on Group IV Photonics (GFP), Vancouver, BC, Canada, 26–28 August 2015; pp. 122–123.

79. Wood, R. ECR-PECVD silicon oxynitride thin films for flat panel displays. In Proceedings of the Opto-Canada: SPIE Regional Meeting on Optoelectronics, Photonics, and Imaging, Ottawa, ON, Canada, 9–10 May 2002; Volume 103133N.
80. Meng, X.S.; Yuan, J.; Ma, Q.S.; Ge, M.Z.; Yang, H. Study on the Preparation Methods of Silicon Oxynitride Thin Films. *Mater. Sci. Eng.* **1999**, *17*, 10–13.
81. Kaghouche, B.; Mansour, F.; Molliet, C.; Rousset, B.; Temple-Boyer, P. Investigation on optical and physico-chemical properties of LPCVD SiO_xN_y thin films. *Eur. Phys. J. Appl. Phys.* **2014**, *66*, 20301. [CrossRef]
82. Gan, Z.; Wang, C.; Chen, Z. Material Structure and Mechanical Properties of Silicon Nitride and Silicon Oxynitride Thin Films Deposited by Plasma Enhanced Chemical Vapor Deposition. *Surfaces* **2018**, *1*, 59–72. [CrossRef]
83. Pernas, P.; Ruiz, E.; Garrido, J.; Piqueras, J.; Pászti, F.; Climent-Font, A.; Lifante, G.; Cantelar, E. Silicon Oxynitride ECR-PECVD Films for Integrated Optics. *Mater. Sci. Forum* **2005**, *480*, 149–154. [CrossRef]
84. Rangarajan, B.; Kovalgin, A.Y.; Schmitz, J. Deposition and properties of silicon oxynitride films with low propagation losses by inductively coupled PECVD at 150 °C. *Surf. Coat. Technol.* **2013**, *230*, 46–50. [CrossRef]
85. Halova, E.; Alexandrova, S.; Szekeres, A.; Modreanu, M. LPCVD-silicon oxynitride films: Interface properties. *Microelectron. Reliab.* **2005**, *45*, 982–985. [CrossRef]
86. Faller, F.; Hurrle, A. High-temperature CVD for crystalline-silicon thin-film solar cells. *IEEE Trans. Electron Devices* **1999**, *46*, 2048–2054. [CrossRef]
87. Lebland, F.; Licoppe, C.; Gao, Y.; Nissim, Y.; Rigo, S. Rapid thermal chemical vapour deposition of SiO_xN_y films. *Appl. Surf. Sci.* **1992**, *54*, 125–129. [CrossRef]
88. Watanabe, J.; Hanabusa, M. Photochemical vapor deposition of silicon oxynitride films by deuterium lamp. *J. Mater. Res.* **1989**, *4*, 882–885. [CrossRef]
89. Bergonzo, P.; Kogelschatz, U.; Boyd, I.W. Photo-CVD of dielectric materials by pseudo-continuous excimer sources. In Proceedings of the Laser-Assisted Fabrication of Thin Films and Microstructures, Quebec City, QC, Canada, 16–20 August 1993; pp. 174–181.
90. Zhang, Z.B.; Luo, Y.M.; Xu, C.H. Research Progress on Silicon Oxynitride Films. *Mater. Rev.* **2009**, *11*, 110–114.
91. Hallam, B.; Tjahjono, B.; Wenham, S. Effect of PECVD silicon oxynitride film composition on the surface passivation of silicon wafers. *Sol. Energy Mater. Sol. Cells* **2012**, *96*, 173–179. [CrossRef]
92. Li, X.B.; Zhang, F.H.; Mou, Q. Effect of Plasma Enhanced Chemical Vapor Deposition Parameters on Characteristics of Silicon nitride Film. *Mater. Prot.* **2006**, *39*, 12–16.
93. Mubarak, A.; Hamzah, E.; Toff, M.R.M. Review of physical vapour deposition (PVD) techniques for hard coating. *J. Mek.* **2005**, *20*, 42–51.
94. Mattox, D.M. Ion plating—Past, present and future. *Surf. Coat. Technol.* **2000**, *133*, 517–521. [CrossRef]
95. Nakao, H.; Hori, K.; Iwazaki, Y.; Ueno, T. Investigation of High Current Tetracene-TFT Using Surface Nitrided SiO_2 Gate Insulator Film. *ECS Trans.* **2016**, *75*, 81–85. [CrossRef]
96. Baptista, A.; Silva, F.; Porteiro, J.; Míguez, J.; Pinto, G. Sputtering Physical Vapour Deposition (PVD) Coatings: A Critical Review on Process Improvement and Market Trend Demands. *Coatings* **2018**, *8*, 402. [CrossRef]
97. Mattox, D.M. *The Foundations of Vacuum Coating Technology*; Noyes Publications: Norwich, UK, 2003; ISBN 978-0-8155-1495-4.
98. Yu, D.H.; Wang, C.Y.; Cheng, X.L.; Song, Y.X. Recent development of magnetron sputtering processes. *Vacuum* **2009**, *46*, 19–25.
99. Tang, C.J.; Jiang, C.C.; Tien, C.L.; Sun, W.C.; Lin, S.C. Optical, structural, and mechanical properties of silicon oxynitride films prepared by pulsed magnetron sputtering. *Appl. Opt.* **2017**, *56*, C168–C174. [CrossRef] [PubMed]
100. Šimurka, L.; Čtvrtlík, R.; Roch, T.; Turutoglu, T.; Erkan, S.; Tomáštík, J.; Bange, K. Effect of deposition conditions on physical properties of sputtered silicon oxynitride thin films on float glass. *Int. J. Appl. Glas. Sci.* **2018**, *9*, 403–412. [CrossRef]
101. Li, J.; Shen, G.; Chen, W.; Li, Z.; Hong, R. Preparation of SiNx multilayer films by mid-frequency magnetron sputtering for crystalline silicon solar cells. *Mater. Sci. Semicond. Process.* **2017**, *59*, 40–44. [CrossRef]
102. Arnell, R.; Kelly, P. Recent advances in magnetron sputtering. *Surf. Coat. Technol.* **1999**, *112*, 170–176. [CrossRef]

103. Sangines, R.; Abundiz-Cisneros, N.; Diliegros-Godines, C.; Machorro-Mejia, R.; Hernández-Utrera, O. Plasma emission spectroscopy and its relation to the refractive index of silicon nitride thin films deposited by reactive magnetron sputtering. *J. Phys. D Appl. Phys.* **2018**, *51*, 095203. [CrossRef]
104. Yount, J.; Lenahan, P. Bridging nitrogen dangling bond centers and electron trapping in amorphous NH3-nitrided and reoxidized nitrided oxide films. *J. Non-Cryst. Solids* **1993**, *164*, 1069–1072. [CrossRef]
105. Hori, T.; Naito, Y.; Iwasaki, H.; Esaki, H. Interface states and fixed charges in nanometer-range thin nitrided oxides prepared by rapid thermal annealing. *IEEE Electron Device Lett.* **1986**, *7*, 669–671. [CrossRef]
106. Itakura, A.; Shimoda, M.; Kitajima, M. Surface stress relaxation in SiO_2 films by plasma nitridation and nitrogen distribution in the film. *Appl. Surf. Sci.* **2003**, *216*, 41–45. [CrossRef]
107. He, S.C.; Jia, Q.M.; Su, H.Y.; Ma, A.H.; Shan, S.Y.; Hu, T.D. Review on the Research Progress of Silicon Oxynitride Materials. *Mater. Rev.* **2016**, *8*, 80–84.
108. Yamamoto, M.; Matsumae, T.; Kurashima, Y.; Takagi, H.; Suga, T.; Itoh, T.; Higurashi, E. Growth Behavior of Au Films on SiO_2 Film and Direct Transfer for Smoothing Au Surfaces. *Int. J. Autom. Technol.* **2019**, *13*, 254–260. [CrossRef]
109. Chen, Z.; Wang, H.; Wang, Y.; Lv, R.; Yang, X.; Wang, J.; Li, L.; Ren, W. Improved optical damage threshold graphene Oxide/SiO_2 absorber fabricated by sol-gel technique for mode-locked erbium-doped fiber lasers. *Carbon* **2019**, *144*, 737–744. [CrossRef]
110. Lai, C.H.; Lin, B.C.; Chang, K.M.; Hsieh, K.Y.; Lai, Y.L. A novel process for forming an ultra-thin oxynitride film with high nitrogen topping. *J. Phys. Chem. Solids* **2008**, *69*, 456–460. [CrossRef]
111. Wong, H.; Yang, B.; Cheng, Y. Chemistry of silicon oxide annealed in ammonia. *Appl. Surf. Sci.* **1993**, *72*, 49–54. [CrossRef]
112. Rebib, F.; Tomasella, E.; Dubois, M.; Cellier, J.; Sauvage, T.; Jacquet, M. SiO_xN_y thin films deposited by reactive sputtering: Process study and structural characterisation. *Thin Solid Films* **2007**, *515*, 3480–3487. [CrossRef]
113. Dasgupta, A.; Takoudis, C.G. Growth kinetics of thermal silicon oxynitridation in nitric oxide ambient. *J. Appl. Phys.* **2003**, *93*, 3615–3618. [CrossRef]
114. Ramírez, J.M.; Ruiz-Caridad, A.; Wojcik, J.; Gutierrez, A.M.; Estradé, S.; Peiró, F.; Sanchís, P.; Mascher, P.; Garrido, B. Luminescence properties of Ce^{3+} and Tb^{3+} co-doped SiO_xN_y thin films: Prospects for color tunability in silicon-based hosts. *J. Appl. Phys.* **2016**, *119*, 113108. [CrossRef]
115. Zhu, M.; Shi, X.; Chen, P.; Liu, W.; Wong, M.; Lin, C.; Chu, P.K. Formation of silicon on plasma synthesized SiO_xN_y and reaction mechanism. *Appl. Surf. Sci.* **2005**, *243*, 89–95. [CrossRef]
116. Lee, H.-I.; Park, J.-B.; Xianyu, W.; Kim, K.; Chung, J.G.; Kyoung, Y.K.; Byun, S.; Yang, W.Y.; Park, Y.Y.; Kim, S.M.; et al. Degradation by water vapor of hydrogenated amorphous silicon oxynitride films grown at low temperature. *Sci. Rep.* **2017**, *7*, 14146. [CrossRef]
117. Dan, Y.P.; Yue, R.F.; Wang, Y.; Liu, L.T. Optical Properties of Silicon Oxynitride Films and Applications in Optical Waveguides. In Proceedings of the 1th National Conference on Nanotechnology and Applications (2008), Xiamen, China, 27–30 November 2000; pp. 98–101.
118. Lin, Z.; Chen, K.; Zhang, P.; Xu, J.; Dong, H.; Li, W.; Ji, Y.; Huang, X. The Role of N-Si-O Defect States in Optical Gain from an a-SiN_xO_y/SiO_2 Waveguide and in Light Emission from an n-a-SiN_xO_y/p-Si Heterojunction LED. *Phys. Status Solidi (A)* **2018**, *215*, 1700750. [CrossRef]
119. Shahpanah, M.; Mehrabian, S.; Abbasi-Firouzjah, M.; Shokri, B. Improving the oxygen barrier properties of PET polymer by radio frequency plasma-polymerized SiO_xN_y thin film. *Surf. Coat. Technol.* **2019**, *358*, 91–97. [CrossRef]
120. Madogni, V.I.; Kounouhéwa, B.; Akpo, A.; Agbomahéna, M.; Hounkpatin, S.A.; Awanou, C.N. Comparison of degradation mechanisms in organic photovoltaic devices upon exposure to a temperate and a subequatorial climate. *Chem. Phys. Lett.* **2015**, *640*, 201–214. [CrossRef]
121. Yu, D.; Yang, Y.-Q.; Chen, Z.; Tao, Y.; Liu, Y.-F. Recent progress on thin-film encapsulation technologies for organic electronic devices. *Opt. Commun.* **2016**, *362*, 43–49. [CrossRef]
122. Oku, T.; Okumura, M.; Totsuka, M.; Shiga, T.; Takemi, M. Surface protection mechanism of insulating films at junctions in compound semiconductor devices. In Proceedings of the 15th International Workshop on Junction Technology (IWJT), Kyoto, Japan, 11–12 June 2015; pp. 71–76.

123. Satoh, R.; Ro, T.; Heo, C.J.; Lee, G.H.; Xianyu, W.; Park, Y.; Park, J.; Lim, S.J.; Leem, D.S.; Bulliard, X.; et al. Bi-layered metal-oxide thin films processed at low-temperature for the encapsulation of highly stable organic photo-diode. *Org. Electron.* **2016**, *41*, 259–265. [CrossRef]
124. Shim, J.; Yoon, H.G.; Na, S.-H.; Kim, I.; Kwak, S. Silicon oxynitride gas barrier coatings on poly(ether sulfone) by plasma-enhanced chemical vapor deposition. *Surf. Coat. Technol.* **2008**, *202*, 2844–2849. [CrossRef]
125. Iwamori, S.; Gotoh, Y.; Moorthi, K. Characterization of silicon oxynitride gas barrier films. *Vacuum* **2002**, *68*, 113–117. [CrossRef]
126. Liu, C.C.; Chang, L.S. Gas permeation properties of silicon oxynitride thin films deposited on polyether sulfone by radio frequency magnetron reactive sputtering in various N_2 contents in atmosphere. *Thin Solid Films* **2015**, *594*, 35–39. [CrossRef]
127. Wen, S.; Xie, X.; Yan, Z.; Huang, T.; Zeng, Z. General memristor with applications in multilayer neural networks. *Neural Netw.* **2018**, *103*, 142–149. [CrossRef]
128. Wang, B.; Zou, F.; Cheng, J. A memristor-based chaotic system and its application in image encryption. *Optik* **2018**, *154*, 538–544. [CrossRef]
129. Liu, S.; Wang, Y.; Fardad, M.; Varshney, P.K. A Memristor-Based Optimization Framework for Artificial Intelligence Applications. *IEEE Circuits Syst. Mag.* **2018**, *18*, 29–44. [CrossRef]
130. Wrazien, S.J.; Zhao, Y.; Krayer, J.D.; White, M.H. Characterization of SONOS oxynitride nonvolatile semiconductor memory devices. *Solid State Electron.* **2003**, *47*, 885–891. [CrossRef]
131. Ielmini, D. Resistive switching memories based on metal oxides: Mechanisms, reliability and scaling. *Semicond. Sci. Technol.* **2016**, *31*, 063002. [CrossRef]
132. Chua, L. Memristor-The missing circuit element. *IEEE Trans. Circuit Theory* **1971**, *18*, 507–519. [CrossRef]
133. Chen, D.; Huang, S.; He, L. Effect of oxygen concentration on resistive switching behavior in silicon oxynitride film. *J. Semicond.* **2017**, *38*, 043002. [CrossRef]
134. Zhang, H.; Ma, Z.; Zhang, X.; Sun, Y.; Liu, J.; Xu, L.; Li, W.; Chen, K.; Feng, D. The Ultra-Low Power Performance of a-SiN_xO_y:H Resistive Switching Memory. *Phys. Status Solidi (A)* **2018**, *215*, 1700753. [CrossRef]
135. Wang, Z.; Joshi, S.; Savel'Ev, S.E.; Jiang, H.; Midya, R.; Lin, P.; Hu, M.; Ge, N.; Strachan, J.P.; Li, Z.; et al. Memristors with diffusive dynamics as synaptic emulators for neuromorphic computing. *Nat. Mater.* **2016**, *16*, 101–108. [CrossRef]
136. Baudzus, L.; Krummrich, P.M. Low Loss Electro-Optic Polymer Based Fast Adaptive Phase Shifters Realized in Silicon Nitride and Oxynitride Waveguide Technology. *Photonics* **2016**, *3*, 49. [CrossRef]
137. Wang, J.Q.; Cheng, Z.; Shu, C.; Tsang, H.K. Optical Absorption in Graphene-on-Silicon Nitride Microring Resonator. *IEEE Photon. Technol. Lett.* **2015**, *27*, 1765–1767. [CrossRef]
138. Jia, Y.; Dai, X.; Xiang, Y.; Fan, D. High quality factor silicon oxynitride optical waveguide ring resonators. *Opt. Mater.* **2018**, *85*, 138–142. [CrossRef]
139. De Ridder, R.; Warhoff, K.; Driessen, A.; Lambeck, P.; Albers, H. Silicon oxynitride planar waveguiding structures for application in optical communication. *IEEE J. Sel. Top. Quantum Electron.* **1998**, *4*, 930–937. [CrossRef]
140. Trenti, A.; Borghi, M.; Biasi, S.; Ghulinyan, M.; Ramiro-Manzano, F.; Pucker, G.; Pavesi, L. Thermo-optic coefficient and nonlinear refractive index of silicon oxynitride waveguides. *AIP Adv.* **2018**, *8*, 025311. [CrossRef]
141. Loka, C.; Lee, K.; Moon, S.W.; Choi, Y.; Lee, K.-S. Enhanced transmittance of sapphire by silicon oxynitride thin films annealed at high temperatures. *Mater. Lett.* **2018**, *213*, 354–357. [CrossRef]
142. Nguyen, H.T.T.; Balaji, N.; Park, C.; Triet, N.M.; Le, A.H.T.; Lee, S.; Jeon, M.; Oh, D.; Dao, V.A.; Yi, J. Al_2O_3 /SiON stack layers for effective surface passivation and anti-reflection of high efficiency n-type c-Si solar cells. *Semicond. Sci. Technol.* **2017**, *32*, 25005. [CrossRef]
143. Sahouane, N.; Necaibia, A.; Ziane, A.; Dabou, R.; Bouraiou, A.; Mostefaoui, M.; Rouabhia, A.; Mostefaoui, M. Realization and modeling of multilayer antireflection coatings for solar cells application. *Mater. Res. Express* **2018**, *5*, 065515. [CrossRef]
144. Parashar, P.K.; Komarala, V.K. Engineered optical properties of silver-aluminum alloy nanoparticles embedded in SiON matrix for maximizing light confinement in plasmonic silicon solar cells. *Sci. Rep.* **2017**, *7*, 12520. [CrossRef]

145. Qu, L.; Tang, L.; Bei, R.; Zhao, J.; Chi, Z.; Liu, S.; Chen, X.; Aldred, M.P.; Zhang, Y.; Xu, J. Flexible Multifunctional Aromatic Polyimide Film: Highly Efficient Photoluminescence, Resistive Switching Characteristic, and Electroluminescence. *ACS Appl. Mater. Interfaces* **2018**, *10*, 11430–11435. [CrossRef]
146. Lam, J.-Y.; Shih, C.-C.; Lee, W.-Y.; Chueh, C.-C.; Jang, G.-W.; Huang, C.-J.; Tung, S.-H.; Chen, W.-C. Bio-Based Transparent Conductive Film Consisting of Polyethylene Furanoate and Silver Nanowires for Flexible Optoelectronic Devices. *Macromol. Rapid Commun.* **2018**, *39*, 1800271. [CrossRef]
147. Lin, Y.-S.; Liao, Y.-H.; Hu, C.-H. Effects of N_2 addition on enhanced scratch resistance of flexible polycarbonate substrates by low temperature plasma-polymerized organo-silicon oxynitride. *J. Non-Cryst. Solids* **2009**, *355*, 182–192. [CrossRef]
148. Lin, Y.S.; Lai, Y.C.; Chen, J.H. Low temperature atmospheric pressure plasma-polymerized organosilicon oxynitride films enhance scratch resistance of flexible carbon fiber-reinforced polymer composites. In Proceedings of the International Symposium on Material Science and Engineering (ISMSE 2018), Seoul, South Korea, 19–21 January 2018; p. 020002.
149. Zhang, Z.B.; Shao, Z.H.; Luo, Y.M.; An, P.Y.; Zhang, M.Y.; Xu, C.H. Hydrophobic, transparent and hard silicon oxynitride coating from perhydropolysilazane. *Polym. Int.* **2015**, *64*, 971–978. [CrossRef]

Review

Towards On-Chip Self-Referenced Frequency-Comb Sources Based on Semiconductor Mode-Locked Lasers

Marcin Malinowski [1,*], Ricardo Bustos-Ramirez [1], Jean-Etienne Tremblay [2], Guillermo F. Camacho-Gonzalez [1], Ming C. Wu [2], Peter J. Delfyett [1,3] and Sasan Fathpour [1,3,*]

1 CREOL, The College of Optics and Photonics, University of Central Florida, Orlando, FL 32816, USA; ricardo.bustos@Knights.ucf.edu (R.B.-R.); gcamacho@Knights.ucf.edu (G.F.C.-G.); delfyett@creol.ucf.edu (P.J.D.)

2 Department of Electrical Engineering and Computer Sciences, University of California, Berkeley, CA 94720, USA; jetremblay@berkeley.edu (J.-E.T.); wu@eecs.berkeley.edu (M.C.W.)

3 Department of Electrical and Computer Engineering, University of Central Florida, Orlando, FL 32816, USA

* Correspondence: marcinmalinowski@knights.ucf.edu (M.M.); fathpour@creol.ucf.edu (S.F.)

Received: 14 May 2019; Accepted: 5 June 2019; Published: 11 June 2019

Abstract: Miniaturization of frequency-comb sources could open a host of potential applications in spectroscopy, biomedical monitoring, astronomy, microwave signal generation, and distribution of precise time or frequency across networks. This review article places emphasis on an architecture with a semiconductor mode-locked laser at the heart of the system and subsequent supercontinuum generation and carrier-envelope offset detection and stabilization in nonlinear integrated optics.

Keywords: frequency combs; heterogeneous integration; second-harmonic generation; supercontinuum; integrated photonics; silicon photonics; mode-locked lasers; nonlinear optics

1. Introduction

The field of integrated photonics aims at harnessing optical waves in submicron-scale devices and circuits, for applications such as transmitting information (communications) and gathering information about the environment (imaging, spectroscopy, etc.). The applications pertaining to the transmission of information include optical transceivers [1], interconnects for high-performance computing [2,3], optical switches [4], and perhaps neural networks [5]. The sensing applications can be long-range, e.g., LiDAR [6], or short-range, e.g., absorption, Raman or florescence spectroscopy [7]. This includes spectroscopy of atomic vapors [8], which is essential in realizing a miniature atomic clock. Somewhere in between these two ranges is the quest to design an on-chip frequency-stabilized comb source. The unprecedented frequency stability, coupled with the broad comb bandwidth, has had such an impact that two of its inventors, John L. Hall and Theodor W. Hänsch, are awarded the 2005 Nobel Prize in Physics for their "contributions to the development of laser-based precision spectroscopy, including the optical frequency-comb technique".

This review paper summarizes efforts in developing an on-chip stabilized broadband supercontinuum source in the context of above sensing goals. The paper discusses in detail the applications, the physics of supercontinuum broadening and finally various integrated-photonic architectures and associated material choices. We concentrate specifically on supercontinuum from waveguides, because the attained spectrum is typically broader and flatter than the competing architecture based on microring resonators [9]. However, this comes at the cost of the need for on-chip narrow-linewidth and high-power mode-locked laser (which is yet to be demonstrated), instead of continuous-wave (CW) pump sources needed for microrings. Irrespective of the specific implementation requirements of such a supercontinuum source, some of the challenges for developing such systems are common to the whole field of integrated photonics. For example, if silicon is chosen as the primary optical material, a.k.a., silicon photonics, there exist fundamental material limitations, such as two-photon absorption and the material's indirect bandgap. The high propagation loss of the III/V compound semiconductor competitors—which can possess direct bandgaps, hence lasing—also renders them a less than ideal alternative. Just like in optical transceivers, the solution is ushered by heterogeneous integration of various materials for different optical functionalities [10,11]. Material heterogeneity is therefore another common feature of the technologies reviewed in this paper.

2. Applications of Miniature Frequency Combs

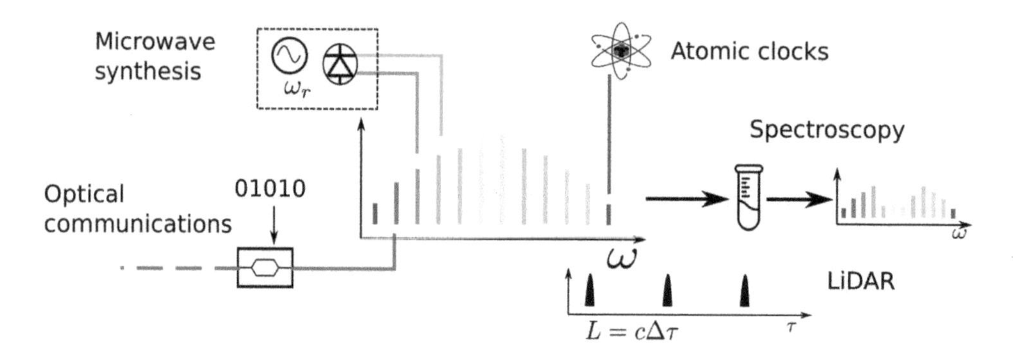

Figure 1. Frequency combs find applications in (i) low-noise microwave signal generation and processing, (ii) gears in atomic clocks that measure the precise frequency of atomic transitions, (iii) broadband light sources in spectroscopy. They can also be viewed as (iv) a source of densely spaced optical channels for optical communications and (v) source of high peak-power pulses in time-of-flight lidar.

The development of mode-locked laser was crucial to the proliferation of frequency combs and has spawned a multitude of applications [12], which are schematically depicted in Figure 1. Among them, spectroscopy is of prime importance. The first spectroscopic experiments concentrated on atoms from the first group of the periodic table. For instance, the Cesium D_1 line (895 nm) was measured, thus its hyperfine constant could be calculated to high precision [13]. Further improvements were made by using self-referenced frequency combs and counter-propagating beams to avoid Doppler broadening [14]. In the initial experiments, only a single line of the frequency comb was used. However, the main advantage of the frequency comb is to use all the available frequency teeth. The next technology leap came with the invention of dual-comb spectroscopy (DCS) [15], where two locked frequency combs, with slightly detuned repetition rates, are used. In DCS, one of the combs is transmitted through the sample and then beat against the other comb producing a radio-frequency (RF) comb. The phase and amplitude information of the probing comb is mapped into the RF comb and thus the sample's absorption spectrum can be recorded. DCS is superior—in terms of resolution, acquisition speed, accuracy and signal-to-noise ratio (SNR)—to other methods, such as high-resolution virtually imaged phased array (VIPA) disperser [16]. Since it does not require a spectrometer, i.e., a grating or an interferometer, it is a perfect candidate for an on-chip source [17]. It should be stressed that for applications in organic chemistry, it is the mid-infrared (mid-IR) region lying approximately between 4000–400 cm^{-1} or 2.5–25 μm is the most spectroscopically interesting window, and is referred to as the fingerprint region. Within this wavelength range, the rich number of rotational and vibrational excitations is sufficient for identification of organic molecules. Examples include the broadband absorption of the O-H bond of alcohols around 2.9 μm or the sharp and strong peak of the carboxyl group (C=O) near 5.5 μm [18]. Should a miniature frequency-comb spectrometer be realized in this regime, one could envision a multitude on biomedical and environmental monitoring applications. However, there are limited laser sources beyond 3 μm, hence alternative light generation methods, e.g., difference-frequency generation and optical-parametric oscillation, are required [19]. Furthermore, mid-IR detectors require cryogenic cooling, in order to limit thermal noise, and even so perform worse than their Si or InGaAs counterparts. Consequently, on-chip stabilization experiments have been primarily performed in the near-IR wavelengths, as summarized in Table 1. In this region, it is still possible to detect the overtones of vibrational resonances of certain molecules, e.g., methane (CH_4) [20]. Additionally, the near-IR range is used in astronomical spectrograms, where frequency combs are used for calibration [21], in order to detect Doppler shifts as small as 1 $cm{\cdot}s^{-1}$. Miniaturization would enable extraterrestrial applications for frequency-comb sources.

Intimately intertwined with the topic of atomic spectroscopy is the subject of atomic clocks. Here, a narrow-linewidth laser is tuned to an atomic transition locked to one of the optical comb teeth. The frequency comb serves as the clockwork that maps the optical frequencies to microwave frequencies that can be counted by electronics. Atomic clocks form the core of international time standard disseminated globally [22]. Simultaneously, this makes them an equally capable frequency standard that would benefit from miniaturization. As the data rates and the number of data channels grow, it becomes increasingly important to synchronize the frequency among devices on the same network [23]. Miniature atomic vapors cells with vertical-cavity surface-emitting lasers (VCSELs), locked to the atomic transition, and Rubidium (Rb) vapor cells integrated with silicon waveguides [8], have already been demonstrated [23]. The next step is integration with an on-chip frequency comb source that would link the optical and microwave frequencies. Miniature atomic clocks would greatly improve the resilience of receivers for global-positioning system (GPS) against jamming [24].

Finally, there are potential applications that stem from the ability to separate individual comb lines and alter their phase and amplitude. In this manner, it is possible to synthesize arbitrary optical pulses [25] and characterize them via multiheterodyne beat [26]. Pulse shaping in the optical domain can also be used to realize programmable and tunable filters for microwave signals encoded on the optical carrier [25]. The individual comb lines can equally well-function as separate channels for coherent terabit-per-second communication [27].

3. Stabilization

Frequency-comb spectra are composed of series of equally spaced lines, hence the name. The frequency comb has two free parameters, namely the repetition rate or the spacing between the comb lines, f_r, and the carrier-envelope offset (CEO) f_0, which is a measure of the phase slippage between the carrier frequency and the peak amplitude of a pulse [12], or alternatively the offset of the comb lines, with respect to zero frequency. Thus, the position of the comb lines is given by the simple relation $f_n = f_0 + nf_r$, where n is an integer as depicted in Figure 2. To take full advantage of the frequency comb as an optical metrology tool, stabilizing both parameters through detection and a feedback loop back to the oscillator is demanded. Stabilization of the repetition rate is easier, as GHz-bandwidth photodetectors are readily available. Some of the repetition stabilization schemes are discussed in the Section 5. Stabilization of the CEO is more challenging, because in absence of an external reference, it is impossible to measure the optical frequency precisely. Therefore, the optical frequencies ought to be mapped into the microwave domain, so that they can be measured electronically. This is accomplished by frequency broadening of the original spectrum, frequency multiplication of one or two portions of the spectrum and subsequent measurement of the heterodyne beat between the two.

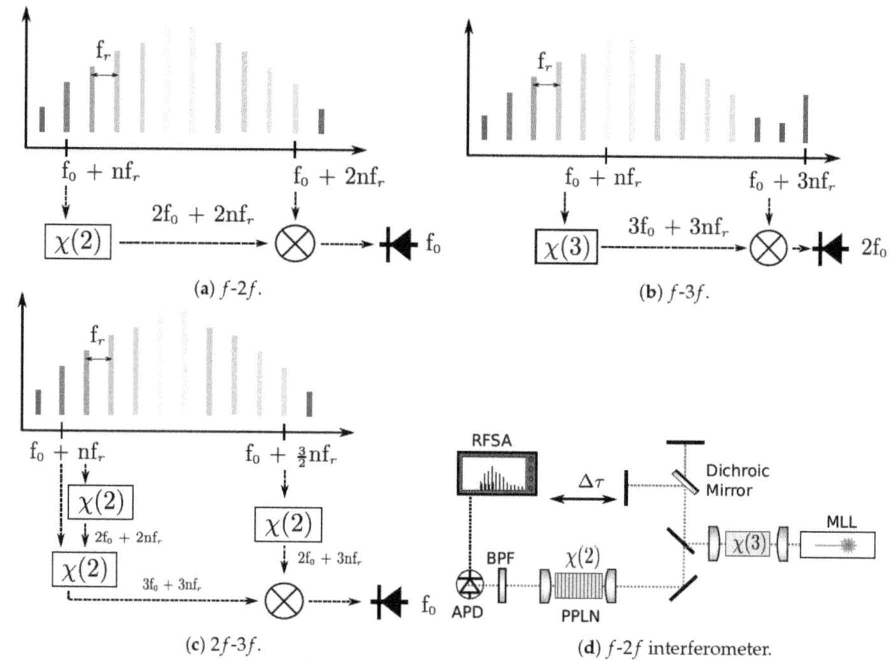

(c) $2f$-$3f$. (d) f-$2f$ interferometer.

Figure 2. Carrier-envelope offset (f_0) detection schemes.

Different frequency-stabilization methods are introduced in the following. Their common challenge is that in general, frequency broadening is a non-trivial task, as the requirements for broad bandwidth and coherence must be met simultaneously. This aspect of frequency broadening is discussed in more detail in the Section 4.

As shown in Figure 2a, the most common scheme is the so-called f-$2f$ self-referencing technique, where a tooth from the long-wavelength portion, $f_0 + nf_r$ is frequency-doubled to $2f_0 + 2nf_r$ and beat against a tooth an octave apart $f_0 + 2nf_r$. In practice, the frequency doubling has a few nanometers of bandwidth, so actually multiple comb lines around these frequencies are used.

Another method is f-$3f$ referencing, which requires two octaves of bandwidth. Here, frequency doubling is replaced with frequency-tripling, which means that the same third-order nonlinear material, i.e., with a strong $\chi(3)$ optical susceptibility, can be simultaneously used for supercontinuum generation and frequency multiplication. In this case, the beat frequency is actually $2f_0$, as evident in Figure 2b. But this attribution does not make a difference for the controllers used in the electronic-feedback loops, because it just affects the proportionality factor.

Finally, there are various fractional schemes, such as $2f$-$3f$ referencing that require shorter bandwidth, in this case 2/3 of an octave, but also frequency multiplication of both sides of the supercontinuum spectrum. The frequency-tripling is done in two stages, through second-harmonic and sum-frequency generations, making the whole process inefficient. The $2f$-$3f$ referencing technique is more prevalent in the case of microring resonators with limited bandwidth [28]. In one report [28], two CW lasers are locked to opposite ends of the spectrum to boost the power for $2f$-$3f$ referencing.

Several CEO detection experiments involving integrated optical waveguides are collected in Table 1. Due to the simplicity of the approach, there is particular allure of using a single, straight waveguide for both frequency doubling and supercontinuum generation. Materials such as AlN and $LiNbO_3$ (LN) or even strained Si_3N_4, possessing both $\chi(3)$ and $\chi(2)$ nonlinear responses, can be considered. However, this comes at the price of increased power consumption for two reasons. First, efficient frequency doubling requires phase matching. In large waveguides used for supercontinuum generation, the phase-matching typically occurs for higher-order modes, as in experiments on $LiNbO_3$ and AlN suggest (Table 1). This results in the CEO signal being generated from beating two different spatial modes, which limits the signal strength. Secondly, the frequency components used in f-$2f$ referencing are separated by an octave, which means that they have substantially different group velocities. Therefore, as they exit the waveguide they are separated in space (or equivalently the arrival times at the detector are different). Thus, in f-$2f$ referencing experiments, it is common to split the spectral components and compensate from the time delay in an interferometer, as depicted in Figure 2d. This feature is not easily available in a single waveguide with simultaneous $\chi(3)$ and $\chi(2)$ nonliterary.

Table 1. On-chip CEO detection experiments. Abbreviations: SHG—second harmonic generation, THG—third harmonic generation, CEO SNR—carrier envelope offset signal to noise ratio.

$\chi(3)$ Material	Si_3N_4	Si_3N_4	Si_3N_4	AlN	$LiNbO_3$
n_2 [$m^2 \cdot W^{-1}$]	2.5×10^{-19}	2.5×10^{-19}	2.5×10^{-19}	2.3×10^{-19}	1.8×10^{-19}
span	600–1700 + nm	520–1700 + nm	600–1900 nm	500–4000 nm	400–2400 nm
pump wavelength	1510 nm	1550 nm	1055 nm	1550 nm	1506 nm
pulse energy	62 pJ	110 pJ	36 pJ	800 pJ	185 pJ
pulse duration	200 fs	80 fs	64 fs	80 fs	160 fs
repetition rate	80 MHz	100 MHz	1 GHz	100 MHz	80 MHz
total insertion loss	7 dB	4 dB	8 dB	8 dB	8.5 dB
$\chi(2)$ Material	**Strained Si_3N_4**	**NA**	**PPLN**	**AlN**	**$LiNbO_3$**
referencing scheme	f-$2f$	f-$3f$	f-$2f$	f-$2f$	f-$2f$
SHG/THG	770 nm	420 nm	680 nm	780 nm	800 nm
variable delay	No	No	Yes	No	No
CEO SNR	27 dB	23 dB	40 dB	37 dB	30 dB
reference	[29]	[30]	[31]	[32]	[33]

Due to above-mentioned reasons, the experiment with the highest CEO SNR and lowest power consumption in Table 1 employs the architecture from Figure 2d, together with periodically poled lithium niobate (PPLN) device to achieve quasi-phase-matching to the preferred fundamental mode. To date, this has only been done in free-space optics, but progress on efforts for on-chip integration are discussed in Section 6.

Once a CEO signal with sufficient SNR (typically 25 dB) is detected, a feedback loop is used to stabilize it. In a semiconductor MLL, there are multiple factors that can affect the magnitude of the CEO. These include the gain current, operating temperature, reverse voltage of the saturable absorber and the current in an integrated phase-shifter. Furthermore, these parameters simultaneously affect the CEO and the repetition rate [34]. It is preferential to choose a free parameter and operating conditions that affect mostly the CEO, while keeping the repetition rate fixed. For semiconductor disk lasers, this has been successfully accomplished via gain current modulation [35]. Another possibility is injection locking [36], as discussed in Section 5.

Strides towards miniaturization have been made that extend beyond CEO detection. A self-referenced frequency-comb source with a fiber oscillator and silicon nitride waveguides, consuming only 5 W of electrical power, was demonstrated [37]. In this work, power-efficient repetition rate control is achieved by resistive thermal fiber heater. In another report [38], the CEO signal of a semiconductor disk laser, operating at 1.8 GHz, is stabilized via supercontinuum generation in photonic-crystal fiber. The whole system consumed 6 W of optical power. Replacing the fiber with integrated silicon nitride waveguides reduced the optical power requirement to 160 mW for a similar system, operating at 1.6 GHz [35]. However, the semiconductor disc lasers still require an external cavity with active feedback to one of the mirrors to limit amplitude and phase noise.

There also exist alternative architectures for self-referenced on-chip frequency combs. They use the spontaneous formation of solitons in microring resonators from CW background. This approach benefits from using a CW laser, instead of a mode-locked laser. An on-chip optical synthesizer has been realized based on this approach [39]. However, the spectrum of frequency combs generated in microrings follows the *sech* envelope function and has a limited bandwidth. Thus, from the spectroscopy viewpoint, photonic waveguides are preferable.

In the future, hybrid approaches could be possible. An example is synchronous pumping of microrings with a mode-locked laser, which should further reduce the system power consumption [40,41].

4. Supercontinuum Generation in Integrated Waveguides

All the self-referencing schemes in Section 3, i.e., $2f$-$3f$, f-$2f$, and f-$3f$, require a broad-bandwidth source that is respectively 2/3 of an octave, an octave and 2 octaves wide. An octave centered around 1550 nm, corresponds to roughly 1000 nm of bandwidth. This is significantly larger than the bandwidth of typical semiconductor lasers, which is around 10 nm (see Table 2). Hence, a crucial element of any on-chip CEO detection scheme is a $\chi(3)$ waveguide, in which frequency broadening, also called supercontinuum generation, occurs. Supercontinuum generation has been thoroughly studied in the context of optical fibers [42]. Here, we review the most important findings that are relevant to obtaining a broadband, coherent supercontinuum in waveguides required for self-referencing.

The master and accurate equation describing the dynamics of fs-range pulses is

$$\frac{\partial A}{\partial z} = -\frac{\alpha}{2}A + \sum_{k\geq 2} \frac{i^{k+1}}{k!}\beta_k \frac{\partial^k A}{\partial T^k} + i\gamma(1 + i\tau_{\text{shock}}\frac{\partial}{\partial T})(A(z,t)\int_{-\infty}^{+\infty} R(T')|A(z,T-T'|^2 dT' + i\Gamma_R(z,T)), \quad (1)$$

where A is the pulse amplitude in $W^{-1/2}$, β_i are dispersion coefficients from the Taylor expansion around the center frequency, ω_0, and γ is the nonlinear coefficient. $\tau_{\text{shock}} \approx 1/\omega_0$ is the shock timescale, although there are more complicated expressions that account for the dispersion of the effective area [42]. $R(T)$ is the Raman response function and various semi-empirical models are developed for it [43]. Finally, $\Gamma_R(z,T)$ is the stochastic noise term arising from spontaneous Raman scattering.

The dynamics of supercontinuum generation from ultrashort (100 fs range) pulse in the anomalous dispersion region, $\beta_2 < 0$, are dominated by soliton fission, as opposed to self-phase modulation in the normal dispersion case. The latter involves amplification of quantum noise via four-wave mixing (FWM) processes, which muddle the coherence of the supercontinuum. Hence, in the context of stabilization we concentrate on the anomalous dispersion regime.

Regarding the dynamics of supercontinuum generation, it is necessary to introduce the soliton number, N, defined as $N^2 = \gamma P_0 T_0^2/|\beta_2|$, where P_0 and T_0 refer to the peak power and duration of the input pulse in Equation (1). The first ejected soliton—which is the shortest and most energetic—has a temporal width of $T_0/(2N-1)$ and a peak power of $(2N-1)^2/N^2 P_0$ [44]. All solitons formed during the fission process are perturbed by the higher-order dispersion terms and Raman scattering, but the perturbation of the first soliton dominates the spectrum. Under phase-matching conditions,

$$D = \sum_{k\geq 2} \frac{\beta_k}{k!} = \frac{1}{2}(2N-1)^2 \frac{|\beta_2|}{T_0^2}, \quad (2)$$

the soliton can transfer energy to a linear dispersive wave [44]. To a good approximation, the right-hand side of the equation can be equated to zero. The dispersive waves are also referred to as Cherenkov radiation [45], due to similarities with radiation emitted by charged particles traveling through a dielectric medium. This equation forms the basis of dispersion engineering. Since semiconductor processing offers tight control of the waveguide geometry, it is possible to control the shape of supercontinuum spectrum in integrated waveguides (such an attribution is difficult to accomplish in largely axially symmetrical fibers). For example, it is possible to introduce a slot into the waveguide structure, effectively flattening the dispersion profile and increasing the bandwidth to two octaves [46]. This design leads the phase-matching condition from Equation (2) to be satisfied at four different wavelengths [47] with relatively flat spectrum, which would be advantageous in spectroscopic applications. Dispersion engineering is especially important in microring resonators [48], where the envelope follows the *sech*-shape of solitons. A simulated spectrum of supercontinuum generated in chalcogenide waveguides together with experimental data from [49] is appended to Figure 4b. The phase-matching condition, shown as a green curve, from Equation (2) predicts the position of dispersive waves to great accuracy.

The second nonlinear effect that perturbs the solitons is Raman scattering. In Raman scattering, some of the energy of impeding photon is transferred to electronic oscillations of molecules. This means that the soliton spectrum is continuously shifted towards red wavelengths. The speed of the frequency drift scales as $d\omega_R/dz \sim |\beta_2|/T_0^4$ [50]. Importantly, spontaneous Raman scattering is an additional source of noise. It has been shown that this noise term leads to the degradation of coherence in the long-wavelength portion of the supercontinuum, hence materials with weak Raman response are preferred [31].

Apart from octave-spanning or wider bandwidth, it is necessary to ensure that the generated supercontinuum is coherent to observe the CEO beat. Coherence is defined as [42]

$$\left|g_{12}^{(1)}\left(\lambda,t_1-t_2\right)\right| = \left|\frac{\left\langle A_1^*\left(\lambda,t_1\right)A_2\left(\lambda,t_2\right)\right\rangle}{\sqrt{\left\langle\left|A_1\left(\lambda,t_1\right)\right|^2\right\rangle\left\langle\left|A_2\left(\lambda,t_2\right)\right|^2\right\rangle}}\right|, \tag{3}$$

where A_1 and A_2 refer to different pulse amplitudes in the pulse train separated by $t_1 - t_2$.

Coherence requires sub-100-fs pulses [42] and the highest degree of coherence is usually observed near the soliton fission point. Also, there are several noise sources that have adverse effect on coherence. As mentioned beforehand, strong Raman effect also leads to undesired additional non-coherent photons generated through spontaneous Raman scattering. Additionally, there is quantum noise (shot noise), which is usually dominant and cannot be eradicated [51]. Furthermore, there are additional sources of noise that come from the oscillator itself, such as phase (optical linewidth), timing jitter (RF linewidth), amplitude noise and technical noise from environmental changes [52]. As the optical-fiber-based oscillators usually do not produce sufficient power for CEO detection experiments, it is necessary to amplify their output, but this leads to additional amplified spontaneous emission (ASE) noise. It has been shown that ASE noise leads to variations in the group velocity of solitons and thus timing jitter [53].

A challenge for supercontinuum generation is that as a nonlinear process, the above noise sources (ASE, Raman and shot noise) are not additive but are actually amplified through FWM processes [54]. Consequently, the FWM gain grows exponentially with pulse energy, whereas the supercontinuum bandwidth increases only linearly [55]. In other words, the minimal pulse energy that produces sufficient bandwidth would also be the point of maximum coherence.

5. Requirements for Semiconductor Lasers

To date, all CEO detection and stabilization experiments using integrated waveguides are performed with either fiber lasers [30], optical-parametric oscillators (OPOs) [29] or pumped vertical external-cavity surface-emitting lasers (VECSEL) [38]. Just like mode-locked fiber lasers were crucial to the development of stabilized frequency combs [13], a semiconductor laser is necessary for on-chip counterparts.

Thus, it would be prudent to compare current state-of-the art semiconductor mode-locked lasers (MLL) with these prior alternatives. As noted in reference [52], the overall frequency-comb performance is much more sensitive to the noise inside the oscillator than the noise added through subsequent amplification. The reason is that the noise inside the cavity (ASE or length variations with temperature) translates into frequency shifts or broader linewidth, whereas the same noise sources outside the cavity would increase the noise floor on the detector that is used for CEO detection.

The comparison presented here is by no means a thorough review of semiconductor MLLs and the reader is directed to references [34,56,57]. Table 2 collects performance of representative lasers. The subsequent discussion compares them against Erbium-doped fiber oscillators used in frequency-comb experiments [37,58–60].

Table 2. Passively mode-locked semiconductor lasers in comparison to Erbium-doped fiber oscillators.

Material	III-V/Si	III-V/Si	III-V/Si	InP	InP	InP	Er fiber (+EDFA)
repetition rate	19 GHz	1 GHz	20 GHz	30 GHz	2.5 GHz	1 GHz	50–200 MHz
pulse duration	1.83 ps	15 ps	900 fs	900 fs	9.8 ps	70 ps	<300 fs (<100 fs)
opt. bandwidth	-	12 nm	3 nm	15 nm	3 nm	5 nm	(~50 nm)
band. threshold.	-	−10 dB	−3 dB	−10 dB	−3 dB	−10 dB	−3 dB
power	9 mW	-	1.8 mW	0.25 mW	80 uW	0.59 mW	~5 mW
	est.			coup.	coup.	coup.	(<200 mW)
3 dB optical lw	-	400 kHz	-	29 MHz	-	80 MHz	10 s kHz
RF lw	6 kHz	0.9 kHz	1.1 kHz	500 kHz	6 kHz/61 kHz	1 MHz	-
RF lw threshold	−3 dB	−10 dB	−3 dB	−20 dB	−3 dB/−20 dB	−20 dB	-
timing jitter	1.2 ps	-	-	4.5 ps	-	4.16 ps	<2 fs
int. range	0.1 MHz–	-	-	100 Hz–	10 kHz–	20 kHz–	10 kHz-
	100 MHz	-	-	30 MHz	10 MHz	80 MHz	10 MHz
reference	[61]	[57]	[56]	[62]	[63]	[64]	[37,58–60]

With regards to the power consumption metric, nonlinear integrated waveguides typically require lower pulse energies than those in photonic-crystal fibers for supercontinuum generation. However, semiconductor-based MLL cavities are shorter and thus the repetition rates are higher than those in fiber cavities. If we take the most optimistic repetition rate value of 1 GHz from Table 2, and the best InGaP waveguides from Table 4 (which require only 2 pJ pulses, and also assume no coupling loss), the average power requirement turns out to be only 2 mW. This power is perfectly reasonable, albeit the value can quickly grow in practice. For instance, a 10-GHz InP cavity and the chalcogenide waveguides discussed later would consume 260 mW of optical power.

The next metric discussed is the pulse width. From the fiber-based comb literature [42], it is generally assumed that coherent supercontinuum generation requires 100 fs pulses, while the best MLL from Table 2 produce 900 fs pulses. However, this is not prohibitive, as it has been shown that some of the pulse compression task can be off-loaded to nonlinear waveguides. It is, for example, shown via simulations that 1 ps pulses can be compressed to 41 fs in 39-cm-long silicon nitride waveguides with ultralow propagation loss of 4.2 dB/m [65]. Similar ideas are proposed for optical fibers [66].

The subtlety lies in the fact that these two requirements are not independent. As seen in Table 2, lasers with long cavities (low repetition rates) tend to produce pulses that span tens of picoseconds. This is because, in general, longer cavities lead to larger net accumulated dispersion, which leads to wider solitons [58], i.e., $\tau_s = 2|\beta_2|/\gamma P_0$ for $A(T) = \sqrt{P_0}/cosh(T/\tau_s)$ in the basic form of Equation (1). In this context, the path forward would probably involve further work on 1 GHz cavities with dispersion engineering or addition of gain flattering filters as in [62], in order to broaden the spectrum and reduce pulse width to picosecond level which is sufficient for further compression in nonlinear waveguides. The issues of quick gain saturation could be addressed via the breathing mode architecture [67], where the pulse is stretched before the gain section and compressed afterwards.

In the elastic-tape model of frequency combs [52,58], there exists a fixed point in the frequency spectrum, f_r, around which the comb teeth breath. In other words, the linewidth of individual comb teeth increases as we move away from f_r [52]. In the case of active/hybrid MLLs, the fixed point is located near the carrier frequency, while in passive MLLs it is close to the zero frequency [34]. In CEO detection experiments, the comb teeth that are beat against one another (after frequency multiplication) are on the extremes of the spectrum. Hence, the linewidth of the CEO is directly proportional to the linewidth of these comb teeth. Reducing the linewidth of the CEO improves its SNR and typically a value of 25 dB is required by the locking electronics for CEO stabilization. For comparison, the free-running CEO linewidth of Erbium-doped fiber lasers is in the range of 10–1000 kHz [58]. Thus, to compare the performance of MLL in the context of frequency-comb generation, several key factors should ideally be known. They include the fixed point and its optical linewidth, as well as the spectral power density of RF phase noise, which dictates the RF linewidth and thus the magnitude of the breathing of the comb lines. However, the fixed point is rarely measured, so the optical linewidth near the carrier frequency is used as an alternative metric.

A clear distinction can be made in Table 2, between the heterogeneously integrated III-V/Si MLL and monolithic InP lasers. The former have narrower optical linewidth (e.g., 400 kHz [57]) than the latter (e.g., tens of MHz [62,64]). This is because the silicon cavities in III-V/Si have longer cavity photon lifetime than the monolithic, high-loss InP counterparts and thus wider optical linewidth through the Schawlow-Townes limit [68]. The narrowest optical linewidth of 400 kHz is still higher than typical values of tens of kHz in Erbium-doped fiber cavities, but it can be further improved with injection-locking techniques mentioned below.

The timing jitter of passive semiconductor MLLs is on the order of picoseconds, as seen in Table 2, in contrast to the femtosecond timing jitter of passive fiber cavities. However, the situation is not as dire as the numbers might imply. The 1.8 GHz VECELS used in CEO stabilization experiments [38], which are closer in repetition rate to semiconductor MLLs, have an integrated timing jitter of 60 fs (1 Hz–100 MHz) after stabilization to an external synthesizer. Therefore, in semiconductor MLLs some feedback system equalizing the repetition rate is necessary before the CEO signal can be detected.

To date, there is no single integrated passively mode-locked laser that can compete with the performance of fiber cavities simultaneously on all metrics. However, there are various methods to augment an oscillator to improve its performance. For example, hybrid-mode locking of cavity is shown to reduce the 10-dB RF linewidth from 0.9 kHz to 1 Hz [57]. In this case, hybrid-mode locking was accomplished by supplying an RF signal to the saturable absorber. We also note that the benefit of hybrid-mode locking is that it provides an access point for repetition rate control in a feedback loop in a fully self-referenced frequency-comb source.

Another possibility is optical feedback, with on-chip external cavity [61]. However, the performance of the feedback is proportional to the cavity length, as it increases the memory of the system. In one report [61], the on-chip external cavity is only twice as long as the oscillator cavity and the 3-dB RF linewidth is 6 kHz. In contrast, in a report based on a much longer (22 m) fiber-based external cavity [69], the RF linewidth is only 192 Hz. However, fabricating equally long waveguides is challenging, due to higher propagation losses of integrated photonics and limited chip space.

Another scheme is to use multitone injection locking [70]. Multitone injection locking is an extension of the injection locking technique. It reduces the optical linewidth through the use of a narrow-linewidth CW seed that is injected into the cavity [71]. The benefit of multitone injection locking is that it simultaneously reduces optical linewidth and also provides a means of controlling the repetition rate by varying the spacing of the tones. We also note that injection locking provides an alternative to the standard way of controlling f_0, via tuning the wavelength of the seed instead of modulation of the current injected into the gain section of the MLL [35].

An example of such an approach is shown in Figure 3a. The cavity used in the experiment is a 10-GHz InP colliding pulse MLL [72]. Here, the multitone injection locking is part of a coupled opto-electronic oscillator loop (COEO), i.e., the repetition rate of the MLL is detected on the electroabsorption modulator (EAM), amplified, and fed back into the Mach-Zehnder modulator, generating sidebands on the injection seed and forming a feedback loop in the system. The resonance condition is ensured by adding a phase-shifter in the loop. Another way to look at COEO is to notice that the MLL acts as a selective filter that amplifies RF modes of the cavity that overlap with the modes of the MLL. In this manner, the performance of the InP MLL is greatly improved. The free-running optical linewidth is reduced by a factor of 6000 to 100 kHz, limited by the optical linewidth of the injection seed, and the COEO loop reduces the RF linewidth by a factor of 70, yielding a 3-dB RF linewidth of 400 Hz [36]. The RF phase noise is further reduced by locking the COEO loop to an external reference, resulting in integrated timing jitter of 500 fs at 1 kHz. Unfortunately, again, the performance of COEO loop is proportional to the length of the optical delay. In this case, the optical delay line is 100 m long.

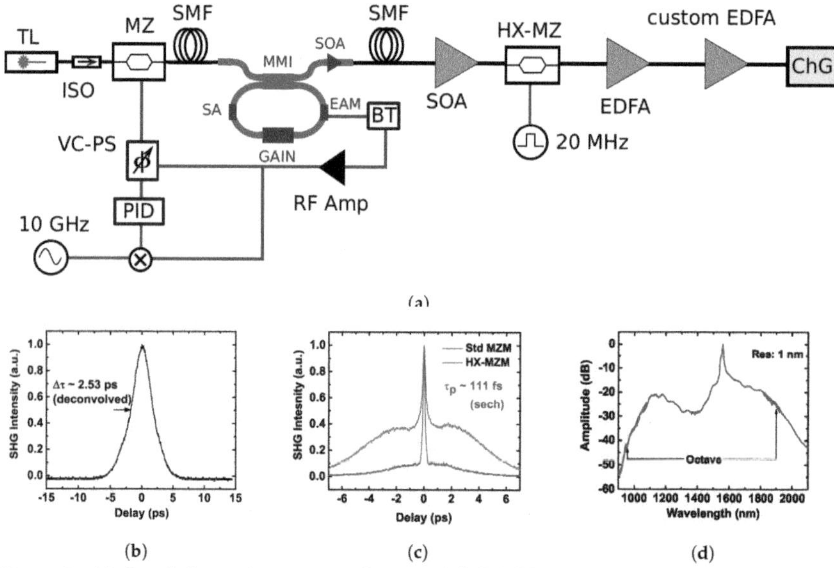

(b) (c) (d)

Figure 3. (**a**) Coupled-optoelectronic oscillator with InP MLL using multitone injection locking. The output pulses are pulse-picked, compressed and amplified to generate supercontinuum in chalcogenide (ChG) waveguides. The experimental data comes from [73]; (**b**) Autocorrelation of pulses coming out of the chip; (**c**) High-extinction modulator is essential for pulse picking, otherwise the residual pulse train is amplified leading to a wide pedestal in the autocorrelation; (**d**) Octave-spanning supercontinuum generated in ChG waveguides using the InP-based source as a seed. Abbreviations: TL—tunable laser, (HX)-MZ—(high-extinction) Mach-Zehnder modulator, MMI—multimode interferometer, EAM—electrooptic amplitude modulator, SA—saturable absorber, SOA—semiconductor optical amplifier, EDFA—Erbium-doped fiber amplifier, ISO—optical isolator, VC-PS voltage-controlled RF phase shifter, BT—bias tee, PID—proportional-integral-derivative controller, SMF—single-mode fiber.

Such augmented cavity acts as a seed for supercontinuum generation in chalcogenide waveguides that are discussed in detail in Section 6. The cavity by itself produces 2.5 ps pulses (see Figure 3b) that are further amplified and compressed to 111 fs. We note that because the carrier dynamics in an EDFA is on the order of μs, every 500th pulse must be picked by a high-extinction modulator. The high-extinction modulator is essential, otherwise the residual 10 GHz pulse train is amplified, leading to a large pedestal in the autocorrelation trace as in Figure 3c.

Finally, an octave-spanning supercontinuum is achieved, as presented in Figure 3d [73]. It is anticipated that the pulse picking and multiple stages of amplification can be integrated on a single InP-based monolithic superchip.

6. Materials for Nonlinear Processes

As mentioned in the introduction, despite significant success of silicon photonics, some of its optical properties have certain limitations. The material has a large third-order nonlinearity, $n_2 = (4.5 \pm 1.5) \times 10^{-18}$ m^2 W^{-1} at 1.55 μm [74], but researchers have been unable to achieve an octave-spanning supercontinuum pumped at 1.55 μm. As noted in [75], where 3/10 of an octave was demonstrated with 100 fJ pulses at 1310 nm, the two-photon and associated free-carrier absorptions clamp the maximum power inside the waveguides and hence limit the spectral broadening.

Nonetheless, the pump does not have to be completely outside the two-photon absorption region supercontinuum to achieve a full octave span as in report [76]) where 1900 nm laser was used. The experimental details are provided in Table 3.

Table 3. On-chip supercontinuum. The n_2 for $Ge_{23}Sb_7S_{70}$ and InGaP are inferred indirectly through measurement of γ coefficient in waveguides. For SiN, AlN, and $LiNbO_3$ refer to Table 1 with CEO detection experiments.

$\chi(3)$ Material	$Ge_{23}Sb_7S_{70}$	As_2S_3	InGaP	Si
n_2 [$m^2 \cdot W^{-1}$]	3.7×10^{-18}	3.8×10^{-18}	2.3×10^{-17}	4.5×10^{-18}
span	1030–2080 nm	1200–1700 + nm	1000–2100 nm	1150–2400 + nm
pump wavelength	1550 nm	1550 nm	1550 nm	1900 nm
pulse energy	26 pJ	60 pJ	2 pJ	4 pJ
pulse duration	240 fs	610 fs	170 fs	50 fs
repetition rate	25 MHz	10 MHz	82 MHz	200 MHz
propagation loss	0.5 dB/cm	0.6 dB/cm	12 dB/cm	1.5 dB/cm
references	[49]	[77]	[78]	[76]

However, once appropriate on-chip sources become available, silicon is an excellent material for supercontinuum generation in the mid-IR. An octave span is already demonstrated with only 5 pJ and 300 fs pulses at 2.5 μm wavelength [79]. The bandwidth could be further improved by removing the buried oxide layer, as in suspended-membrane air-clad silicon [80], silicon-germanium [81], or silicon-on-sapphire waveguides [82].

In the nearest future, a near-IR frequency comb is the most viable path, and therefore the $\chi(3)$ limitations of silicon must be addressed by heterogeneous integration with other materials. Additionally, silicon has a centrosymmetric lattice structure, hence does not possess an intrinsic second-order nonlinear optical susceptibility, $\chi(2)$, needed for frequency doubling.

When comparing the waveguides used for CEO detection supercontinuum generation, i.e., experiments from Tables 1 and 3, respectively, it is evident that materials with a strong $\chi(3)$ response are preferred, as it can lead to lower pulse energies required for octave-spanning supercontinuum. This is despite the fact that the Kerr nonlinearity, n_2, scales as $1/E_g^4$ with bandgap, E_g, and is inherently tied to the two-photon absorption via Kramers–Kroning relationship [83]. This means that materials with high n_2 possess a bandgap that is relatively close to the pump. For example, the transparency range of Si_3N_4 is 0.4–4.6 μm, whereas for silicon the transmission window is 1.2–8 μm. Due to the proximity of the bandgap, the high n_2 materials are likely to suffer from two-photon absorption. Nevertheless, within the tables, the best performing material is InGaP, which has the highest nonlinearity, despite having the highest propagation losses of 12 dB/cm [78]. Octave span is achieved with only 2 pJ of pulse energy in short (2 mm long) waveguides, which is why the high loss is not prohibitive.

The second functionality, required for the f-$2f$ referencing scheme, is efficient SHG. For a thorough review of SHG in heterogeneous integrated photonics, the reader is directed to [84] and for a discussion of appropriate figures of merit to [85]. Table 4 collects the most important demonstrations in integrated photonics. First, we note the efficiency of SHG can be greatly improved through the power enhancement of resonant cavities such as microrings. Consequently, the highest efficiency from Table 4, as expressed in $\%W^{-1}$, is for low-loss AlN microrings, when compared to LN and GaAs waveguides, despite the AlN's smaller $\chi(2)$ nonlinearity. However, the bandwidth of these resonant structures is limited by their quality factor, Q, and thus they have limited applicability to the f-$2f$ referencing schemes, because for SHG to occur one of the comb lines would have to be tuned to the microring resonance, and tuning would amount to having an already stabilized frequency-comb source. Secondly, it is advantageous to use all comb lines that fall within the phase-matching bandwidth (usually couple of nanometers) for highest CEO SNR. Thus, in terms of architecture, straight waveguides are a better choice.

Table 4. Second-harmonic generation in integrated waveguides.

$\chi(2)$ Material	$LiNbO_3$	GaAs	AlN	GaN	Strained Si_3N_4
d $[\frac{1}{2}\chi(2)]$	30 pmV^{-1}	119 pmV^{-1}	1 pmV^{-1}	8 pmV^{-1}	eff. 0.02 pmV^{-1}
efficiency	17% W^{-1}	255% W^{-1}	2300% W^{-1}	0.015% W^{-1}	-
efficiency	4600% $W^{-1}cm^{-2}$	13,000% $W^{-1}cm^{-2}$	-	-	-
SHG mode order	fundamental	fundamental	5th	6th	6th
architecture	waveguide	waveguide	microring	microring	microring
reference	[85]	[86]	[87]	[88]	[89]

Efficient SHG requires phase matching between the pump and the signal. Since the two are separated by an octave, in terms of frequency, this leads to the phase-matching condition being satisfied for a higher-order mode at the signal wavelength as for Si_3N_4, GaN and AlN cases in Table 4. This is suboptimal, as it compromises the mode-overlap between the pump and signal and thus overall conversion efficiency [84], as well as the mode-overlap between the signal short-wavelength part of supercontinuum (which is usually in the fundamental mode).

As shown in Table 4, record efficiencies of 13,000% $W^{-1}cm^{-2}$ have been demonstrated in thin-film GaAs [86]. This is possible due to GaAs's highest $\chi(2)$ response among the materials in Table 4 and a large refractive index contrast leading to a small mode area. However, this has been achieved by modal-phase matching between fundamental modes of different polarizations, i.e., the pump being in the transverse-electric (TE) and the signal in the transverse-magnetic (TM mode). This attribution has potential hurdles in f-$2f$ referencing scheme polarization rotators must be included, because the generated SHG would be in an orthogonal polarization state to the short end of the supercontinuum.

Finally, there is the thin-film equivalent of the commercially successful PPLN technology, where the phase matching is achieved through periodic-poling of the ferroelectric LN crystals [84,90,91]. Here, efficiencies of up to 4600% $W^{-1}cm^{-2}$ have been recently demonstrated for 1550 nm pump [85] and functionality for 2 μm pump has also been shown [92,93]. Similar to the above GaAs report, operation in the fundamental mode at both pump and signal wavelengths is feasible. An advantage of the above PPLN devices is that the pump and signal can have the same TE polarization modes, hence polarization rotation is not needed.

To summarize, the most optimal on-chip CEO detection scheme, for sufficient CEO signal with the lowest pulse energy, would incorporate (a) strong $\chi(3)$ material, (b) integrated spectral splitters and delay lines, and (c) a strong $\chi(2)$ material with a device geometry leading to SHG generated in the fundamental mode and same polarization as the pump.

A sensible approach to this task is the integration of chalcogenide glass, e.g., $Ge_{23}Sb_7S_{70}$, with thin-film LN. As discussed earlier, LN is one of the prime contenders for efficient SHG and ChG was chosen as the $\chi(3)$ material, because it is a glass that can be easily deposited on LN, without heating the substrate that would damage the thin-film LN.

Another material alternative, with strong $\chi(3)$ nonliterary, is silicon nitride. Low-loss material is typically achieved by low-pressure chemical-vapor deposition (LPCVD), which requires deposition temperature of ~700 °C. Hence, in this case, it is preferred to bond LN onto patterned silicon nitride layer to avoid thermal damage to the LN thin films [94].

The fabrication flow for integration of ChG and LN has already been demonstrated together with efficient mode-conversion from the ChG layer and rib-loaded LN with only 1.5 dB loss [95]. The strongest $\chi(2)$ response of LN is along the z axis. Since the demonstrated thin-film PPLN devices use y-cut crystals, so the electrodes could be placed in the plane of the wafer, the pump field for the SHG process has to be oriented along the z-axis. This in turn requires the geometry of the ChG waveguide to support broadband TE supercontinuum generation, which is shown to be feasible [96].

The grand vision of the final device to be demonstrated is depicted in Figure 4, together with the measured performance of discrete components. Figure 4b shows an octave-spanning TM supercontinuum generated with 240-fs-wide pulses, carrying 26 pJ of energy [49]. Figure 4c shows the performance of spectral splitters, as measured using a fiber-based supercontinuum source. Spectral separation is achieved through intermediate coupling to a higher-order mode (TE_1), which permits definition of the whole structure in a single lithography step and avoids tiny gaps between waveguides or sharp terminations [97].

As mentioned, second harmonic was demonstrated on thin-film PPLN, reviews of which can be found elsewhere [84,85]. The first PPLN device on silicon substrates showed only 3 dB difference between the signal and the pump [90] and the recorded spectrum is shown in Figure 4d. We note that this early experiment was performed using a pulsed source and therefore there is a significant contribution from sum-frequency signal generation in the signal. Highly efficient devices, under CW pumping, were later reported [85].

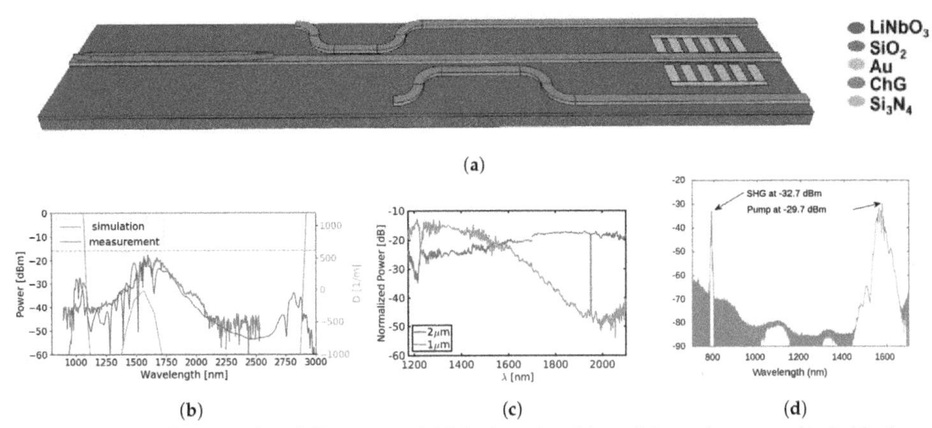

Figure 4. (**a**) The vision for a fully integrated f-$2f$ referencing chip and the performance of individual components; (**b**) measured (red) octave-spanning TM supercontinuum requiring only 26 pJ of pulse energy and the simulated spectrum (blue) together with the phase-matching condition (green) that predicts the position of dispersive waves; (**c**) spectral splitters showing 30 dB extinction ratio at 2 μm; (**d**) second-harmonic generation in thin-film PPLN.

Lastly, Figure 5 shows an experiment used to detect a second harmonic generated from supercontinuum from integrated waveguides. The long end of the supercontinuum is amplified using a custom-made Thulium-doped fiber amplifier (TDFA). The pump wavelength at 1984 nm is amplified by 20 dB to compensate for the high, 12 dB, coupling losses of the ChG chip. This is sufficient to observe a SHG signal at 992 nm at −71 dBm that falls within the short end of the supercontinuum [98]. A similar experiment with a highly nonlinear fiber and thin-film PPLN was also performed and is shown in Figure 5c. It is anticipated that integration would reduce the coupling losses rendering the TDFA unnecessary.

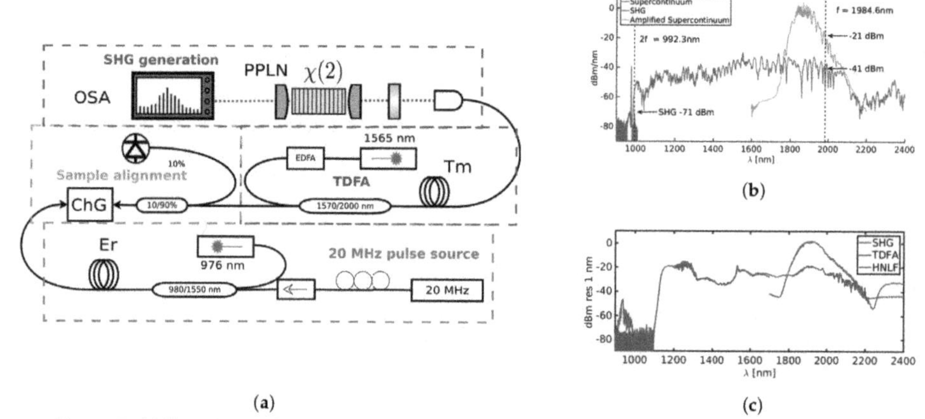

(a) (b) (c)

Figure 5. (**a**) Experimental setup for generating a second harmonic from supercontinuum, in turn, generated in chalcogenide (ChG) waveguides. A Thulium-doped amplifier (TDFA) is inserted between the ChG chip and bulk periodically poled lithium niobate (PPLN) to compensate for insertion losses. (**b**) The measured spectra in the experiment: blue—supercontinuum generated from ChG, green—amplified supercontinuum, red—second harmonic. (**c**) The same experiment performed with highly nonlinear fiber and thin-film lithium niobate.

7. Future Outlook

Great progress has been made towards realizing on-chip frequency-stabilized supercontinuum-based comb sources. The integrated nonlinear components outperform the bulk counterparts, due to their smaller effective mode areas that increase the strength of interaction. The $\chi(3)$ waveguides require less power than optical fibers, in order to produce an octave-spanning supercontinuum, and in some cases the supercontinuum extends far enough to even enable f-$3f$ referencing. Some materials, such as AlN, exhibit both $\chi(3)$ and $\chi(2)$ responses, allowing the detection of the CEO signal straight out of the waveguides. Record-high second-harmonic generation efficiencies have been demonstrated in thin-film PPLN and GaAs. The missing component is a semiconductor mode-locked laser that would serve as the heart of an on-chip frequency-comb source. In this case, heterogeneous integration of III-V gain media with long low-loss cavities, and possibly further stages of amplification and pulse compression to achieve low noise, sub-ps, 1-GHz oscillators seems to be the path forward. Subpicosecond pulses would be sufficient, provided that the nonlinear waveguides could shoulder compression to 100 fs pulses. Injection locking and hybrid-mode locking could be used to reduce the optical and RF linewidth of passive cavities. In the nearest future, an stabilized frequency combs in the near-IR is the next milestone, with potential applications in microwave synthesis, miniature atomic clocks, astronomy, and LiDAR. In the distant future, the technology could be extended to encompass the mid-IR range, where spectroscopic identification of organic molecules is possible. This is where a flat-, broadband- and stabilized-supercontinuum source would shine and open plethora of possibilities for inexpensive biomedical diagnostics and environmental monitoring.

Funding: Large portion work presented here was carried out under the Defense Advanced Research Project Agency (DARPA) DODOS project, grant number HR0011-15-C-0057. The views, opinions, and/or findings expressed are those of the authors and should not be interpreted as representing the official views or policies of the Department of Defense or the U.S. Government.

Conflicts of Interest: The authors declare no conflict of interest.

References

1. Li, H.; Balamurugan, G.; Sakib, M.; Sun, J.; Driscoll, J.; Kumar, R.; Jayatilleka, H.; Rong, H.; Jaussi, J.; Casper, B. A 112 Gb/s PAM4 transmitter with silicon photonics microring modulator and CMOS driver. In Proceedings of the Optical Fiber Communication Conference Postdeadline Papers 2019, San Diego, CA, USA, 3–7 March 2019.
2. Krishnamoorthy, A.V.; Ho, R.; Zheng, X.; Schwetman, H.; Lexau, J.; Koka, P.; Li, G.; Shubin, I.; Cunningham, J.E. Computer systems based on silicon photonic interconnects. *Proc. IEEE* **2009**, *97*, 1337–1361. [CrossRef]
3. Sun, C.; Wade, M.T.; Lee, Y.; Orcutt, J.S.; Alloatti, L.; Georgas, M.S.; Waterman, A.S.; Shainline, J.M.; Avizienis, R.R.; Lin, S.; et al. Single-chip microprocessor that communicates directly using light. *Nature* **2015**, *528*, 534–538. [CrossRef] [PubMed]
4. Nikolova, D.; Rumley, S.; Calhoun, D.; Li, Q.; Hendry, R.; Samadi, P.; Bergman, K. Scaling silicon photonic switch fabrics for data center interconnection networks. *Opt. Express* **2015**, *23*, 1159–1175. [CrossRef] [PubMed]
5. Shen, Y.; Harris, N.C.; Skirlo, S.; Prabhu, M.; Baehr-Jones, T.; Hochberg, M.; Sun, X.; Zhao, S.; Larochelle, H.; Englund, D.; et al. Deep learning with coherent nanophotonic circuits. *Nat. Photonics* **2017**, *11*, 441–446. [CrossRef]
6. Sun, J.; Timurdogan, E.; Yaacobi, A.; Hosseini, E.S.; Watts, M.R. Large-scale nanophotonic phased array. *Nature* **2013**, *493*, 195–199. [CrossRef] [PubMed]
7. Subramanian, A.Z.; Ryckeboer, E.; Dhakal, A.; Peyskens, F.; Malik, A.; Kuyken, B.; Zhao, H.; Pathak, S.; Ruocco, A.; De Groote, A.; et al. Silicon and silicon nitride photonic circuits for spectroscopic sensing on-a-chip. *Photonics Res.* **2015**, *3*, B47–B59. [CrossRef]
8. Yang, W.; Conkey, D.B.; Wu, B.; Yin, D.; Hawkins, A.R.; Schmidt, H. Atomic spectroscopy on a chip. *Nat. Photonics* **2007**, *1*, 331–335. [CrossRef]
9. Okawachi, Y.; Saha, K.; Levy, J.S.; Wen, Y.H.; Lipson, M.; Gaeta, A.L. Octave-spanning frequency comb generation in a silicon nitride chip. *Opt. Lett.* **2011**, *36*, 3398–3400. [CrossRef] [PubMed]
10. Fathpour, S. Emerging heterogeneous integrated photonic platforms on silicon. *Nanophotonics* **2015**, *4*, 143–164. [CrossRef]
11. Jones, R.; Doussiere, P.; Driscoll, J.B.; Lin, W.; Yu, H.; Akulova, Y.; Komljenovic, T.; Bowers, J.E. Heterogeneously Integrated InP/Silicon Photonics: Fabricating Fully Functional Transceivers. *IEEE Nanotechnol. Mag.* **2019**, *13*, 17–26. [CrossRef]
12. Diddams, S.A. The evolving optical frequency comb. *JOSA B* **2010**, *27*, B51–B62. [CrossRef]
13. Udem, T.; Reichert, J.; Holzwarth, R.; Hänsch, T. Absolute optical frequency measurement of the cesium D 1 line with a mode-locked laser. *Phys. Rev. Lett.* **1999**, *82*, 3568–3571. [CrossRef]
14. Stalnaker, J.E.; Mbele, V.; Gerginov, V.; Fortier, T.M.; Diddams, S.A.; Hollberg, L.; Tanner, C.E. Femtosecond frequency comb measurement of absolute frequencies and hyperfine coupling constants in cesium vapor. *Phys. Rev. A* **2010**, *81*, 043840. [CrossRef]
15. Coddington, I.; Swann, W.C.; Newbury, N.R. Coherent multiheterodyne spectroscopy using stabilized optical frequency combs. *Phys. Rev. Lett.* **2008**, *100*, 013902. [CrossRef] [PubMed]
16. Diddams, S.A.; Hollberg, L.; Mbele, V. Molecular fingerprinting with the resolved modes of a femtosecond laser frequency comb. *Nature* **2007**, *445*, 627–630. [CrossRef]
17. Coddington, I.; Newbury, N.; Swann, W. Dual-comb spectroscopy. *Optica* **2016**, *3*, 414–426. [CrossRef]
18. Colthup, N. *Introduction to Infrared and Raman Spectroscopy*; Elsevier: Amsterdam, The Netherlands, 2012.
19. Schliesser, A.; Picqué, N.; Hänsch, T.W. Mid-infrared frequency combs. *Nat. Photonics* **2012**, *6*, 440–449. [CrossRef]
20. Tombez, L.; Zhang, E.; Orcutt, J.; Kamlapurkar, S.; Green, W. Methane absorption spectroscopy on a silicon photonic chip. *Optica* **2017**, *4*, 1322–1325. [CrossRef]
21. Li, C.H.; Benedick, A.J.; Fendel, P.; Glenday, A.G.; Kärtner, F.X.; Phillips, D.F.; Sasselov, D.; Szentgyorgyi, A.; Walsworth, R.L. A laser frequency comb that enables radial velocity measurements with a precision of 1 $cm{\cdot}s^{-1}$. *Nature* **2008**, *452*, 610–612. [CrossRef]
22. Ludlow, A.D.; Boyd, M.M.; Ye, J.; Peik, E.; Schmidt, P.O. Optical atomic clocks. *Rev. Mod. Phys.* **2015**, *87*, 637–701. [CrossRef]

23. Knappe, S.; Shah, V.; Schwindt, P.D.D.; Hollberg, L.; Kitching, J.; Liew, L.A.; Moreland, J. A microfabricated atomic clock. *Appl. Phys. Lett.* **2004**, *85*, 1460–1462. [CrossRef]
24. Fruehauf, H. Fast "Direct-P(Y)" GPS signal acquisition using a special portable clock. In Proceedings of the 33rd Annual Precise Time and Time Interval Meeting, Long Beach, CA, USA, 27–29 November 2001; p. 359.
25. Song, M.; Long, C.M.; Wu, R.; Seo, D.; Leaird, D.E.; Weiner, A.M. Reconfigurable and tunable flat-top microwave photonic filters utilizing optical frequency combs. *IEEE Photonics Technol. Lett.* **2011**, *23*, 1618–1620. [CrossRef]
26. Delfyett, P.J.; Ozdur, I.; Hoghooghi, N.; Akbulut, M.; Davila-Rodriguez, J.; Bhooplapur, S. Advanced ultrafast technologies based on optical frequency combs. *IEEE J. Sel. Top. Quantum Electron.* **2012**, *18*, 258–274. [CrossRef]
27. Pfeifle, J.; Brasch, V.; Lauermann, M.; Yu, Y.; Wegner, D.; Herr, T.; Hartinger, K.; Schindler, P.; Li, J.; Hillerkuss, D.; et al. Coherent terabit communications with microresonator Kerr frequency combs. *Nat. Photonics* **2014**, *8*, 375–380. [CrossRef] [PubMed]
28. Brasch, V.; Lucas, E.; Jost, J.D.; Geiselmann, M.; Kippenberg, T.J. Self-referenced photonic chip soliton Kerr frequency comb. *Light. Sci. Appl.* **2017**, *6*, e16202. [CrossRef] [PubMed]
29. Okawachi, Y.; Yu, M.; Cardenas, J.; Ji, X.; Klenner, A.; Lipson, M.; Gaeta, A.L. Carrier envelope offset detection via simultaneous supercontinuum and second-harmonic generation in a silicon nitride waveguide. *Opt. Lett.* **2018**, *43*, 4627–4630. [CrossRef]
30. Carlson, D.R.; Hickstein, D.D.; Lind, A.; Droste, S.; Westly, D.; Nader, N.; Coddington, I.; Newbury, N.R.; Srinivasan, K.; Diddams, S.A.; et al. Self-referenced frequency combs using high-efficiency silicon-nitride waveguides. *Opt. Lett.* **2017**, *42*, 2314–2317. [CrossRef]
31. Klenner, A.; Mayer, A.S.; Johnson, A.R.; Luke, K.; Lamont, M.R.; Okawachi, Y.; Lipson, M.; Gaeta, A.L.; Keller, U. Gigahertz frequency comb offset stabilization based on supercontinuum generation in silicon nitride waveguides. *Opt. Express* **2016**, *24*, 11043–11053. [CrossRef]
32. Hickstein, D.D.; Jung, H.; Carlson, D.R.; Lind, A.; Coddington, I.; Srinivasan, K.; Ycas, G.G.; Cole, D.C.; Kowligy, A.; Fredrick, C.; et al. Ultrabroadband supercontinuum generation and frequency-comb stabilization using on-chip waveguides with both cubic and quadratic nonlinearities. *Phys. Rev. Appl.* **2017**, *8*, 014025. [CrossRef]
33. Yu, M.; Desiatov, B.; Okawachi, Y.; Gaeta, A.L.; Loncar, M. Coherent two-octave-spanning supercontinuum generation in lithium-niobate waveguides. *Opt. Lett.* **2019**, *44*, 1222–1225. [CrossRef]
34. Delfyett, P.J.; Klee, A.; Bagnell, K.; Juodawlkis, P.; Plant, J.; Zaman, A. Exploring the limits of semiconductor-laser-based optical frequency combs. *Appl. Opt.* **2019**, *58*, D39–D49. [CrossRef] [PubMed]
35. Waldburger, D.; Mayer, A.; Alfieri, C.; Nürnberg, J.; Johnson, A.; Ji, X.; Klenner, A.; Okawachi, Y.; Lipson, M.; Gaeta, A.; et al. Tightly locked optical frequency comb from a semiconductor disk laser. *Opt. Express* **2019**, *27*, 1786–1797. [CrossRef] [PubMed]
36. Ramirez, R.B.; Plascak, M.E.; Bagnell, K.; Bhardwaj, A.; Ferrara, J.; Hoefler, G.E.; Kish, F.A.; Wu, M.C.; Delfyett, P.J. Repetition rate stabilization and optical axial mode linewidth reduction of a chip-scale MLL using regenerative multitone injection locking. *J. Light. Technol.* **2018**, *36*, 2948–2954. [CrossRef]
37. Manurkar, P.; Perez, E.F.; Hickstein, D.D.; Carlson, D.R.; Chiles, J.; Westly, D.A.; Baumann, E.; Diddams, S.A.; Newbury, N.R.; Srinivasan, K.; et al. Fully self-referenced frequency comb consuming 5 Watts of electrical power. *OSA Contin.* **2018**, *1*, 274–282. [CrossRef]
38. Jornod, N.; Gürel, K.; Wittwer, V.J.; Brochard, P.; Hakobyan, S.; Schilt, S.; Waldburger, D.; Keller, U.; Südmeyer, T. Carrier-envelope offset frequency stabilization of a gigahertz semiconductor disk laser. *Optica* **2017**, *4*, 1482–1487. [CrossRef]
39. Spencer, D.T.; Drake, T.; Briles, T.C.; Stone, J.; Sinclair, L.C.; Fredrick, C.; Li, Q.; Westly, D.; Ilic, B.R.; Bluestone, A.; et al. An optical-frequency synthesizer using integrated photonics. *Nature* **2018**, *557*, 81–85. [CrossRef] [PubMed]
40. Obrzud, E.; Lecomte, S.; Herr, T. Temporal solitons in microresonators driven by optical pulses. *Nat. Photonics* **2017**, *11*, 600–607. [CrossRef]
41. Malinowski, M.; Rao, A.; Delfyett, P.; Fathpour, S. Optical frequency comb generation by pulsed pumping. *APL Photonics* **2017**, *2*, 066101. [CrossRef]
42. Dudley, J.M.; Genty, G.; Coen, S. Supercontinuum generation in photonic crystal fiber. *Rev. Mod. Phys.* **2006**, *78*, 1135–1184. [CrossRef]

43. Stolen, R.H.; Gordon, J.P.; Tomlinson, W.; Haus, H.A. Raman response function of silica-core fibers. *JOSA B* **1989**, *6*, 1159–1166. [CrossRef]
44. Roy, S.; Bhadra, S.K.; Agrawal, G.P. Effects of higher-order dispersion on resonant dispersive waves emitted by solitons. *Opt. Lett.* **2009**, *34*, 2072–2074. [CrossRef] [PubMed]
45. Akhmediev, N.; Karlsson, M. Cherenkov radiation emitted by solitons in optical fibers. *Phys. Rev. A* **1995**, *51*, 2602–2607. [CrossRef] [PubMed]
46. Zhang, L.; Yan, Y.; Yue, Y.; Lin, Q.; Painter, O.; Beausoleil, R.G.; Willner, A.E. On-chip two-octave supercontinuum generation by enhancing self-steepening of optical pulses. *Opt. Express* **2011**, *19*, 11584–11590. [CrossRef] [PubMed]
47. Zhang, L.; Lin, Q.; Yue, Y.; Yan, Y.; Beausoleil, R.G.; Willner, A.E. Silicon waveguide with four zero-dispersion wavelengths and its application in on-chip octave-spanning supercontinuum generation. *Opt. Express* **2012**, *20*, 1685–1690. [CrossRef] [PubMed]
48. Brasch, V.; Geiselmann, M.; Herr, T.; Lihachev, G.; Pfeiffer, M.H.P.; Gorodetsky, M.L.; Kippenberg, T.J. Photonic chip–based optical frequency comb using soliton Cherenkov radiation. *Science* **2016**, *351*, 357–360. [CrossRef] [PubMed]
49. Tremblay, J.É.; Malinowski, M.; Richardson, K.A.; Fathpour, S.; Wu, M.C. Picojoule-level octave-spanning supercontinuum generation in chalcogenide waveguides. *Opt. Express* **2018**, *26*, 21358–21363. [CrossRef] [PubMed]
50. Gordon, J.P. Theory of the soliton self-frequency shift. *Opt. Lett.* **1986**, *11*, 662–664. [CrossRef] [PubMed]
51. Corwin, K.L.; Newbury, N.R.; Dudley, J.M.; Coen, S.; Diddams, S.A.; Weber, K.; Windeler, R. Fundamental noise limitations to supercontinuum generation in microstructure fiber. *Phys. Rev. Lett.* **2003**, *90*, 113904:1–113904:4. [CrossRef]
52. Newbury, N.R.; Swann, W.C. Low-noise fiber-laser frequency combs. *JOSA B* **2007**, *24*, 1756–1770. [CrossRef]
53. Gordon, J.P.; Haus, H.A. Random walk of coherently amplified solitons in optical fiber transmission. *Opt. Lett.* **1986**, *11*, 665–667. [CrossRef]
54. Cavalcanti, S.B.; Agrawal, G.P.; Yu, M. Noise amplification in dispersive nonlinear media. *Phys. Rev. A* **1995**, *51*, 4086–4092. [CrossRef] [PubMed]
55. Newbury, N.R.; Washburn, B.; Corwin, K.L.; Windeler, R. Noise amplification during supercontinuum generation in microstructure fiber. *Opt. Lett.* **2003**, *28*, 944–946. [CrossRef] [PubMed]
56. Davenport, M.L.; Liu, S.; Bowers, J.E. Integrated heterogeneous silicon/III–V mode-locked lasers. *Photonics Res.* **2018**, *6*, 468–478. [CrossRef]
57. Wang, Z.; Van Gasse, K.; Moskalenko, V.; Latkowski, S.; Bente, E.; Kuyken, B.; Roelkens, G. A III-V-on-Si ultra-dense comb laser. *Light. Sci. Appl.* **2017**, *6*, e16260. [CrossRef] [PubMed]
58. Droste, S.; Ycas, G.; Washburn, B.R.; Coddington, I.; Newbury, N.R. Optical frequency comb generation based on erbium fiber lasers. *Nanophotonics* **2016**, *5*, 196–213. [CrossRef]
59. Sinclair, L.C.; Deschênes, J.D.; Sonderhouse, L.; Swann, W.C.; Khader, I.H.; Baumann, E.; Newbury, N.R.; Coddington, I. Invited Article: A compact optically coherent fiber frequency comb. *Rev. Sci. Instrum.* **2015**, *86*, 081301. [CrossRef] [PubMed]
60. Kim, J.; Song, Y. Ultralow-noise mode-locked fiber lasers and frequency combs: Principles, status, and applications. *Adv. Opt. Photonics* **2016**, *8*, 465–540. [CrossRef]
61. Liu, S.; Komljenovic, T.; Srinivasan, S.; Norberg, E.; Fish, G.; Bowers, J.E. Characterization of a fully integrated heterogeneous silicon/III-V colliding pulse mode-locked laser with on-chip feedback. *Opt. Express* **2018**, *26*, 9714–9723. [CrossRef]
62. Parker, J.S.; Bhardwaj, A.; Binetti, P.R.; Hung, Y.J.; Coldren, L.A. Monolithically integrated gain-flattened ring mode-locked laser for comb-line generation. *IEEE Photonics Technol. Lett.* **2012**, *24*, 131–133. [CrossRef]
63. Latkowski, S.; Moskalenko, V.; Tahvili, S.; Augustin, L.; Smit, M.; Williams, K.; Bente, E. Monolithically integrated 2.5 GHz extended cavity mode-locked ring laser with intracavity phase modulators. *Opt. Lett.* **2015**, *40*, 77–80. [CrossRef]
64. Cheung, S.; Baek, J.H.; Scott, R.P.; Fontaine, N.K.; Soares, F.M.; Zhou, X.; Baney, D.M.; Yoo, S.B. 1-GHz monolithically integrated hybrid mode-locked InP laser. *IEEE Photonics Technol. Lett.* **2010**, *22*, 1793–1795. [CrossRef]

65. Johnson, A.R.; Ji, X.; Lamont, M.R.; Okawachi, Y.; Lipson, M.; Gaeta, A.L. Coherent supercontinuum generation with picosecond pulses. In Proceedings of the CLEO: Science and Innovations. Optical Society of America, San Jose, CA, USA, 14–19 May 2017.
66. Li, F.; Li, Q.; Yuan, J.; Wai, P.K.A. Highly coherent supercontinuum generation with picosecond pulses by using self-similar compression. *Opt. Express* **2014**, *22*, 27339–27354. [CrossRef] [PubMed]
67. Resan, B.; Delfyett, P.J. Dispersion-managed breathing-mode semiconductor mode-locked ring laser: Experimental characterization and numerical simulations. *IEEE J. Quantum Electron.* **2004**, *40*, 214–221. [CrossRef]
68. Takushima, Y.; Sotobayashi, H.; Grein, M.E.; Ippen, E.P.; Haus, H.A. Linewidth of mode combs of passively and actively mode-locked semiconductor laser diodes. In Proceedings of the Active and Passive Optical Components for WDM Communications IV. International Society for Optics and Photonics, Philadelphia, PA, USA, 25–28 October 2004; Volume 5595, pp. 213–228.
69. Haji, M.; Hou, L.; Kelly, A.E.; Akbar, J.; Marsh, J.H.; Arnold, J.M.; Ironside, C.N. High frequency optoelectronic oscillators based on the optical feedback of semiconductor mode-locked laser diodes. *Opt. Express* **2012**, *20*, 3268–3274. [CrossRef] [PubMed]
70. Lee, W.; Delfyett, P.J. Dual-mode injection locking of two independent modelocked semiconductor lasers. *Electron. Lett.* **2004**, *40*, 1182–1183. [CrossRef]
71. Takada, A.; Imajuku, W. Linewidth narrowing and optical phase control of mode-locked semiconductor ring laser employing optical injection locking. *IEEE Photonics Technol. Lett.* **1997**, *9*, 1328–1330. [CrossRef]
72. Bhardwaj, A.; Ferrara, J.; Ramirez, R.B.; Plascak, M.; Hoefler, G.; Lal, V.; Kish, F.; Delfyett, P.; Wu, M. An integrated racetrack colliding-pulse mode-locked laser with pulse-picking modulator. In *CLEO: Science and Innovations*; Optical Society of America: Washington, DC, USA, 2017.
73. Plascak, M.E.; Ramirez, R.B.; Malinowski, M.; Tremblay, J.E.; Bhardwaj, A.; Hoefler, G.C.; Fathpour, S.; Wu, M.C.; Delfyett, P.J. Progress towards full stabilization of an injection locked 10 GHz chip-scale mode-locked laser on InP. In Proceedings of the 2018 Conference on Lasers and Electro-Optics (CLEO), San Jose, CA, USA, 13–18 May 2018.
74. Tsang, H.; Liu, Y. Nonlinear optical properties of silicon waveguides. *Semicond. Sci. Technol.* **2008**, *23*, 064007. [CrossRef]
75. Hsieh, I.W.; Chen, X.; Liu, X.; Dadap, J.I.; Panoiu, N.C.; Chou, C.Y.; Xia, F.; Green, W.M.; Vlasov, Y.A.; Osgood, R.M. Supercontinuum generation in silicon photonic wires. *Opt. Express* **2007**, *15*, 15242–15249. [CrossRef]
76. Singh, N.; Xin, M.; Vermeulen, D.; Shtyrkova, K.; Li, N.; Callahan, P.T.; Magden, E.S.; Ruocco, A.; Fahrenkopf, N.; Baiocco, C.; et al. Octave-spanning coherent supercontinuum generation in silicon on insulator from 1.06 mm to beyond 2.4 mm. *Light. Sci. Appl.* **2018**, *7*, 17131. [CrossRef] [PubMed]
77. Lamont, M.R.; Luther-Davies, B.; Choi, D.Y.; Madden, S.; Eggleton, B.J. Supercontinuum generation in dispersion engineered highly nonlinear ($\gamma = 10$ /W/m) As_2S_3 chalcogenide planar waveguide. *Opt. Express* **2008**, *16*, 14938–14944. [CrossRef]
78. Dave, U.D.; Ciret, C.; Gorza, S.P.; Combrie, S.; Rossi, A.D.; Raineri, F.; Roelkens, G.; Kuyken, B. Dispersive-wave-based octave-spanning supercontinuum generation in InGaP membrane waveguides on a silicon substrate. *Opt. Lett.* **2015**, *40*, 3584–3587. [CrossRef] [PubMed]
79. Lau, R.K.; Lamont, M.R.; Griffith, A.G.; Okawachi, Y.; Lipson, M.; Gaeta, A.L. Octave-spanning mid-infrared supercontinuum generation in silicon nanowaveguides. *Opt. Lett.* **2014**, *39*, 4518–4521. [CrossRef] [PubMed]
80. Chiles, J.; Gai, X.; Luther-Davies, B.; Fathpour, S. Mid-infrared supercontinuum generation in high-contrast, fusion-bonded silicon membrane waveguides. In Proceedings of the 2017 IEEE Photonics Conference (IPC), Orlando, FL, USA, 1–5 October 2017; pp. 313–314. [CrossRef]
81. Sinobad, M.; Monat, C.; Luther-Davies, B.; Ma, P.; Madden, S.; Moss, D.J.; Mitchell, A.; Allioux, D.; Orobtchouk, R.; Boutami, S.; et al. Mid-infrared octave spanning supercontinuum generation to 8.5 μm in silicon-germanium waveguides. *Optica* **2018**, *5*, 360–366. [CrossRef]
82. Singh, N.; Hudson, D.D.; Yu, Y.; Grillet, C.; Jackson, S.D.; Casas-Bedoya, A.; Read, A.; Atanackovic, P.; Duvall, S.G.; Palomba, S.; et al. Midinfrared supercontinuum generation from 2 to 6 μm in a silicon nanowire. *Optica* **2015**, *2*, 797–802. [CrossRef]
83. Sheik-Bahae, M.; Hagan, D.J.; Van Stryland, E.W. Dispersion and band-gap scaling of the electronic Kerr effect in solids associated with two-photon absorption. *Phys. Rev. Lett.* **1990**, *65*, 96–99. [CrossRef] [PubMed]

84. Rao, A.; Fathpour, S. Second-Harmonic Generation in Integrated Photonics on Silicon. *Phys. Status Solidi A* **2018**, *215*, 1700684. [CrossRef]
85. Fathpour, S. Heterogeneous Nonlinear Integrated Photonics. *IEEE J. Quantum Electron.* **2018**, *54*, 1–16. [CrossRef]
86. Chang, L.; Boes, A.; Guo, X.; Spencer, D.T.; Kennedy, M.; Peters, J.D.; Volet, N.; Chiles, J.; Kowligy, A.; Nader, N.; et al. Heterogeneously integrated GaAs waveguides on insulator for efficient frequency conversion. *Laser Photonics Rev.* **2018**, *12*, 1800149. [CrossRef]
87. Guo, X.; Zou, C.L.; Tang, H.X. Second-harmonic generation in aluminum nitride microrings with 2500%/W conversion efficiency. *Optica* **2016**, *3*, 1126–1131. [CrossRef]
88. Xiong, C.; Pernice, W.; Ryu, K.K.; Schuck, C.; Fong, K.Y.; Palacios, T.; Tang, H.X. Integrated GaN photonic circuits on silicon (100) for second harmonic generation. *Opt. Express* **2011**, *19*, 10462–10470. [CrossRef]
89. Levy, J.S.; Foster, M.A.; Gaeta, A.L.; Lipson, M. Harmonic generation in silicon nitride ring resonators. *Opt. Express* **2011**, *19*, 11415–11421. [CrossRef] [PubMed]
90. Rao, A.; Malinowski, M.; Honardoost, A.; Talukder, J.R.; Rabiei, P.; Delfyett, P.; Fathpour, S. Second-harmonic generation in periodically-poled thin film lithium niobate wafer-bonded on silicon. *Opt. Express* **2016**, *24*, 29941–29947. [CrossRef] [PubMed]
91. Wang, C.; Langrock, C.; Marandi, A.; Jankowski, M.; Zhang, M.; Desiatov, B.; Fejer, M.M.; Lončar, M. Ultrahigh-efficiency wavelength conversion in nanophotonic periodically poled lithium niobate waveguides. *Optica* **2018**, *5*, 1438–1441. [CrossRef]
92. Khan, S.; Malinowski, M.; Tremblay, J.E.; Rao, A.; Camacho-González, G.F.; Ramirez, R.B.; Plascak, M.; Richardson, K.A.; Delfyett, P.; Wu, M.C.; et al. Integrated thin-film lithium-niobate waveguides on silicon for second-harmonic generation pumped at 1875 nm. In Proceedings of the Conference on Lasers and Electro-Optics, San Jose, CA, USA, 13–18 May 2018.
93. Chang, L.; Volet, N.; Li, Y.; Peters, J.; Bowers, J.E. A thin-film PPLN waveguide for second-harmonic generation at 2-μm. In Proceedings of the 2016 IEEE Photonics Conference (IPC), Waikoloa, HI, USA, 2–6 October 2016; pp. 587–588.
94. Chang, L.; Pfeiffer, M.H.; Volet, N.; Zervas, M.; Peters, J.D.; Manganelli, C.L.; Stanton, E.J.; Li, Y.; Kippenberg, T.J.; Bowers, J.E. Heterogeneous integration of lithium niobate and silicon nitride waveguides for wafer-scale photonic integrated circuits on silicon. *Opt. Lett.* **2017**, *42*, 803–806. [CrossRef] [PubMed]
95. Honardoost, A.; Gonzalez, G.F.C.; Khan, S.; Malinowski, M.; Rao, A.; Tremblay, J.; Yadav, A.; Richardson, K.; Wu, M.C.; Fathpour, S. Cascaded integration of optical waveguides with third-order nonlinearity with lithium niobate waveguides on silicon substrates. *IEEE Photonics J.* **2018**, *10*, 1–9. [CrossRef]
96. Gonzalez, G.F.C.; Malinowski, M.; Honardoost, A.; Fathpour, S. Design of a hybrid chalcogenide-glass on lithium-niobate waveguide structure for high-performance cascaded third- and second-order optical nonlinearities. *Appl. Opt.* **2019**, *58*, D1–D6. [CrossRef]
97. Tremblay, J.E.; Malinowski, M.; Camacho-Gonzalez, G.; Fathpour, S.; Wu, M.C. Large Bandwidth Waveguide Spectral Splitters Using Higher-Order Mode Evolution. In Proceedings of the Conference on Lasers and Electro-Optics, San Jose, CA, USA, 13–18 May 2018.
98. Malinowski, M.; Tremblay, J.; Gonzalez, G.F.C.; Rao, A.; Khan, S.; Hsu, P.; Yadav, A.; Richardson, K.A.; Delfyett, P.; Wu, M.C.; et al. Amplified octave-spanning supercontinuum from chalcogenide waveguides for second-harmonic generation. In Proceedings of the 2017 IEEE Photonics Conference (IPC), Orlando, FL, USA, 1–5 October 2017; pp. 261–262.

 *micromachines*

Article

Effect of Nitrogen Doping on the Photoluminescence of Amorphous Silicon Oxycarbide Films

Jie Song, Rui Huang, Yi Zhang, Zewen Lin, Wenxing Zhang *, Hongliang Li, Chao Song, Yanqing Guo and Zhenxu Lin *

School of Materials Science and Engineering, Hanshan Normal University, Chaozhou 521041, China; songjie@hstc.edu.cn (J.S.); rhuang@hstc.edu.cn (R.H.); cavtor@126.com (Y.Z.); 2636@hstc.edu.cn (Z.L.); lhl4@hstc.edu.cn (H.L.); chaosong@hstc.edu.cn (C.S.); yqguo126@126.com (Y.G.)

* Correspondence: wenxingzhang@hstc.edu.cn (W.Z.); linzhenxu2013@163.com (Z.L.)

Received: 26 August 2019; Accepted: 26 September 2019; Published: 27 September 2019

Abstract: The effect of nitrogen doping on the photoluminescence (PL) of amorphous SiC_xO_y films was investigated. An increase in the content of nitrogen in the films from 1.07% to 25.6% resulted in red, orange-yellow, white, and blue switching PL. Luminescence decay measurements showed an ultrafast decay dynamic with a lifetime of ~1 ns for all the nitrogen-doped SiC_xO_y films. Nitrogen doping could also widen the bandgap of SiC_xO_y films. The microstructure and the elemental compositions of the films were studied by obtaining their Raman spectra and their X-ray photoelectron spectroscopy, respectively. The PL characteristics combined with an analysis of the chemical bonds configurations present in the films suggested that the switching PL was attributed to the change in defect luminescent centers resulting from the chemical bond reconstruction as a function of nitrogen doping. Nitrogen doping provides an alternative route for designing and fabricating tunable and efficient SiC_xO_y-based luminescent films for the development of Si-based optoelectronic devices.

Keywords: photoluminescence; amorphous silicon oxycarbide; nitrogen doping; defect; plasma enhanced chemical vapor deposition

1. Introduction

Efficient silicon (Si)-based luminescent materials are indispensable components to realize a cheap and complementary metal oxide semiconductor (CMOS) optical integration. Thus far, different systems of Si-based luminescent materials, such as SiO_x, SiN_x, SiC_x, and SiN_xO_y, have been developed, and efforts have been devoted to understanding and ameliorating the light emission of Si-based materials [1–9]. Silicon oxycarbide (SiC_xO_y) has been widely explored because of its strong light emission and high solid solubility for rare earths [10–13]. SiC_xO_y also features a tunable band gap. As such, it is beneficial to obtaining strong white electroluminescence at a low driving voltage in SiC_xO_y-based light-emitting diodes [14]. In the recent reference, Gallis et al. systematically studied the white photoluminescence (PL) dynamics from SiC_xO_y film, where the band tail states related to the Si–O–C and/or the Si–C bonds were suggested as the sources of the luminescence [11]. Recently, optical gain was demonstrated in a-SiC_xO_y under ultraviolet excitation, which was attributed to the formation of a three-level luminescence model with the intermediate level related to Si dangling bond (DB) defects radiative state [15]. Furthermore, an increase in C content in SiC_xO_y films can cause a strong light emission ranging from near-infrared to orange regions [15]. Although performance is enhanced in SiC_xO_y films, progress remains slow. The main obstacle lies in the fact that the light emission efficiency generally remains too low to allow the fabrication of efficient light-emitter devices. To date, studies on the effect of doping on the optical properties of SiC_xO_y films have mainly focused on rare earth (RE) doping, such as Er and Eu doping [16–18]. However, up to now, the effect of other elements on the optical properties of SiC_xO_y films is still unclear.

In this letter, the effect of nitrogen doping on the PL of amorphous SiC_xO_y film was investigated. Interestingly, an increase in nitrogen content in the films induced strong red, orange-yellow, white, and blue switching PL. Combining the PL results with the analysis of the microstructure and the chemical bonding configurations within the films, it suggests that the rearrangement of chemical bonds with varying nitrogen plays an important role in the evolution of PL characteristics in the films.

2. Materials and Methods

Nitrogen-doped SiC_xO_y films with the thickness of 550 nm were grown at 250 °C on Si substrates and quartz by radio frequency (RF) glow-discharge decomposition of SiH_4, CH_4, O_2, and NH_3 mixtures in the very high frequency plasma enhanced chemical vapor deposition (VHF-PECVD) system. The flow rates of SiH_4, CH_4, and O_2 were kept at 3.5, 5, and 1.2 sccm, respectively, whereas the flow rate of NH_3 varied from 0.5 sccm to 5 sccm to control the N content in the films. The films were named S_x (x = 1, 2, 3, 4) for the NH_3 flow rates at 0.5, 1, 3, and 5 sccm, respectively. The RF power and the deposition pressure for the growth were maintained at 30 W and 20 Pa, respectively. The optical band gaps of the films were calculated using the Tauc method, which were determined by spectrophotometer (Shimadzu UV-3600, Shimadzu Corporation, Kyoto, Japan). The PL spectra of the films were measured at room temperature by use of a fluorescence spectrophotometer (Jobin Yvon fluorolog-3, Horiba, Ltd., Kyoto, Japan). Time resolved PL were measured by use of an Edinburgh FLS1000 spectrometer (Edinburgh Instruments Ltd., Livingston, UK) equipped with a 600 mW 375 nm-laser beam with the repetition rate of 20 MHz. The microstructures of the films were characterized by Raman spectra. The chemical bonds of the films were examined by Fourier transform infrared (FTIR) absorption, and Si, O, C, and N contents in the films were identified through X-ray photoelectron spectroscopy.

3. Results and Discussion

Figure 1a shows the PL spectra of the SiCO:N films prepared at different NH_3 flow rates. Strong visible light emission could be tuned from red to blue regions at room temperature by adjusting the NH_3 flow rates increased from 0.5 sccm to 5 sccm. Red, orange-yellow, white, and blue switching luminescence were strong enough to be seen with naked eyes even at 325 nm Xe lamp light excitation. Figure 1b illustrates the optical band gap energy of the films as a function of the NH_3 flow rate. The optical band gap E_{opt} of the films was obtained in accordance with the Tauc plot Equation (1):

$$(\alpha h\nu)^{1/2} = A^{1/2}(h\nu - E_{opt}) \quad (1)$$

where α is the absorption coefficient, A is a coefficient quantifying the slope of the absorption edge, and $h\nu$ is the photon energy [19]. The calculated E_{opt} increases linearly from 2.83 eV to 3.66 eV as the NH_3 flow rate increases from 0.5 sccm to 5 sccm. This finding demonstrates that N doping can widen the bandgap of SiC_xO_y films, which may result from the substitution of stronger Si–N bonds for weak Si–Si bonds or Si–C bonds. The comparison of PL with E_{opt} results indicated that the value of PL peak energy of all the films was obviously smaller than the corresponding E_{opt}, suggesting that the origin of PL was not from the band-to-band recombination.

The microstructure of the SiCO:N films was examined using Raman spectra (Figure 2a) to further understand the origin of PL characteristics. All the SiCO:N films exhibited similar line shape characteristics typical of amorphous silicon-based materials. A broad Raman band, which was ascribed to the transverse optical (TO) vibration mode of amorphous silicon, peaked at ~470 cm^{-1} for all the SiCO:N films. These results showed that all the SiCO:N films had a uniform amorphous structure without the presence of Si nanocrystals [20]. Furthermore, there was no obvious change in the surface morphology of the films prepared at different NH_3 flow rates, as was revealed by atomic force microscopy (Figure 2b).

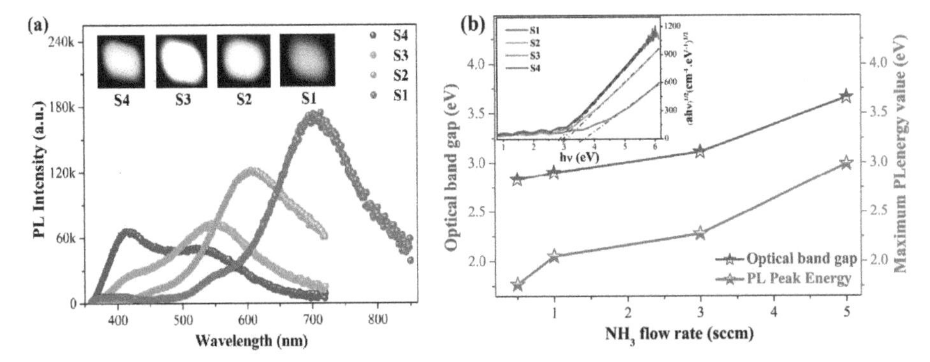

Figure 1. (**a**) Photoluminescence (PL) spectra of the SiCO:N films prepared by different NH_3 flow rates: S1 (0.5 sccm), S2 (1 sccm), S3 (3 sccm), and S4 (5 sccm). The inset is the optical images of PL from the films under 325 nm Xe lamp light excitation. (**b**) The optical band gap of the SiCO:N films vs. the NH_3 flow rates. The inset shows the ($(\alpha h\nu)^{1/2}$-vs-$h\nu$) plot of the SiCO:N film S_x (x = 1, 2, 3, 4).

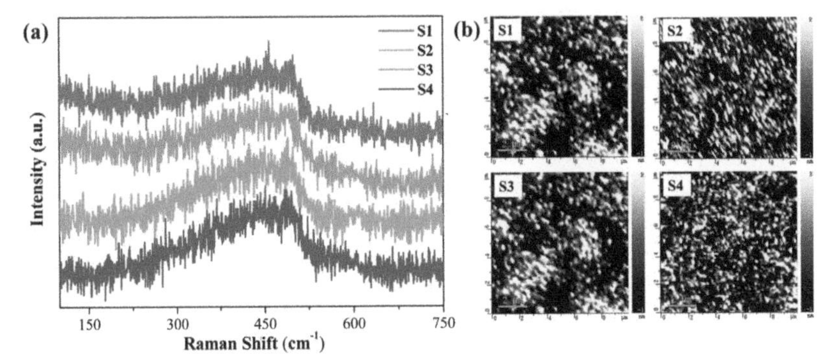

Figure 2. (**a**) Raman spectra of the SiCO:N films with various NH_3 flow rates, (**b**) atomic force microscopic images of the SiCO:N films prepared by different NH_3 flow rates: S1 (0.5 sccm), S2 (1 sccm), S3 (3 sccm), and S4 (5 sccm).

The films were measured by time-resolved PL to obtain further insights into the PL mechanism of the SiCO:N films (Figure 3). The decay curves could be fitted with a double exponential function:

$$I(t) = A_1 \exp(\frac{-t}{\tau_1}) + A_2 \exp(\frac{-t}{\tau_2}) \quad (2)$$

where A_i and τ_i (i = 1, 2) are the normalized amplitudes of the components and the time constants, respectively [21]. The obtained average lifetime of the SiCO:N films was about 1 ns. The luminescent dynamic behavior was similar to that observed in defect-related luminescent Si-based materials, such as SiN_xO_y and SiC_xO_y [19,21]. Furthermore, it was also found that the luminescence decay lifetimes in our case were shorter than those in the band-tail recombination model where a broader band-tail brought a longer lifetime, as the photogenerated carriers could be thermalized into deeper localized states [22]. Therefore, the results suggested that the light emission of the SiCO:N films originated from the defect luminescent centers in the films.

The FTIR spectra of the SiCO:N films were obtained to study the local bonding changes in the films grown at different NH_3 flow rates (Figure 4). In the S1 film, the vibration modes related to Si–C, Si–N, C–Si–O, Si–H, and C–H bonds could be clearly observed. The bands centered at 860 and 1039 cm^{-1} could be ascribed to Si–N and C–Si–O stretching modes, respectively [23,24]. Additionally, a band at 1265 cm^{-1} was assigned to the Si–CH_3 stretching vibration [25]. A small band shoulder at

800 cm^{-1} was observed and was assigned to the Si–C stretching vibration [24]. A distinct absorption peak at 2170 cm^{-1} and a weak band located at 2965 cm^{-1} were attributed to the Si–H and the C–H stretching vibrations, respectively [26]. A weak band around 3375 cm^{-1} was associated with the N–H stretching mode [23]. The most important feature for the FTIR spectra was the strong dependence of major bands on NH_3 flow rates. As the NH_3 flow rates increased, the intensity of C–Si–O bonds gradually decreased, and the peak gradually became red shifted. As the NH_3 flow rate increased to 3 sccm, this band broadened and red shifted to ~1010 cm^{-1} with a shoulder at ~940 cm^{-1}, which was assigned to the N–Si–O vibration [27]. As the NH_3 flow rate further increased to 5 sccm, the band of the N–Si–O vibration became dominant, indicating that the silicon oxycarbide-dominant phase of the film transformed into silicon oxynitride. Apparently, the increase in the NH_3 flow rate resulted in chemical bond reconstruction in the films. Based on the FTIR spectra, the evolution of PL characteristics could be suggested from the chemical bond reconstruction in SiCO:N films.

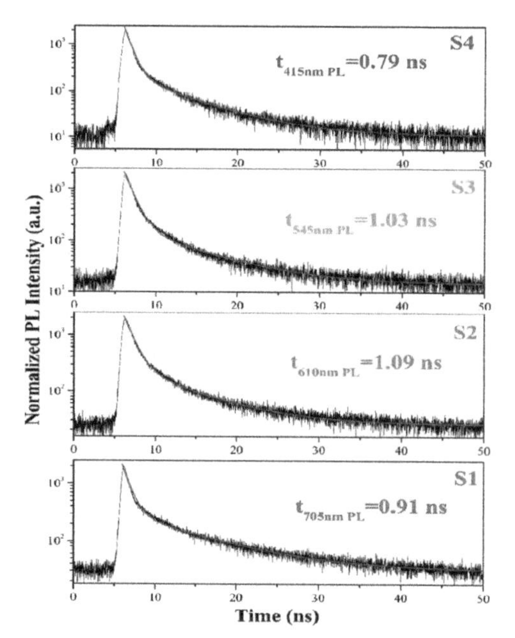

Figure 3. Room temperature time resolved photoluminescence for the SiCO:N films with various NH_3 flow rates.

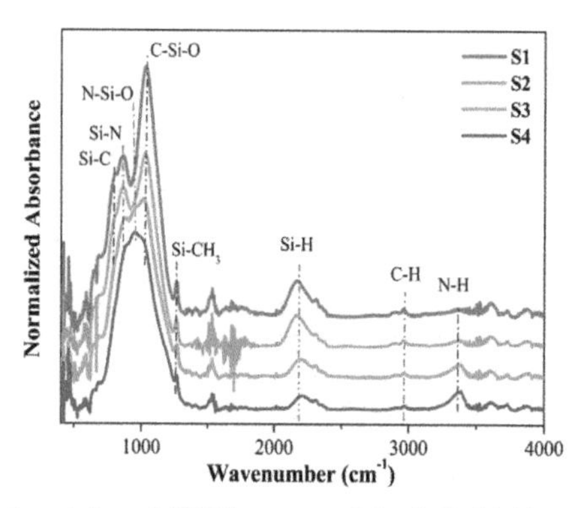

Figure 4. Fourier transform infrared (FTIR) spectra of the SiCO:N films grown at different NH_3 flow rates.

The composition of the SiCO:N films was examined through X-ray photoelectron spectroscopy (XPS) (Figure 5). The atomic percentages of Si, C, O, and N in the SiCO:N film fabricated at an NH_3 flow rate of 0.5 sccm were 51.89%, 19.82%, 27.22%, and 1.07%, respectively. This finding indicated that the Si-rich silicon oxycarbide phase was dominant in the S1 film. The change in the XPS spectra was the gradual decrease in Si and C concentrations with the increase in N concentration and NH_3 flow rates (Figure 5). As the NH_3 flow rate increased to 5 sccm, the N concentration rapidly increased to 25.6%, whereas the Si and the C concentrations decreased to 40.8% and 8.22%, respectively. This finding was consistent with the observed results in the FTIR spectra shown in Figure 4, that is, the N–Si–O vibration band became dominant, while the C–Si–O vibration band significantly weakened as the NH_3 flow rate increased to 5 sccm. This result indicated that the dominant phase in the films changed from silicon oxycarbide to silicon oxynitride when the NH_3 flow rate increased to 5 sccm.

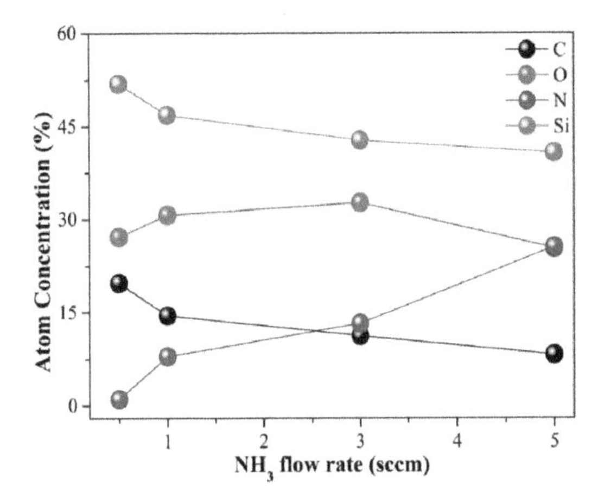

Figure 5. The atom concentration of Si, C, O, and N of the SiCO:N films against the NH_3 flow rates.

The PL decay analysis (Figure 3) revealed that the luminescent dynamic behavior in the nitrogen doped SiC_xO_y films featured a defect-related luminescent characteristic, as observed in SiN_xO_y and SiC_xO_y films. Previous studies clarified that C-related nonbridging oxygen hole centers (NBOHC) are the principal radiative recombination centers in silicon oxycarbide, and they are responsible for light emission ranging from the green region to the red region [28]. In our case, the PL intensity in the film decreased as the NH_3 flow rate increased. This change was similar to that of the intensity of the C–Si–O bonds (Figure 4). Therefore, the observed tunable light emissions from green to red may have originated from recombination through C-related NBOHC defects in SiC_xO_y films. The PL spectra of the S3 film could be deconvoluted into a strong green band and a weak blue band. As the NH_3 flow rate increased to 5 sccm, the intensity of the green PL band decreased dramatically, whereas the blue PL band of the film S4 became dominant. This behavior could be attributed to the change in the dominant phase of the films from silicon oxycarbide to silicon oxynitride as a result of the increase in NH_3 flow rate to 5 sccm (Figure 4). In the case of amorphous SiN_xO_y films, the blue PL could be ascribed to the radiative recombination between N–Si–O defect states and the valence band tail states [27]. Thus, the blue PL from S3 and S4 was likely from N–Si–O defect luminescent centers.

4. Conclusions

In summary, we report the effect of nitrogen doping on the PL of amorphous SiC_xO_y films. Nitrogen doping can induce strong red, orange-yellow, white, and blue switching PL with a recombination lifetime in nanoseconds. This process can also widen the band gap of SiC_xO_y films. The PL results and the FTIR analyses reveal that the switching characteristics in PL originate from the variation in defect luminescent centers resulting from the chemical bond re-construction as a function of nitrogen doping. Apparently, nitrogen doping provides an alternative route for designing and fabricating tunable and efficient SiC_xO_y-based luminescent films for the development of Si-based optoelec-tronic devices.

Author Contributions: Data curation, J.S. and Z.L. (Zhenxu Lin); Investigation, Y.Z., Z.L. (Zewen Lin), and H.L.; Methodology, C.S., and Y.G., All authors analyzed the experimental data; Writing–review & editing, J.S., Z.L. (Zhenxu Lin), R.H. and W.Z.; All authors participated in discussions and knew implications of the work.

Funding: This work was supported by National Natural Science Foundation of China (Nos. 61274140), Natural Science Foundation of Guangdong Province (2015A030313871), Young Talents in Higher Education of Guangdong (2017KQNCX129), the Distinguished Young Teacher Training Program in Higher Education of Guangdong (YQ2015112), Science and Technology Planning Project of Guangdong Province (2017B090921002) and Science and Technology Planning Project of Chaozhou (2018SS24).

Conflicts of Interest: The authors declare no conflict of interest.

References

1. Pavesi, L.; Negro, L.D.; Mazzoleni, C.; Franzo, G.; Priolo, F. Optical gain in silicon nanocrystals. *Nature* **2000**, *408*, 440–444. [CrossRef] [PubMed]
2. Cheng, C.H.; Lien, Y.C.; Wu, C.L.; Lin, G.R. Mutlicolor electroluminescent Si quantum dots embedded in SiOx thin film MOSLED with 24% external quantum efficiency. *Opt. Express* **2013**, *21*, 391–403. [CrossRef] [PubMed]
3. Li, D.; Wang, F.; Yang, D. Evolution of electroluminescence from silicon nitride light-emitting devices via nanostructural silver. *Nanoscale* **2013**, *5*, 3435–3440. [CrossRef] [PubMed]
4. Huang, R.; Lin, Z.; Lin, Z.; Song, C. Suppression of hole overflow and enhancement of light emission efficiency in Si quantum dots based silicon nitride light emitting diodes. *IEEE J. Sel. Top. Quantum. Electron.* **2014**, *20*, 212–217. [CrossRef]
5. Limpens, R.; Luxembourg, S.L.; Weeber, A.W.; Gregorkiewicz, T. Emission efficiency limit of Si nanocrystals. *Sci. Rep.* **2016**, *6*, 19566. [CrossRef] [PubMed]
6. Yao, L.; Yu, T.; Ba, L.; Meng, H.; Fang, X.; Wang, Y. Efficient silicon quantum dots light emitting diodes with an inverted device structure. *J. Mater. Chem. C* **2016**, *4*, 673–677. [CrossRef]
7. Lin, Z.; Chen, K.; Zhang, P.; Xu, J.; Li, W.; Yang, H. Improved power efficiency in phosphorus doped na-SiN_xO_y/p-Si heterojunction light emitting diode. *Appl. Phys. Lett.* **2017**, *110*, 081109. [CrossRef]
8. Li, M.; Jiang, L.; Sun, Y.; Xiao, T.; Xiang, P.; Tan, X. Silicon content influence on structure and photoluminescence properties of carbon rich hydrogenated amorphous silicon carbide thin films. *J. Alloys Compd.* **2018**, *753*, 320–328. [CrossRef]
9. Zhang, X.; Chen, R.; Wang, P.; Gan, Z.; Zhang, Y.; Jin, H.; Jian, J.; Xu, J. Investigation of energy transfer mechanisms in rare earth ions and SnO_2 quantum dots co-doped amorphous silica films. *Opt. Express* **2019**, *27*, 2783–2791. [CrossRef]
10. Ding, Y.; Shirai, H. White light emission from silicon oxycarbide films prepared by using atmospheric pressure microplasma jet. *J. Appl. Phys.* **2009**, *105*, 043515. [CrossRef]
11. Tabassum, N.; Nikas, V.; Ford, B.; Huang, M.; Kaloyeros, A.E.; Gallis, S. Time-resolved analysis of the white photoluminescence from chemically synthesized SiC_xO_y thin films and nanowires. *Appl. Phys. Lett.* **2016**, *109*, 043104. [CrossRef]
12. Bellocchi, G.; Iacona, F.; Miritello, M.; Cesca, T.; Franzò, G. SiOC thin films: An efficient light source and an ideal host matrix for Eu^{2+} ions. *Opt. Express* **2013**, *21*, 20280–20290. [CrossRef] [PubMed]
13. Lin, Z.; Huang, R.; Wang, H.; Wang, Y.; Zhang, Y.; Guo, Y. Dense nanosized europium silicate clusters induced light emission enhancement in Eu-doped silicon oxycarbide films. *J. Alloys Compd.* **2017**, *694*, 946–951. [CrossRef]
14. Ding, Y.; Shirai, H.; He, D. White light emission and electrical properties of silicon oxycarbide-based metal-oxide-semiconductor diode. *Thin Solid Films* **2011**, *519*, 2513–2515. [CrossRef]
15. Lin, Z.; Huang, R.; Zhang, Y.; Song, J.; Li, H.; Guo, Y.; Song, C. Defect emission and optical gain in SiC_xO_y: H films. *ACS Appl. Mater. Interfaces* **2017**, *9*, 22725–22731. [CrossRef]
16. Gallis, S.; Huang, M.; Kaloyeros, A.E. Efficient energy transfer from silicon oxycarbide matrix to Er ions via indirect excitation mechanisms. *Appl. Phys. Lett.* **2007**, *90*, 161914. [CrossRef]

17. Bellocchi, G.; Franzò, G.; Miritello, M.; Iacona, F. White light emission from Eu-doped SiOC films. *Appl. Phys. Express* **2013**, *7*, 012601. [CrossRef]
18. Bellocchi, G.; Fabbri, F.; Miritello, M.; Iacona, F.; Franzò, G. Multicolor depth-resolved cathodoluminescence from Eu-doped SiOC thin films. *ACS Appl. Mater. Interfaces* **2015**, *7*, 18201–18205. [CrossRef]
19. Lin, Z.; Guo, Y.; Song, J.; Zhang, Y.; Song, C.; Wang, X. Effect of thermal annealing on the blue luminescence of amorphous silicon oxycarbide films. *J. Non-Cryst. Solids* **2015**, *428*, 184–188. [CrossRef]
20. Rui, Y.; Li, S.; Xu, J.; Song, C.; Jiang, X.; Li, W. Size-dependent electroluminescence from Si quantum dots embedded in amorphous SiC matrix. *J. Appl. Phys.* **2011**, *110*, 064322. [CrossRef]
21. Huang, R.; Lin, Z.; Guo, Y.; Song, C.; Wang, X.; Lin, H. Bright red, orange-yellow and white switching photoluminescence from silicon oxynitride films with fast decay dynamics. *Opt. Mater. Express* **2014**, *4*, 205–212. [CrossRef]
22. Kato, H.; Kashio, N.; Ohki, Y.; Seol, K.S.; Noma, T. Band-tail photoluminescence in hydrogenated amorphous silicon oxynitride and silicon nitride films. *J. Appl. Phys.* **2003**, *93*, 239–244. [CrossRef]
23. Huang, R.; Wang, X.; Song, J.; Guo, Y.; Ding, H.; Wang, D. Strong orange–red light emissions from amorphous silicon nitride films grown at high pressures. *Scripta Mater.* **2010**, *62*, 643–645. [CrossRef]
24. Grill, A.; Neumayer, D.A. Structure of low dielectric constant to extreme low dielectric constant SiCOH films: Fourier transform infrared spectroscopy characterization. *J. Appl. Phys.* **2003**, *94*, 6697–6707. [CrossRef]
25. Yao, R.; Feng, Z.; Zhang, B.; Zhao, H.; Yu, Y.; Chen, L. Blue photoluminescence from continuous freestanding β-SiC/SiO_xC_y/Cfree nanocomposite films with polycarbosilane (pcs) precursor. *Opt. Mater.* **2011**, *33*, 635–642. [CrossRef]
26. Wang, J.; Suendo, V.; Abramov, A.; Yu, L.; Pere, R.I.C. Strongly enhanced tunable photoluminescence in polymorphous silicon carbon thin films via excitation-transfer mechanism. *Appl. Phys. Lett.* **2010**, *97*, 221113. [CrossRef]
27. Zhang, P.; Chen, K.; Lin, Z.; Tan, D.; Dong, H.; Li, W. Dynamics of high quantum efficiency photoluminescence from N-Si-O bonding states in oxygenated amorphous silicon nitride films. *Appl. Phys. Lett.* **2016**, *118*, 111103. [CrossRef]
28. Gallis, S.; Nikas, V.; Suhag, H.; Huang, M.; Kaloyeros, A.E. White light emission from amorphous silicon oxycarbide (a-SiC_xO_y) thin films: Role of composition and postdeposition annealing. *Appl. Phys. Lett.* **2010**, *97*, 081905. [CrossRef]

Article

Programmable SCOW Mesh Silicon Photonic Processor for Linear Unitary Operator

Liangjun Lu *, Linjie Zhou and Jianping Chen

State Key Laboratory of Advanced Optical Communication Systems and Networks, Shanghai Institute for Advanced Communication and Data Science, Shanghai Key Lab of Navigation and Location Services, Department of Electronic Engineering, Shanghai Jiao Tong University, Shanghai 200240, China; ljzhou@sjtu.edu.cn (L.Z.); jpchen62@sjtu.edu.cn (J.C.)

* Correspondence: luliangjun@sjtu.edu.cn; Tel.: +86-21-6493-2761

Received: 27 August 2019; Accepted: 24 September 2019; Published: 26 September 2019

Abstract: Universal unitary multiport interferometers (UMIs) can perform any arbitrary unitary transformation to a vector of input optical modes, which are essential for a wide range of applications. Most UMIs are realized by fixed photonic circuits with a triangular or a rectangular architecture. Here, we present the implementation of an $N \times N$ rectangular UMI with a programmable photonic processor based on two-dimensional meshes of self-coupled optical waveguide (SCOW) resonant structures. Our architecture shows a high tolerance to the unbalanced loss upon interference. This work enriches the functionality of the SCOW mesh photonic processors, which are promising for field-programmable photonic arrays.

Keywords: photonic processors; unitary transformation; silicon photonics

1. Introduction

Photonic integration has attracted increasing interest in the potential of extensive reduction of size, weight, and power (SWaP). Silicon photonics is a promising solution for complex integrated photonic systems with low cost and high integration density, due to the advantages of complementary metal-oxide-semiconductor (CMOS) compatible processing and high refractive index contrast. Lots of silicon-based photonic integrated circuits (PICs) have been demonstrated in a wide variety of applications. Among them, unitary multiport interferometers (UMIs) are now receiving more and more attention. UMIs can implement various arbitrary unitary transformation, addressing a wide range of applications including mode-division (de)multiplexing [1,2], photonic deep machine learning [3], quantum particle simulation [4], and quantum signal processing [5,6]. Besides, a UMI can be extended to any $M \times N$ linear (non-unitary) transformation. The most common architecture for UMIs is based on a specific triangular mesh of 2×2 couplers and phase shifters, proposed by Reck et al. [7]. Recently, a silicon 4×4 photonic UMI has been experimentally demonstrated [8]. An optimal design with a more symmetrical rectangular mesh for the UMIs shows the superior performance of less optical depth and more robustness to optical losses compared with Reck's architecture [9].

Inspired by the concept of Field Programmable Gate Arrays (FPGAs), general photonic processors, which can be programmed to perform multiple photonic processing functions by using the same hardware configuration, have the merits of higher flexibility and more cost-effectiveness compared with application-specific PICs (ASPICs) [10–12]. Several elegant programmable photonic processors have been proposed and demonstrated with square [13], hexagonal [14], or triangular [15] mesh networks. These processors show versatile functionalities in optical and microwave photonics signal processing [16–18]. The method and algorithm for setting up mesh networks composed of non-ideal components have also been proposed [19–21]. Recently, Perez et al. realized a rectangular UMI with the hexagonal mesh processor by properly setting tunable waveguide couplers. This work, in turn, verifies the reconfiguration capability of the hexagonal meshes [22].

In our previous work, we proposed a reconfigurable photonic processor based on a mesh of self-coupled optical waveguide (SCOW) resonators [23]. This architecture has the advantages of high scalability and versatile configurations, which can be programmed to various optical components like ring resonators, Mach–Zehnder interferometers (MZIs), Fabry–Perot (FP) cavities, and also more complicated structures composed of these basic components. These structures are key components in various photonic processing. In this work, we extend the functionality of the SCOW-mesh photonic processors to an arbitrary unitary operator. We present, in detail, the implementation method as well as the synthesis algorithm for an $N \times N$ rectangular UMI. These results show the capability of the programmable SCOW mesh processors.

2. Materials and Methods

The SCOW resonator is formed by bending a single waveguide twice to form two directional couplers (DCs) at the input and output ports [24,25]. A single-stage SCOW resonator has versatile spectral responses from both transmission and reflection sides, depending on the coupling coefficients of these two DCs. We can also realize a two-dimensional array of SCOW resonators as illustrated in Figure 1a. In the vertical direction, two contiguous SCOWs share two DCs, while in the lateral direction, the two adjacent SCOWs are placed closely to form an additional DC.

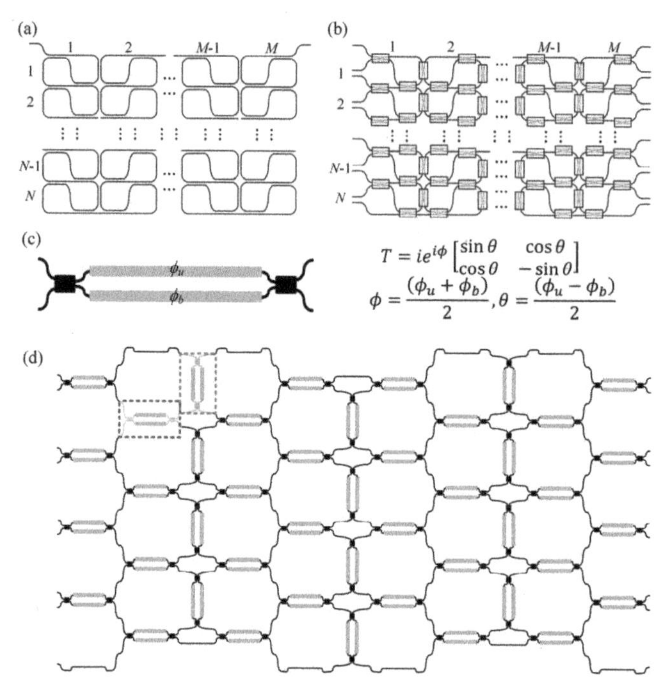

Figure 1. (**a**) Schematic of an array of $N \times M$ SCOW resonators placed in a two-dimensional mesh. (**b**) Schematic of an $N \times M$ SCOW mesh photonic processor with tunable couplers. (**c**) Structure of a tunable coupler based on a 2×2 MZI. ϕ_u and ϕ_b are phases of the upper and bottom arms of the MZI, respectively. T is the transfer matrix of the MZI. (**d**) Schematic implementation of a 4×4 SCOW mesh processor constructed by 2×2 MZI tunable couplers. The green dashed circle represents a SCOW unit composed of six MZIs placed in the vertical and horizontal directions. All the vertical MZIs (blue dashed rectangle) and the horizontal MZIs (red dashed rectangle) are designed with the same equivalent optical path length.

To realize a reconfigurable photonic processor, we replace the passive couplers with tunable couplers (TCs), as shown in Figure 1b [26,27]. By adjusting the coupling states of the TCs, the SCOW mesh processor can be reconfigured to various optical structures with versatile spectral responses, which can accomplish multiple signal processing tasks. For an $N \times M$ photonic processor, there are totally $3MN - N$ TCs. As shown in Figure 1b, we break the connecting waveguides at the right and left edges of the processor to form multiple input and output ports. The TC can be formed by a 2×2 MZI, as depicted in Figure 1c. According to the transfer matrix, the differential phase between the two arms of the MZI, $\theta = (\phi_u - \phi_b)/2$, decides the coupling coefficient of the TC, while the common phase, $\phi = (\phi_u + \phi_b)/2$, decides the phases of the output ports. When the two MZI arms are in phase ($\theta = 0$), light from the input port is fully transmitted to the cross port, which is referred to as cross-state. In the case where the phase difference of two arms is equal to π ($\theta = \pi/2$), the MZI is changed to the bar-state. For these two specific cases, the MZI works as parallel or crossed waveguides with their phases controlled by ϕ, which can be recognized as phase shifters.

For universal unitary transformation, any two interference optical paths must have an equal path length. Therefore, the MZI elements should be designed with the same equivalent optical path length. Figure 1d shows an example of a 4×4 SCOW mesh processor for UMI application. The green dashed circle represents a SCOW unit consisting of six directly-connected MZIs without any extra connecting waveguide. The MZIs are arranged in either vertical or horizontal directions. In order to get an equal optical path length, these two kinds of MZIs are slightly different at the input and output ports as shown the red and blue dashed rectangles in Figure 1d. At the top and bottom edges, one of the MZIs in the SCOW unit is replaced by a propagation waveguide with an equal optical length. In this case, we can easily construct a rectangular UMI for arbitrary unitary transformation applications even when used with a pulsed signal.

3. Results

Here, we present the implementation of the SCOW mesh photonic processor as a rectangular UMI. Figure 2a shows an example of the 9×9 rectangular UMI proposed by Clements et al. [9]. There are 36 crossings in the UMI. Each crossing represents a two-path interferometer. It corresponds to a variable beam splitter described by an optical field transfer matrix, $T_{m,n}(\theta, \phi)$, as depicted in Figure 2b. It can be implemented by an MZI with an additional phase shifter at one input port. In our proposed processor, the variable beam splitter can be constructed by three MZIs as illustrated in Figure 2b. The left two MZIs are considered as input ports of the variable beam splitter, which are at the bar-state. These two MZIs also serve as two phase shifters with target phases of $\phi + \phi_0$ and ϕ_0. To be mentioned, for an MZI at the bar-state, the bottom path has an additional π phase shift compared with the upper path, as shown in Figure 1c. Therefore, the applied phase (ϕ_u, ϕ_b) of the upper MZI has an extra common phase shift of π for the target phase of $\phi + \phi_0$. That is very important in the implementation of the rectangular UMI, where we share the bar-state MZI as the phase shifters for two adjacent variable beam splitters in the same column, as indicated by the dashed circles in Figure 2c. The right MZI works as the TC. Therefore, at the output ports of the right MZI, the transfer matrix is:

$$M = ie^{i\phi_0}\begin{bmatrix} e^{i\phi}\sin\theta & \cos\theta \\ e^{i\phi}\cos\theta & -\sin\theta \end{bmatrix} \quad (1)$$

From Equation (1), we can see that M equals to $i\exp(i\phi_0)T_{m,n}(\theta, \phi)$. The common phase item $i\exp(i\phi_0)$ only brings a phase to the final diagonal matrix, which can be omitted.

Figure 2c shows the implementation of a 9×9 rectangular UMI using the proposed photonic processor. The processor is based on a 5×9 SCOW mesh, which incorporates 130 MZIs in total. We use simple rectangles to represent all MZIs and different colors for MZI working states. The MZI design parameters are the same as in Figure 1d. The optical path length of one MZI is defined as L_{mzi}. There are four variable beam splitters in every column, which are all constructed by three MZIs. Each two adjacent beam splitters share a bar-sate MZI. Therefore, in every column, there are five bar-state MZIs and 4 MZIs working as the TCs. The rest of the MZIs are implemented as cross-connecting waveguides once they are tuned to the cross-state. The target phases ϕ and θ of the variable beam splitter in the dth row and the qth column of the two-dimensional SCOW mesh are denoted as ϕ_{dq} and θ_{dq} (d = 1, 2, 3, 4; q = 1, 2, . . . , 9). These target phases can be calculated by the decomposition method proposed by Clements et al. [9], which will be presented later. The phases of the bar-state MZIs are denoted as φ_{pq} (p = 1, 2, . . . , 5), as labeled in Figure 2c. As the phases of all the 5 bar-state MZIs in one column can be tuned independently, it is sufficient to determine the 4 independent phases in the variable beam splitters as required for practical application.

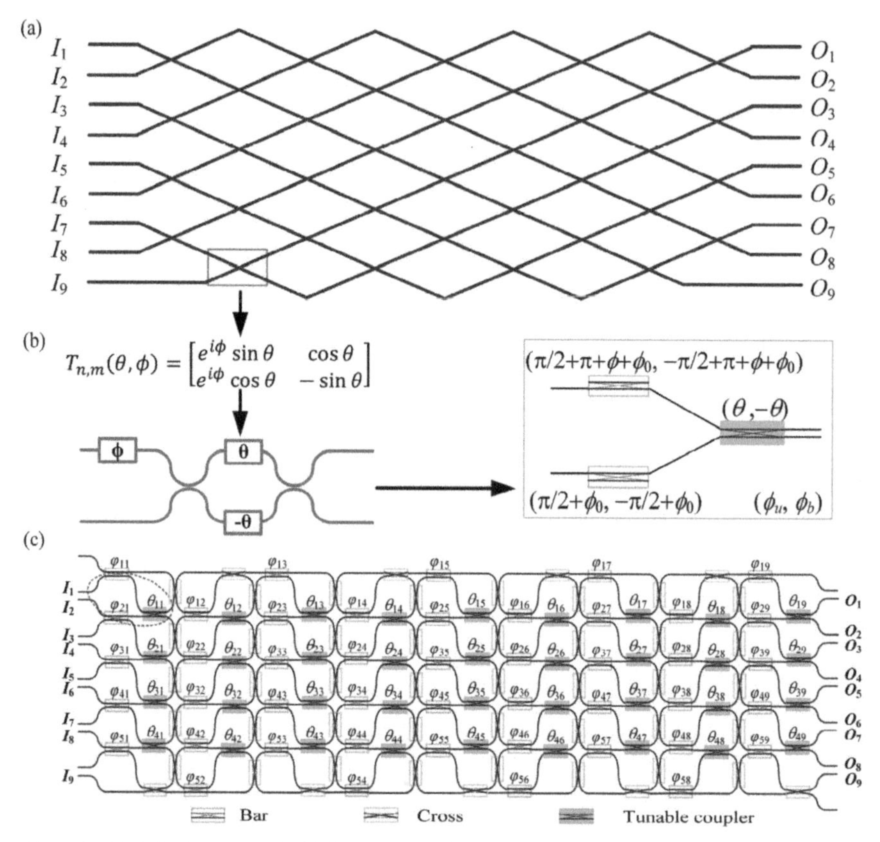

Figure 2. (**a**) A 9 × 9 rectangular UMI as proposed in [9]. (**b**) The crossing in the rectangular UMI is represented by a transfer matrix, $T_{m,n}(\theta, \phi)$, implemented by a 2 × 2 coupler and a phase shifter. It can be equivalently implemented by three MZIs in the SCOW-based mesh network. (**c**) Equivalent implementation of the 9 × 9 rectangular UMI using the SCOW-based photonic processor.

For universal unitary applications, any interferometer in the UMI must have an equal path length. We can see that in the first column, each interferometer has an equal path length of $2L_{mzi}$. In the second column, the first three interferometers have two interferential optical paths with one shared cross-state MZI and one independent bar-state MZI. The optical path length is $3L_{mzi}$. For the fourth interferometer, light from I_9 passes two cross-state MZIs, a bar-state MZI and a propagation waveguide with a length of L_{mzi} before interfering with another path. The other optical path includes the first stage interferometer, a cross-state MZI, and a bar-state MZI, which has the same path length as that for interference. For the first interferometer of the third column, the upper optical path originates from the first column interferometer and then passes one propagation waveguide and three MZIs with

$4L_{mzi}$ equivalent path, which is the same as the bottom optical path. It is easy to verify that the rest of the interferometers also have equal path lengths, which proves the functionality of our SCOW-mesh photonic processor.

Our SCOW-based processors can be easily programmed to an arbitrary rectangular UMI with such an implementation method. For an $N \times N$ rectangular UMI, the minimum number of variable beam splitters is $N(N-1)/2$. Figure 3 shows the schematics of various rectangular UMIs and the corresponding implementations with SCOW-based processors. The number of columns in the SCOW processor linearly increases with N, while the number of rows equals to *mod* $((N+2)/2)$. In each SCOW, the left MZI is configured to the bar-state. When N is odd, $N-1$ of the right MZIs in each column work as TCs, while the other MZI is at the cross-state. When N is even, the number of cross-state MZIs increases to two for each even column. The rest of the MZIs are implemented to the cross-state. The required numbers of MZIs are listed in Table 1.

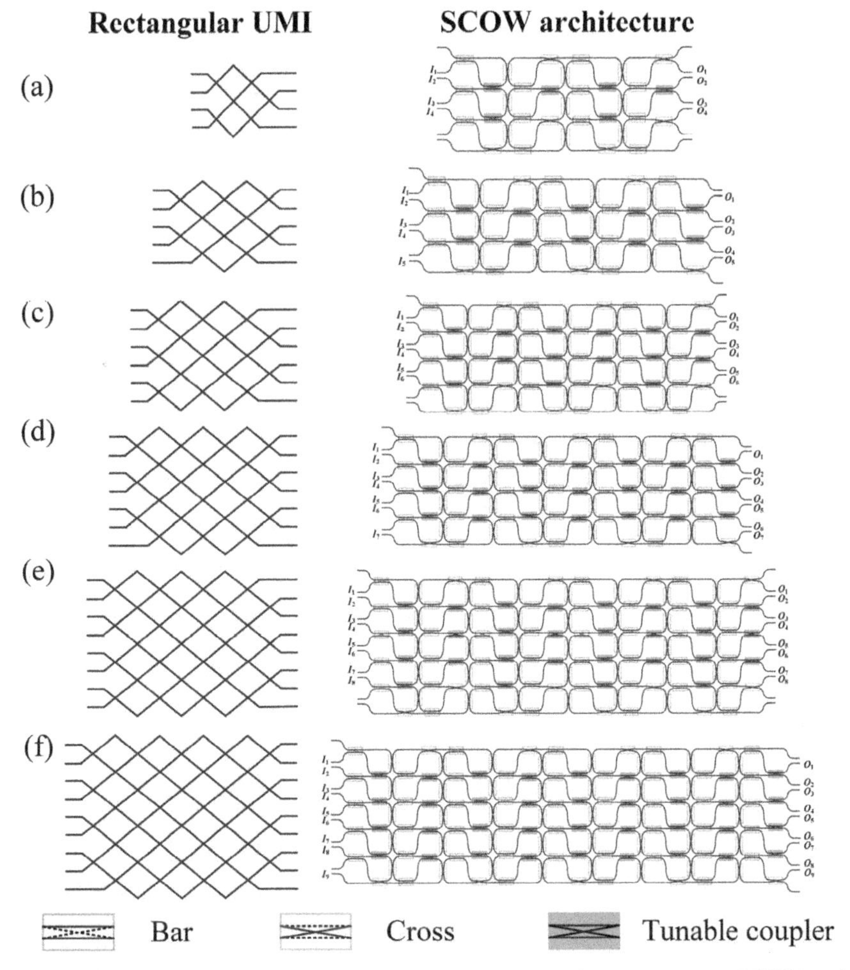

Figure 3. Schematics of the rectangular UMIs and the equivalent implementations using SCOW-based photonic processors. The scales of the UMIs are (**a**) 4 × 4, (**b**) 5 × 5, (**c**) 6 × 6, (**d**) 7 × 7, (**e**) 8 × 8, and (**f**) 9 × 9.

Table 1. Required number of MZIs to construct an $N \times N$ rectangular UMI.

UMI Scale N	SCOW Architecture
$2n + 1$	$6n^2 + 8n + 2$
$2n$	$6n^2 + 5n - 1$

The synthesis procedure of the rectangular UMI with the hexagonal mesh processor has been proposed by Perez et al. [22]. The implementation procedure of an $N \times N$ UMI for the SCOW-based processor is similar to that of hexagonal meshes. The algorithm proceeds by nulling successive matrix elements starting from the targeted unitary $N \times N$ matrix U. Depending on the location of the element $U(n, m)$ to be canceled, a row or column combination of the matrix is required [9,22]. If $N - n - m$ is odd, any target element in column n or m of U can be nulled by multiplying U from the right by a $T_{m,n-1}$ matrix. While if $N - n - m$ is even, then, any element in rows n or m can be nulled by $T_{m,n}U$. For the case of the proposed processor, $T_{m,n}$ represents:

$$T_{m,n} = \begin{bmatrix} 1 & 0 & \cdots & \cdots & & \cdots & \cdots & \cdots & 0 \\ 0 & 1 & & & & & & & \cdot \\ \cdot & & \cdot & & & & & & \cdot \\ \cdot & & & e^{i\phi}\sin\theta & \cos\theta & & & & \cdot \\ \cdot & & & e^{i\phi}\cos\theta & -\sin\theta & & & & \cdot \\ \cdot & & & & & & \cdot & & \cdot \\ \cdot & & & & & & & 1 & 0 \\ 0 & \cdots & \cdots & \cdots & & \cdots & \cdots & 0 & 1 \end{bmatrix} \tag{2}$$

According to [22], the values of θ and ϕ of the corresponding matrix T are given as:

$$\begin{aligned} \theta &= a\sin\left|\sqrt{\frac{|U(n,m)|^2}{|U(n,m)|^2+|U(n,m+1)|^2}}\right| \quad (\textit{when } N-n-m \textit{ is odd}) \\ \phi &= \angle U(n,m) - \angle U(n,m+1) - \pi \end{aligned} \tag{3}$$

$$\begin{aligned} \theta &= a\sin\left|\sqrt{\frac{|U(n,m)|^2}{|U(n,m)|^2+|U(n-1,m)|^2}}\right| \quad (\textit{when } N-n-m \textit{ is even}) \\ \phi &= \angle U(n,m) - \angle U(n-1,m) \end{aligned} \tag{4}$$

With the calculated θ and ϕ of each variable beam splitter, we can deduce the applied phases to the MZI arms. The target phases ϕ and θ of the beam splitter in the dth row and the qth column of the two-dimensional SCOW mesh are defined as ϕ_{dq} and θ_{dq} ($d = 1, 2, 3$, mod $((N-1)/2)$, $q = 1, 2, \ldots, N$). The applied phases of the MZI TCs are θ_{dq} and $-\theta_{dq}$ for the two MZI arms, respectively. Each bar-state MZI is shared by two beam splitters, which is different from that of hexagonal meshes. The bar-state MZIs are defined as φ_{pq} ($p = 1, 2, \ldots$, mod $((N-1)/2) + 1$). It means that the applied phases on the two arms of the bar-state MZIs are $\pi/2 + \varphi_{pq}$ and $-\pi/2 + \varphi_{pq}$, respectively. As there are enough bar-state MZIs in each column independently determining the target phases of the variable beam splitters, we can easily decide the phases of all the bar-state MZIs one by one from the former bar-state MZI with the relation of $\varphi_{p+1,q} = \varphi_{pq} + \pi - \phi_{dq}$. Once we fix the first bar-sate MZI in each column, all the other bar-state MZIs can be decided. Therefore, with the implementation method, the SCOW mesh photonic processors can be easily programmed to an arbitrary rectangular UMI, which shows the reconfiguration capability of the SCOW architecture.

In practice, both the tunable MZIs and the connection waveguides are lossy, which degrades the performance of the UMIs. There are two types of loss, namely, the balanced loss and the unbalanced loss. They depend on the path difference of all the paths in the interferometers. In our design, the propagation loss in an MZI is expected to mainly contribute to the balanced loss. For an $N \times N$ UMI, each path incorporates $3N - 1$ MZIs to the maximum. Therefore, the balanced loss of our architecture increases as $3N-1$. The unbalanced loss in the interferometers degrades the fidelity of the transformations. As the insertion loss of the connection waveguide and the MZI is not necessarily the same, the path loss may be different for the interferometers. Therefore, we calculate the fidelity of the SCOW mesh-based UMI using the transfer matrix method. In each MZI, we add identical extra insertion loss to the two output ports. To compare the fidelity of our design with the triangular and rectangular designs, we use a similar procedure as mentioned in [9]. For a given N, we generate 500 groups of ϕ_{dq} and θ_{dq} with random values. With all the phase parameters of the MZIs, we can quantify the fidelity of the unitary transformation by our SCOW-based processor using the following equation [9],

$$F(U_{\text{exp}}, U) = \left| \frac{tr(U^{\dagger} U_{\text{exp}})}{\sqrt{N \times tr(U_{\text{exp}}^{\dagger} U_{\text{exp}})}} \right|^2 \quad (5)$$

where U_{exp} represents the transformation of a lossy SCOW processor and U represents the intended lossless unitary transformation. Figure 4a shows the average fidelity for a UMI changing with the size N, when the loss of the MZIs is set to 0.2 dB. We can see that the triangular architecture has the least tolerance to the unbalanced loss of the interferometers. As our design is originated from the rectangular architecture, the fidelity is only slightly lower than that of the rectangular design. The inset shows the close-up view of the fidelity of our design and the rectangular design. We can see that the fidelity values of our SCOW architecture are divided into two groups depending on the port numbers of the interferometers. The fidelity for odd number ports is lower than that for even number ports, because of the larger unbalanced loss in the interference paths. Figure 4b shows the fidelity as a function of MZI loss when the UMI size N is fixed to 20. As the insertion loss of one MZI is typically less than 0.5 dB, the fidelity of a 20 × 20 UMI based on the SCOW architecture is around 0.999 which is better than that of the triangular design. These simulation results show that our architecture has a high tolerance to the path loss difference in the interferometers.

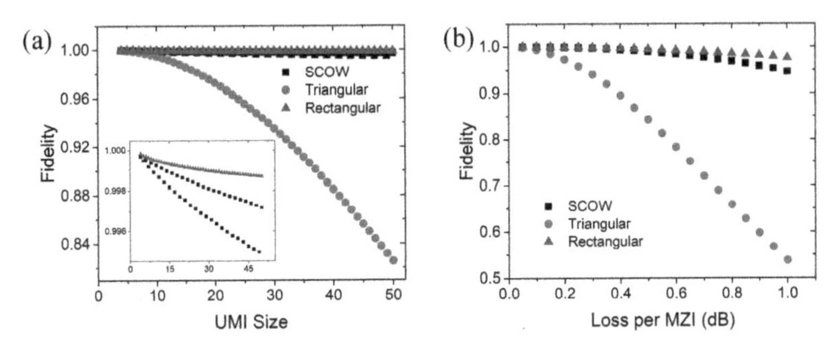

Figure 4. (**a**) Simulated average fidelity as a function of UMI size when the MZIs have a constant loss of 0.2 dB. The inset shows the close-up view of the fidelity of our design and the rectangular design. (**b**) Simulated average fidelity as a function of MZI loss when the UMIs perform 20 × 20 unitary transformations. The fidelity values of the triangular and rectangular architectures are from [9].

4. Conclusions

In summary, we proposed a two-dimensional SCOW mesh photonic processor to work as a rectangular UMI, which can accomplish arbitrary optical unitary transformation. Arbitrary unitary transformation has wide applications in quantum signal processing, photonic deep machine learning, etc. We presented the detailed adaptation synthesis algorithm for constructing an $N \times N$ rectangular UMI using the SCOW mesh-based photonic processor. The simulated fidelity shows that our architecture has a high tolerance to the path loss of the interferometers. This work extends the functionality of the SCOW mesh photonic processor, making it more powerful for multitude signal processing tasks.

Author Contributions: L.L. proposed the idea, analyzed the results and wrote the manuscript; L.Z. and J.C. supervised the project and edited the manuscript; all authors discussed the results and commented on the manuscript.

Funding: This work was supported in part by the National Natural Science Foundation of China (NSFC) (61705129, 61535006), the National Key R&D Program of China (2018YFB2201702), and Shanghai Municipal Science and Technology Major Project (2017SHZDZX03).

Conflicts of Interest: The authors declare no conflict of interest.

References

1. Annoni, A.; Guglielmi, E.; Carminati, M.; Ferrari, G.; Sampietro, M.; Miller, D.A.; Melloni, A.; Morichetti, F. Unscrambling light-automatically undoing strong mixing between modes. *Light Sci. Appl.* **2017**, *6*, e17110. [CrossRef] [PubMed]
2. Miller, D.A. All linear optical devices are mode converters. *Opt. Express* **2012**, *20*, 23985–23993. [CrossRef]
3. Shen, Y.; Harris, N.C.; Skirlo, S.; Prabhu, M.; Baehr-Jones, T.; Hochberg, M.; Sun, X.; Zhao, S.; Larochelle, H.; Englund, D.; et al. Deep learning with coherent nanophotonic circuits. *Nat. Photon.* **2017**, *11*, 441. [CrossRef]
4. Harris, N.C.; Steinbrecher, G.R.; Prabhu, M.; Lahini, Y.; Mower, J.; Bunandar, D.; Chen, C.; Wong, F.N.C.; Baehr-Jones, T.; Hochberg, M.; et al. Quantum transport simulations in a programmable nanophotonic processor. *Nat. Photon.* **2017**, *11*, 447. [CrossRef]
5. Qiang, X.; Zhou, X.; Wang, J.; Wilkes, C.M.; Loke, T.; O'Gara, S.; Kling, L.; Marshall, G.D.; Santagati, R.; Ralph, T.C.; et al. Large-scale silicon quantum photonics implementing arbitrary two-qubit processing. *Nat. Photon.* **2018**, *12*, 534–539. [CrossRef]
6. Wang, J.; Paesani, S.; Ding, Y.; Santagati, R.; Skrzypczyk, P.; Salavrakos, A.; Tura, J.; Augusiak, R.; Mancinska, L.; Bacco, D.; et al. Multidimensional quantum entanglement with large-scale integrated optics. *Science* **2018**, *360*, 285–291. [CrossRef] [PubMed]
7. Reck, M.; Zeilinger, A.; Bernstein, H.J.; Bertani, P. Experimental realization of any discrete unitary operator. *Phys. Rev. Lett.* **1994**, *73*, 58–61. [CrossRef] [PubMed]
8. Ribeiro, A.; Ruocco, A.; Vanacker, L.; Bogaerts, W. Demonstration of a 4×4-port universal linear circuit. *Optica* **2016**, *3*, 1348. [CrossRef]
9. Clements, W.R.; Humphreys, P.C.; Metcalf, B.J.; Kolthammer, W.S.; Walsmley, I.A. Optimal design for universal multiport interferometers. *Optica* **2016**, *3*, 1460. [CrossRef]
10. Birth of the programmable optical chip. *Nat. Photon.* **2015**, *10*, 1.
11. Miller, D.A.B. Silicon photonics: Meshing optics with applications. *Nat. Photon.* **2017**, *11*, 403–404. [CrossRef]
12. Liu, W.; Li, M.; Guzzon, R.S.; Norberg, E.J.; Parker, J.S.; Lu, M.; Coldren, L.A.; Yao, J. A fully reconfigurable photonic integrated signal processor. *Nat. Photon.* **2016**, *10*, 190–195. [CrossRef]
13. Zhuang, L.; Roeloffzen, C.G.H.; Hoekman, M.; Boller, K.-J.; Lowery, A.J. Programmable photonic signal processor chip for radiofrequency applications. *Optica* **2015**, *2*, 854–859. [CrossRef]
14. Perez, D.; Gasulla, I.; Crudgington, L.; Thomson, D.J.; Khokhar, A.Z.; Li, K.; Cao, W.; Mashanovich, G.Z.; Capmany, J. Multipurpose silicon photonics signal processor core. *Nat. Commun.* **2017**, *8*, 636. [CrossRef] [PubMed]
15. Perez, D.; Gasulla, I.; Capmany, J.; Soref, R.A. Reconfigurable lattice mesh designs for programmable photonic processors. *Opt. Express* **2016**, *24*, 12093–12106. [CrossRef] [PubMed]
16. Xie, Y.; Geng, Z.; Zhuang, L.; Burla, M.; Taddei, C.; Hoekman, M.; Leinse, A.; Roeloffzen, C.G.H.; Boller, K.-J.; Lowery, A.J. Programmable optical processor chips: Toward photonic RF filters with DSP-level flexibility and MHz-band selectivity. *Nanophotonics* **2017**, *7*, 421–454. [CrossRef]
17. Pérez-López, D.; Sánchez, E.; Capmany, J. Programmable True Time Delay Lines Using Integrated Waveguide Meshes. *J. Lightw. Technol.* **2018**, *36*, 4591–4601. [CrossRef]
18. Pérez, D.; Gasulla, I.; Capmany, J. Toward Programmable Microwave Photonics Processors. *J. Lightw. Technol.* **2018**, *36*, 519–532. [CrossRef]
19. Miller, D.A.B. Self-configuring universal linear optical component. *Photonics Res.* **2013**, *1*, 1–15. [CrossRef]
20. Miller, D.A.B. Perfect optics with imperfect components. *Optica* **2015**, *2*, 747. [CrossRef]
21. Miller, D.A.B. Setting up meshes of interferometers-reversed local light interference method. *Opt. Express* **2017**, *25*, 29233–29248. [CrossRef]

22. Perez, D.; Gasulla, I.; Fraile, F.J.; Crudgington, L.; Thomson, D.J.; Khokhar, A.Z.; Li, K.; Cao, W.; Mashanovich, G.Z.; Capmany, J. Silicon Photonics Rectangular Universal Interferometer. *Laser Photonics Rev.* **2017**, *11*, 1700219. [CrossRef]
23. Lu, L.; Shen, L.; Zhou, L.; Chen, J. Reconfigurable Silicon Photonic Signal Processor Based on the SCOW Resonant Structure. In Proceedings of the CLEO: Science and Innovations 2018, San Jose, CA, USA, 13–18 May 2018; p. STh3B.4.
24. Zhou, L.; Ye, T.; Chen, J. Coherent interference induced transparency in self-coupled optical waveguide-based resonators. *Opt. Lett.* **2011**, *36*, 13–15. [CrossRef] [PubMed]
25. Zou, Z.; Zhou, L.; Sun, X.; Xie, J.; Zhu, H.; Lu, L.; Li, X.; Chen, J. Tunable two-stage self-coupled optical waveguide resonators. *Opt. Lett.* **2013**, *38*, 1215. [CrossRef] [PubMed]
26. Sun, X.; Zhou, L.; Xie, J.; Zou, Z.; Lu, L.; Zhu, H.; Li, X.; Chen, J. Investigation of Coupling Tuning in Self-Coupled Optical Waveguide Resonators. *IEEE Photon. Technol. Lett.* **2013**, *25*, 936–939. [CrossRef]
27. Tang, H.X.; Zhou, L.J.; Xie, J.Y.; Lu, L.J.; Chen, J.P. Electromagnetically Induced Transparency in a Silicon Self-Coupled Optical Waveguide. *J. Lightw. Technol.* **2018**, *36*, 2188–2195. [CrossRef]

Article

Active On-Chip Dispersion Control Using a Tunable Silicon Bragg Grating

Charalambos Klitis [1], **Marc Sorel** [1] **and Michael J. Strain** [2,*]

[1] School of Engineering, University of Glasgow, Glasgow G12 8LT, UK
[2] Institute of Photonics, Department of Physics, University of Strathclyde, Glasgow G1 1RD, UK
* Correspondence: michael.strain@strath.ac.uk

Received: 26 July 2019 ; Accepted: 24 August 2019; Published: 28 August 2019

Abstract: Actively controllable dispersion in on-chip photonic devices is challenging to implement compared with free space optical components where mechanical degrees of freedom can be employed. Here, we present a method by which continuously tunable group delay control is achieved by modulating the refractive index profile of a silicon Bragg grating using thermo-optic effects. A simple thermal heater element is used to create tunable thermal gradients along the grating length, inducing chirped group delay profiles. Both effective blue and red chirp are realised using a single on-chip device over nanometre scale bandwidths. Group delay slopes are continuously tunable over a few ps/nm range from red to blue chirp, compatible with on-chip dispersion compensation for telecommunications picosecond pulse systems.

Keywords: silicon photonics; dispersion control; Bragg gratings

---

1. Introduction

Integrated Bragg grating filters are an established and widely used technology on the Silicon-on-Insultor (SOI) material platform. They are applied in a large variety of applications, including, optical filtering [1,2], sensing [3], laser cavity feedback [4] and all-optical signal processing [5,6]. A wide range of device geometries have been demonstrated in order to exercise control over the grating optical characteristics, namely the filter bandwidth [7], ripple [8], extinction and dispersion [2,9,10]. In turn, the optical characteristics of the grating can be designed through the coupling coefficient, κ, and the grating Bragg wavelength, λ_B, as a function of the propagation length [10,11]. Active control of Bragg grating devices has also been demonstrated using multi-section p-n junction elements to create tunable notch features in the device stopband [12].

The control afforded by the Bragg grating device over the dispersion of a propagating signal is crucial in applications such as all-optical signal processing [5] and pulse shaping [13]. Commonly, the group delay profile of a device is varied through the local Bragg wavelength function. Therefore, by simply varying the grating period [14], or effective waveguide index, devices can be created with linear and higher order dispersion profiles [9] or even complex multi-band filter designs [15]. Filter ripple effects have been effectively minimised by suitably apodising the grating κ along its length [9,10,16]. The dispersion profile of a grating device is generally defined during the fabrication procedure as a modulation of the period [14] or effective index [17]. Some rigid tunability of the device spectrum has been demonstrated through global thermal [18] or electronic tuning [19]. Narrowband delay has also been demonstrated using multi-section gratings at the band-edge [20,21]. However, for future Photonic Integrated Circuits (PICs), true tunable dispersion on-chip will be a valuable component functionality. The principle of reconfigurable silicon photonic circuits has been well established in devices including ring resonators [22], Mach–Zehnder interferometers [23,24], and modulators [25], using both thermo-optic and carrier injection methods.

In this work, we present the design considerations for realising tunable dispersion gratings on the silicon photonics platform and present measured performance of fabricated devices. Control is demonstrated producing both red and blue chirp in a single device over bandwidths and group delay ranges compatible with picosecond pulse compensation schemes.

2. Results

2.1. Actively Tunable Bragg Grating Design

The Bragg grating device is implemented with a constant Bragg wavelength, $\lambda_B(z)$, along its length. To create a chirp in the grating response, the Bragg wavelength function is controlled by creating a thermal gradient along the device length. A schematic of the device is shown in Figure 1.

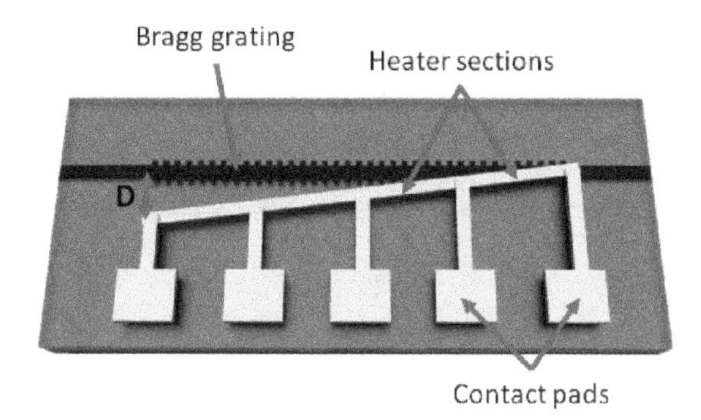

Figure 1. Schematic of a Bragg grating with an integrated heater element. The displacement between the waveguide axis and heater central position is a function of propagation length, with a maximum displacement *D*.

It has been demonstrated by a number of groups that by placing a resistive element over a waveguide, the local index of refraction can be tuned through the thermo-optic effect [26–28]. Furthermore, due to the large mismatch in thermal conductivity of silicon and silica, the heat dissipation as a function of lateral displacement from the element allows differential heating of closely spaced structures, e.g., for tuning the coupling fraction of an evanescent field coupler [29]. Thus, by making the heater element displacement from the waveguide a function of the propagation length, local variations in refractive index, and hence Bragg wavelength, can be induced in the grating. Figure 2a shows a cross-sectional, thermal finite element model of the heater element above a silicon-on-insulator waveguide in the steady state. By varying the offset between the central axis of the heater element and the silicon waveguide, the temperature of the silicon guide can be controlled. In this example, the heater element is dissipating 11 mW of power. For power dissipation levels similar to the results presented here, the thermal control element has a temperature of tens of degrees above room temperature. The heater element then dominates the thermal state of the silicon waveguide with respect to environmental temperature variations. No active temperature stabilisation of the silicon substrate was used in this work. Figure 2b shows the relationship between the offset of the heater axis to the waveguide and the resultant waveguide temperature. The curve fits well to a quadratic function. There is a variation from the curve when the waveguide and heater are co-linear. This is due to the finite nature of the heater width, resulting in saturation of the temperature as a reducing function of offset. The thermal impedance of the silica layers also results in a high thermal gradient over micron scales, and therefore allows longitudinal chirping of the grating structure. Thus, given the quadratic relationship between offset and waveguide temperature, and the proportional relationship between temperature and material refractive index, a refractive index chirp function can be designed as a function of heater displacement from the waveguide axis.

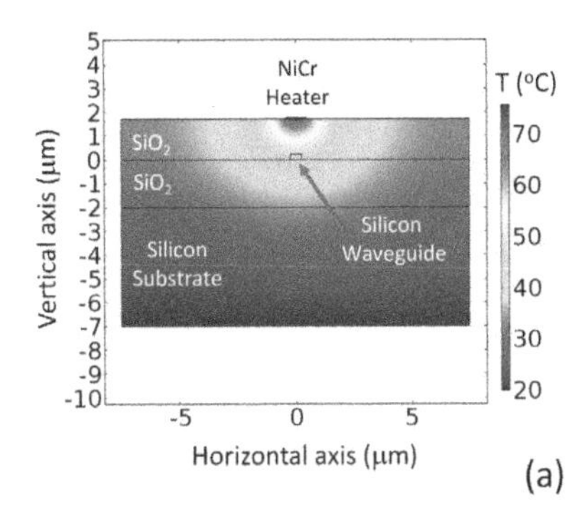

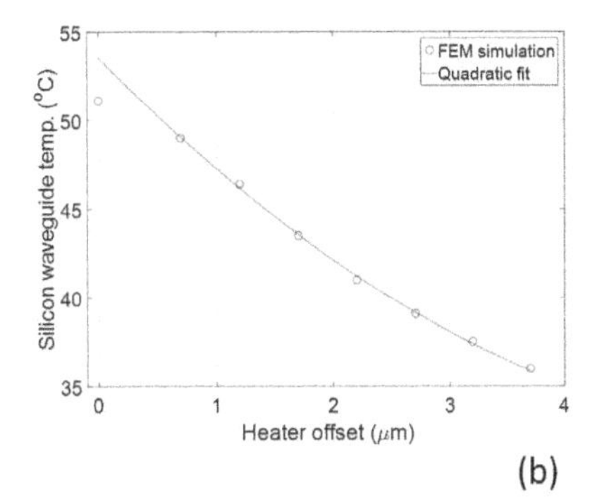

Figure 2. (**a**) FEM thermal model of a heater element co-axial with a silicon-on-insulator waveguide; and (**b**) calculated temperature of the silicon waveguide as a function of offset of the heater central axis from the central axis of the waveguide.

In Figure 1, a signal injected from the left hand side of the device would see an increasing refractive index with propagation length through the grating, and therefore an effective red chirp. Alternatively, light coupled in from the right hand side of the device would see decreasing refractive index with propagation length and, therefore, effective blue chirp. The magnitude of the induced chirp is dependent on the grating length, coupling coefficient and thermal gradient. The difference in temperature and, therefore, refractive index, along the grating can be controlled electronically through the power dissipated in the heater.

To demonstrate the tunable dispersion concept, a grating response was chosen to cover bandwidths of $\approx$1 nm, and group delays in the picosecond range, corresponding to typical dispersion compensation requirements for picosecond telecommunications pulses. The gratings were designed on a silicon-on-insulator platform with a core layer thickness of 220 nm and a mean waveguide width of 500 nm. The grating period, Λ_0, was set at 318 nm, to give a Bragg wavelength around 1550 nm. The grating was designed with a length of 250 µm and a sinusoidally varying sidewall amplitude modulation, d, of 6 nm. The grating perturbation is realised as a sinusoidal variation around an average waveguide width, w. A schematic of the sidewall perturbation grating is shown in Figure 3 along with a SEM image of the fabricated device before deposition of the upper-cladding.

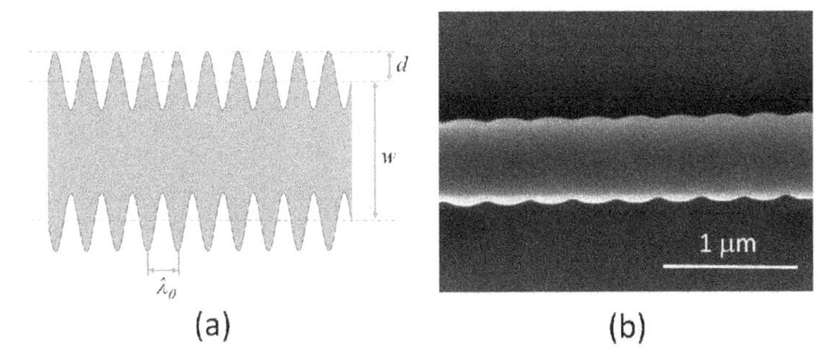

Figure 3. (**a**) Schematic of a sinusoidal sidewall perturbation Bragg grating, and (**b**) SEM image of a fabricated device.

A Gaussian apodisation of the sidewall perturbation was applied to minimise ripples in the grating amplitude and group delay response. The grating response can simulated using a Transfer Matrix Method (TMM) model with $\lambda_B(z)$ as a parameter [30]. Since $\lambda_B(z) = 2n_{\mathrm{eff}}(z)\Lambda_0$, where $n_{\mathrm{eff}}(z)$

is the local modal refractive index and Λ_0 is the grating period, the grating chirp can be modelled using a variation in refractive index. The modal index is in turn dependent on the temperature at the waveguide generated by the local heater element. The heater element is fabricated using a 900 nm wide track of 50 nm thick NiCr. To create a linear chirp profile, the displacement, $D(z)$, between the centre lines of the waveguide and heater element, is varied using a $z^{0.5}$ relationship. It is worth noting that the waveguide dispersion of silicon nanowire waveguides is significant [31], and therefore must be taken into consideration when designing and modelling Bragg gratings in silicon.

Figure 4a shows a measured transmission spectrum of the fabricated device with no power dissipated on the heater, i.e., the device is unchirped. Measured device spectra can be fitted to the TMM model using a least squares curve fitting method. In this case, the form of the grating chirp is assumed, e.g., unchirped or linear blue-chirp. The coupling coefficient and the maximum and minimum effective index values at the grating ends are then left as free parameters to perform the fit. The extracted modal index from the fitting of the unchirped grating is 2.4386 and the coupling coefficient is 323 cm^{-1}. Furthermore, since an apodisation scheme based on variation of the grating sidewall amplitude was employed, residual ripples in the grating reflectivity and group delay spectra are still apparent. Grating coupling coefficient apodisation schemes that more effectively reduce the spectral ripple of the device, and can be fabricated within nanometric tolerances, have recently been demonstrated using phase based methods [10,16].

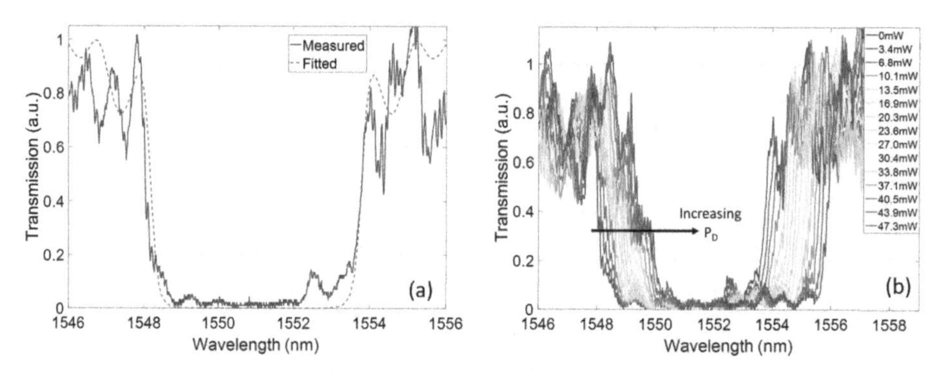

Figure 4. (**a**) Measured and fitted Bragg grating transmission spectra of the grating device in the unchirped state; and (**b**) measured grating spectra for increasing power dissipation, P_D, on the thermal heating element.

To assess the effects of the heater on the z-dependent modal index, the grating transmission spectrum was measured as a function of total electrical power dissipated on the heater element, P_D. Figure 4b shows the measured transmission spectra, linearly varying P_D up to 47.3 mW. The grating fitting was carried out in each case, assuming a linearly varying modal index as a function of the propagation length, with extreme modal index values as free parameters. The local shift of the Bragg grating along the device can be decomposed into two components: a baseline shift of the full device since the full length sees some minimum thermal increase, and a relative temperature change along the propagation direction due to the displacement of the heater section. This can be approximated as:

$$\lambda_B(z) = 2(n_{\text{eff},0} + dn_{\text{min}} + dn(z))\Lambda_0 \tag{1}$$

where $n_{\text{eff},0}$ is the effective modal index of the grating waveguide at room temperature, dn_{min} is the increase in modal effective index at point of greatest displacement between the waveguide and the heater element, and $dn(z)$ is the additional thermal increase along the grating length above dn_{min} as the displacement of the heater and waveguide reduces. Figure 5 shows the extracted values of

$dn_{\min}$ and $dn(z)_{\max}$ from the transmission spectra. As s expected, $dn(z)_{\max}$ has a higher gradient as a function of P_D than $dn_{\min}$, indicating the effective chirping of the grating response.

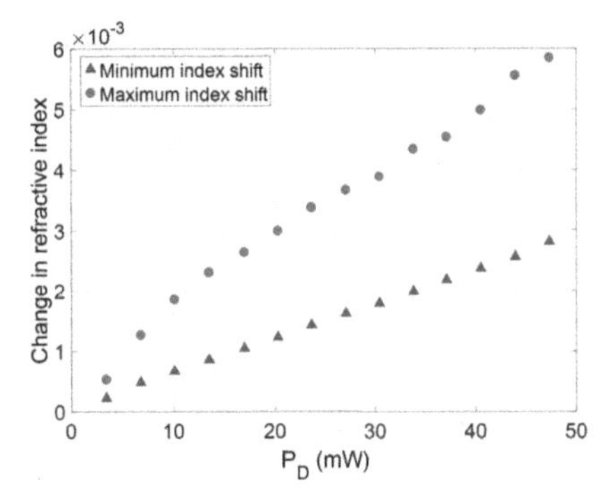

Figure 5. Variation in minimum and maximum thermally induced modal refractive index shifts of the device extracted from fitting of the transmission spectra.

2.2. Group Delay Chirp Control

Figure 6 shows the calculated reflectivity spectra and group delay profiles for the grating, with the extracted values of $dn_{\min}$ and $dn(z)_{\max}$ as parameters. The group delay is only calculated for the portion of the reflectivity spectrum above 90% of its maximum value.

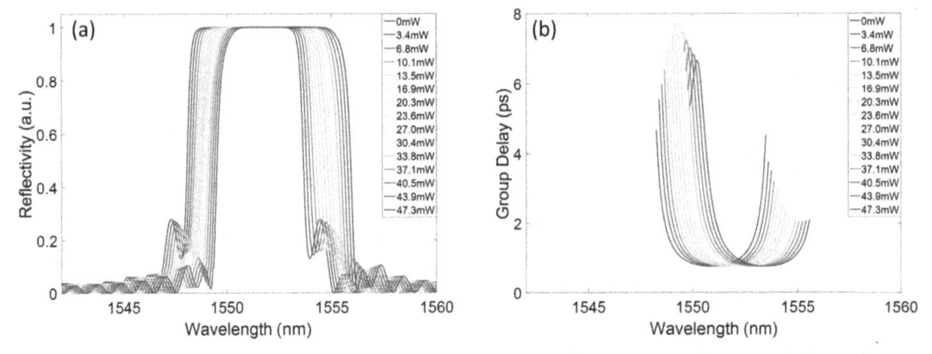

Figure 6. (**a**) Simulated reflectivity; and (**b**) group delay of a Bragg grating device with thermally induced refractive index differences as parameters.

The group delay profile of the grating response shows a sensitivity to the induced refractive index profile, with an increasing blue chirp developing across the central region of the grating reflectivity band. The wide stopband, due to the high grating coupling coefficient, means that, even for chirped gratings, there is a portion of the reflection band with flat group delay. As noted previously, 1 nm bandwidths are considered here, corresponding to typical telecommunications picosecond pulse sources. By selecting a particular central wavelength within the grating stopband, controllable chirp can be exhibited as a function of power dissipated on the heater. For example, sub-bands centred at 1551 nm and 1552.2 nm are shown in Figure 7a,b, respectively. In Figure 7a, increasing the dissipated power on the heater induces a variation in the blue chirped gradient of the group delay in the few ps/nm range. Alternatively, by selecting a central wavelength towards the red tuned side of the grating reflection spectrum, a range of chirp values spanning from blue to red chirp can be accessed, as shown in Figure 7b. As this change from blue to red chirp crosses a zero dispersion condition, the absolute chirp rate is less than the blue chirp rate achievable at similar power dissipation levels, at bands further detuned from this wavelength.

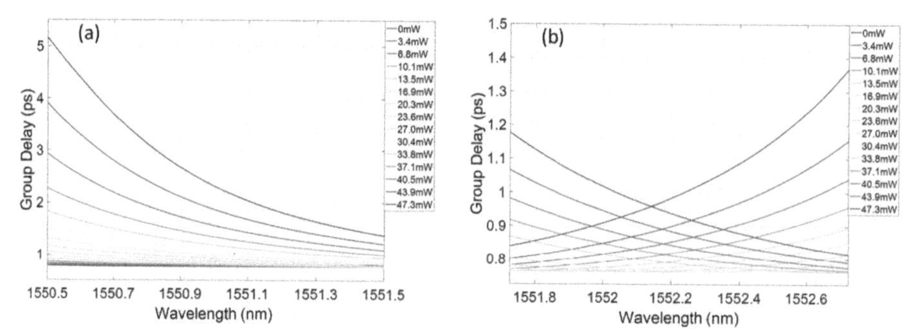

Figure 7. Simulated group delay bands as a function of dissipated power on the heater centred at: (**a**) 1551 nm; and (**b**) 1552.2 mn.

The group delay of the fabricated tunable grating was measured as detailed in Section 5. Figure 8 shows the measured group delay as a function of power dissipated on the heater.

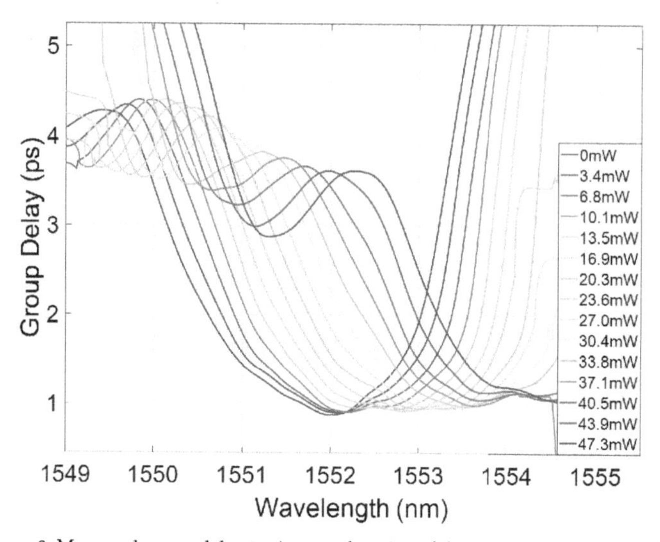

Figure 8. Measured group delay tuning as a function of dissipated power on the heater.

As expected, there is a clear increase of the group delay in the centre of the band, indicating increasing blue chirp of the spectrum. Evidence of the residual ripple in the grating reflectivity is also apparent in the group delay kink developing for increasing dispersion. Figure 9a,b shows sub-bands of the grating response centred around 1551.7 and 1553 nm, respectively. As predicted, for the lower wavelength range, increasing the heater power creates an increasing group delay slope, with effective blue chirp. In the longer wavelength band, both blue and red chirp responses can be accessed, passing through a flat group delay characteristic.

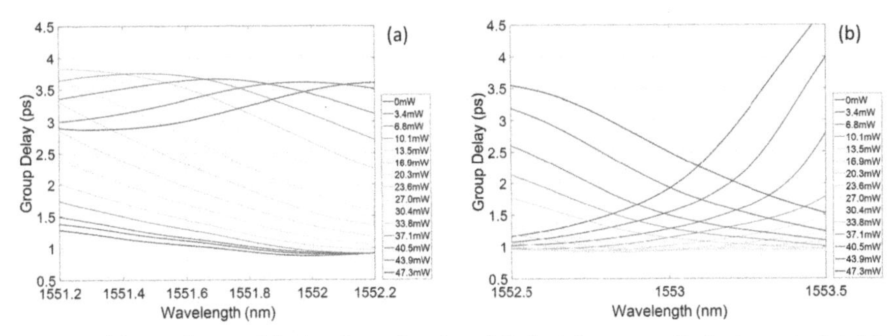

Figure 9. Measured group delay bands as a function of dissipated power on the heater centred at: (**a**) 1551.7 nm; and (**b**) 1553 mn.

Figure 10 presents the average group delay slope for both sub-bands in Figure 9. In both cases, quasi-continuous tuning of the group delay slope is demonstrated in the ps/nm range, with both blue and red chirp accessible from a single device and from a single injection direction.

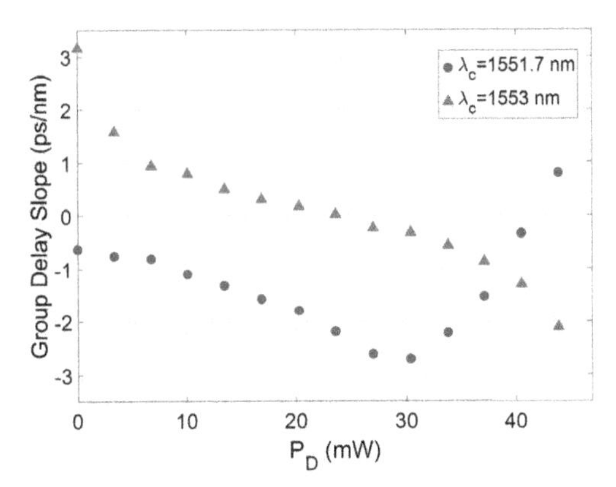

Figure 10. Measured group delay slope as a function of dissipated power on the heater corresponding to the dispersion bands centred at 1551.7 nm; and 1553 nm.

3. Discussion

By integrating a thermal heater element with a varying displacement along the length of a uniform Bragg grating device, thermal gradients and therefore local modal refractive index profiles have been imposed onto the device. The strong grating coupling coefficient means that the full group delay profile is complex, with both blue and red chirp regions occurring over a bandwidth of a few nanometres. However, this can be used to sub-select regions of interest for dispersion control over nanometre scale bands, suitable for picosecond pulse systems. In this way, both blue and red chirp control have been demonstrated in a single device. The group delay slope is also related to the grating length, i.e., longer physical distances between grating sections tuned to different spectral components will induce larger group delay shifts between those spectral components [32]. Therefore, to tailor the accessible maximum group delay slope, the grating length can be defined as a design parameter. Figure 11 shows TMM simulated grating reflectivity and group delay spectra, dependent on the grating length. In these simulations, the thermal index gradient was assumed to be linear and corresponding to the measurement conditions for the maximum power dissipated on the thermal heater of 47.3 mW. The apodisation profile is Gaussian.

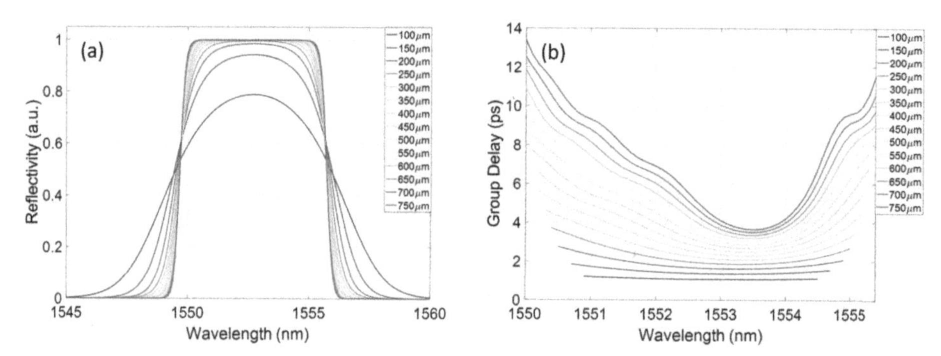

Figure 11. Simulated grating: (**a**) reflectivity; and (**b**) group delay spectra with grating length as a parameter.

The group delay range across the spectrum is, as expected, dependent on the grating length. Therefore, devices can be designed based on the required application requirements.

In addition to the relative group delay dispersion induced across the device, the absolute spectral position of the grating response is dependent on two main factors. The first is related to fabrication tolerances in the manufacture of the device. Since the absolute Bragg wavelength of a fixed period grating is proportional to its effective modal index, any variation from the design geometry will lead to a shift in the resultant Bragg spectrum. However, in high resolution fabrication schemes such as optimised electron beam or deep ultraviolet (UV) lithography, such variations are global rather than local. That is, the average waveguide width may vary from run to run, or even between spatially separated devices, but is unlikely to see significant variation across a single device. Typical geometrical variations can be in the nanometre range, leading to effective index variations in the order of 10^{-3}–10^{-2} [33]. The second factor affecting the absolute spectral position of the Bragg grating response is the global thermal shift experienced by the full grating when the heater is electrically driven. The thermal element induces both a global wavelength shift of the grating characteristic and a local gradient to induce the chirped profile, as detailed in Figure 5, The global, or minimum index, shift experienced by the full grating is in the order of 10^{-3}, comparable with any potential geometric effects. To implement control over both the group delay slope and central wavelength, a second heater element could be implemented parallel to the Bragg grating. This element would produce a rigid index shift of the full grating response, shifting the demonstrated group delay curves in wavelength. Therefore, using both heater controls, varying group delay slopes in both chirp directions could be accessed at a single central wavelength, as long as the un-heated grating central Bragg wavelength was designed to be at a shorter wavelength than the required application space.

4. Conclusions

Control over the group delay slope of an integrated silicon Bragg grating is achieved using a simple thermal tuning element. Spatial displacement of the heater along the propagation length of the grating produces a longitudinal variation in modal effective index, and, therefore, dispersion. Variation of the electrical power dissipated by the heater allows for control over the group delay slope, with fabricated devices covering the range of 0–3 ps/nm over nanometre range bandwidths. This range is ultimately limited by the grating length and can be designed for specific applications before fabrication. Finally, due to the strong coupling coefficient of the gratings presented in this work, both blue and red chirp have been demonstrated in a single device, with the possibility to tune across this full range for a particular central wavelength condition.

5. Materials and Methods

5.1. Device Fabrication

The devices were fabricated on a 220 nm thick silicon core on a buried oxide lower cladding layer of 2 µm. The patterns were written using e-beam lithography into hydrogen silsesquioxane (HSQ) resist that was subsequently used as a hardmask for Inductively Coupled Plasma (ICP)-Reactive Ion Etching (RIE) etching of the silicon. Polymer spot-size convertors [34] were fabricated to match the input lensed fibre mode to the waveguides. The final devices were coated with a 900 nm thick silica uppercladding before definition of the heater elements and electrical transmission lines. The heaters were fabricated as 900 nm wide strips of 50 nm thick NiCr. Multiple point contacts on the heater lines were defined to reduce the necessary voltage required to drive the device, as shown in the schematic of Figure 1. The heaters were divided into 50 µm long sections, each with a resistance of $\approx$1.5 kΩ, and were driven in an alternating positive/ground, arrangement.

5.2. Transmission and Group Delay Measurements

The Bragg grating devices were probed using a tunable laser coupled through fibre polarisation optics and a lensed fibre tip to the chip, as shown in Figure 12.

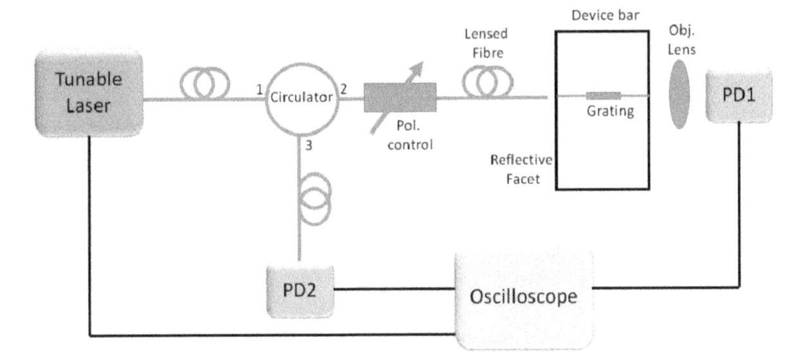

Figure 12. Transmission and group delay measurement setup. (Pol., polarisation; PD, photodiode).

The transmission spectra were measured by coupling the light from the output facet of the device bar to an objective lens and InGaAs photodiode. The reflected spectra were captured by coupling the back reflected light from the chip through a circulator to an InGaAs photodiode. Both photodiode signals were measured using an oscilloscope, triggered from the swept tunable laser source. The reflected spectra exhibit Fabry–Perot interference fringes created by the cavity formed between the reflective facet of the device bar and the Bragg grating, as shown in Figure 13. While the device bar facet is a broadband reflector with a spatially fixed position, the grating group delay profile creates a reflector with an effective spatial position that is wavelength dependent. Therefore, the effective Fabry–Perot cavity length is wavelength dependent and will create interference fringes that are directly related to the group delay response of the grating. This interference allows the direct extraction of the Fabry–Perot cavity phase, and hence group delay, introduced by the grating element as detailed in [35]. Figure 8 shows the measured group delay curves for the device measured using this Fabry–Perot technique.

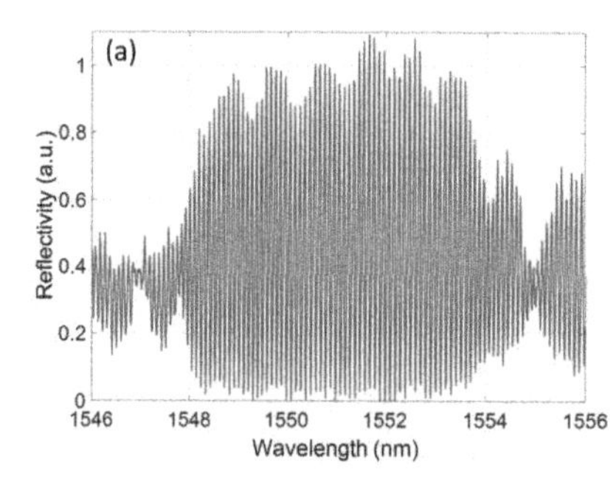

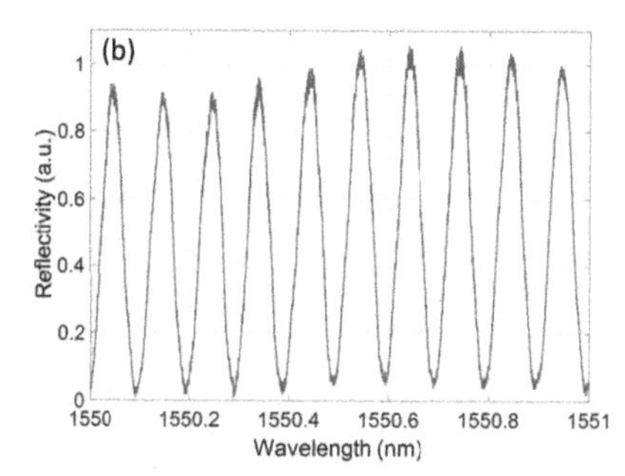

Figure 13. (**a**) Measured reflectivity spectrum of the grating with no induced chirp; and (**b**) detailed spectrum of the Fabry–Perot fringes created by the intereference between the device bar facet and grating reflector.

Author Contributions: Conceptualisation, M.S. and M.J.S.; Funding acquisition, M.S. and M.J.S.; Methodology, C.K. and M.J.S.; Software, M.J.S.; Writing—original draft, C.K. and M.J.S.; and Writing—review and editing, C.K., M.S. and M.J.S.

Funding: This research was funded by EPSRC grant numbers EP/P013597/1, EP/P013570/1, and EP/L021129/1.

Acknowledgments: The authors acknowledge the support of the James Watt Nanofabrication Centre staff.

Conflicts of Interest: The authors declare no conflict of interest.

References

1. Harris, N.C.; Grassani, D.; Simbula, A.; Pant, M.; Galli, M.; Baehr-jones, T.; Hochberg, M.; Englund, D.; Bajoni, D.; Galland, C. Integrated Source of Spectrally Filtered Correlated Photons for Large-Scale Quantum Photonic Systems. *Phys. Rev. X* **2014**, *4*, 041047. [CrossRef]
2. Oser, D.; Mazeas, F.; Le Roux, X.; Pérez-Galacho, D.; Alibart, O.; Tanzilli, S.; Labonté, L.; Marris-Morini, D.; Vivien, L.; Cassan, É.; et al. Coherency-Broken Bragg Filters: Overcoming On-Chip Rejection Limitations. *Laser Photonics Rev.* **2019**, *13*, 1800226. [CrossRef]
3. Prabhathan, P.; Murukeshan, V.M.; Jing, Z.; Ramana, P.V. Compact SOI nanowire refractive index sensor using phase shifted Bragg grating. *Opt. Express* **2009**, *17*, 15330–15341. [CrossRef] [PubMed]
4. Zhang, C.; Srinivasan, S.; Tang, Y.; Heck, M.J.R.; Davenport, M.L.; Bowers, J.E. Low threshold and high speed short cavity distributed feedback hybrid silicon lasers. *Opt. Express* **2014**, *22*, 10202–10209. [CrossRef] [PubMed]
5. Burla, M.; Cortés, L.R.; Li, M.; Wang, X.; Chrostowski, L.; Azaña, J. Integrated waveguide Bragg gratings for microwave photonics signal processing. *Opt. Express* **2013**, *21*, 25120–25147. [CrossRef] [PubMed]
6. Rutkowska, K.A.; Duchesne, D.; Strain, M.J.; Morandotti, R.; Sorel, M.; Azaña, J. Ultrafast all-optical temporal differentiators based on CMOS-compatible integrated-waveguide Bragg gratings. *Opt. Express* **2011**, *19*, 19514–19522. [CrossRef]
7. Wang, X.; Shi, W.; Yun, H.; Grist, S.; Jaeger, N.A.; Chrostowski, L. Narrow-band waveguide Bragg gratings on SOI wafers with CMOS-compatible fabrication process. *Opt. Express* **2012**, *20*, 15547–15558. [CrossRef]
8. Veerasubramanian, V.; Beaudin, G.; Giguere, A.; Le Drogoff, B.; Aimez, V.; Kirk, A.G. Design and Demonstration of Apodized Comb Filters on SOI. *IEEE Photonics J.* **2012**, *4*, 1133–1139. [CrossRef]
9. Chen, G.F.R.; Wang, T.; Donnelly, C.; Tan, D.T.H. Second and third order dispersion generation using nonlinearly chirped silicon waveguide gratings. *Opt. Express* **2013**, *21*, 29223–29230. [CrossRef]
10. Cheng, R.; Han, Y.; Chrostowski, L. Characterization and compensation of apodization phase noise in silicon integrated Bragg gratings. *Opt. Express* **2019**, *27*, 9516–9535. [CrossRef]
11. Strain, M.J.; Sorel, M. Design and fabrication of integrated chirped Bragg gratings for on-chip dispersion control. *Quantum Electron. IEEE J.* **2010**, *46*, 774–782. [CrossRef]
12. Zhang, W.; Yao, J. A fully reconfigurable waveguide Bragg grating for programmable photonic signal processing. *Nat. Commun.* **2018**, *9*, 1396. [CrossRef] [PubMed]

13. Rivas, L.M.; Strain, M.J.; Duchesne, D.; Carballar, A.; Sorel, M.; Morandotti, R.; Azaña, J. Picosecond linear optical pulse shapers based on integrated waveguide Bragg gratings. *Opt. Lett.* **2008**, *33*, 2425–2427. [CrossRef] [PubMed]
14. Tan, D.T.H.; Ikeda, K.; Saperstein, R.E.; Slutsky, B.; Fainman, Y. Chip-scale dispersion engineering using chirped vertical gratings. *Opt. Lett.* **2008**, *33*, 3013–3015. [CrossRef] [PubMed]
15. Strain, M.J.; Thoms, S.; MacIntyre, D.S.; Sorel, M. Multi-wavelength filters in silicon using superposition sidewall Bragg grating devices. *Opt. Lett.* **2014**, *39*, 413–416. [CrossRef] [PubMed]
16. Cheng, R.; Chrostowski, L. Apodization of Silicon Integrated Bragg Gratings through Periodic Phase Modulation. *IEEE J. Sel. Top. Quantum Electron.* **2019**, *26*. [CrossRef]
17. Kim, M.s.; Ju, J.J.; Park, S.K.; Lee, M.H.; Kim, S.H.; Lee, K.D. Tailoring Chirp Characteristics of Waveguide Bragg Gratings Using Tapered Core Profiles. *IEEE Photonics Technol. Lett.* **2006**, *18*, 2413–2415. [CrossRef]
18. Giuntoni, I.; Stolarek, D.; Kroushkov, D.I.; Bruns, J.; Zimmermann, L.; Tillack, B.; Petermann, K. Continuously tunable delay line based on SOI tapered Bragg gratings. *Opt. Express* **2012**, *20*, 11241–11246. [CrossRef]
19. Khan, S.; Baghban, M.A.; Fathpour, S. Electronically tunable silicon photonic delay lines. *Opt. Express* **2011**, *19*, 11780–11785. [CrossRef] [PubMed]
20. Jiang, L.; Huang, Z.R. Integrated Cascaded Bragg Gratings for On-Chip Optical Delay Lines. *IEEE Photonics Technol. Lett.* **2018**, *30*, 499–502. [CrossRef]
21. Chung, C.J.; Xu, X.; Wang, G.; Pan, Z.; Chen, R.T. On-chip optical true time delay lines featuring one-dimensional fishbone photonic crystal waveguide. *Appl. Phys. Lett.* **2018**, *112*, 071104. [CrossRef]
22. Zheng, X.; Patil, D.; Lexau, J.; Liu, F.; Li, G.; Thacker, H.; Luo, Y.; Shubin, I.; Li, J.; Yao, J.; et al. Ultra-efficient 10 Gb/s hybrid integrated silicon photonic transmitter and receiver. *Opt. Express* **2011**, *19*, 5172–5186. [CrossRef] [PubMed]
23. Orlandi, P.; Morichetti, F.; Strain, M.J.; Sorel, M.; Bassi, P.; Melloni, A. Photonic Integrated Filter with Widely Tunable Bandwidth. *J. Light. Technol.* **2014**, *32*, 897–907. [CrossRef]
24. Pérez, D.; Gasulla, I.; Crudgington, L.; Thomson, D.J.; Khokhar, A.Z.; Li, K.; Cao, W.; Mashanovich, G.Z.; Capmany, J. Multipurpose silicon photonics signal processor core. *Nat. Commun.* **2017**, *8*, 636. [CrossRef] [PubMed]
25. Reed, G.T.; Mashanovich, G.; Gardes, F.Y.; Thomson, D.J. Silicon optical modulators. *Nat. Photonics* **2010**, *4*, 518–526. [CrossRef]
26. Atabaki, A.H.; Eftekhar, A.A.; Yegnanarayanan, S.; Adibi, A. Sub-100-nanosecond thermal reconfiguration of silicon photonic devices. *Opt. Express* **2013**, *21*, 18312–18323. [CrossRef] [PubMed]
27. Densmore, A.; Janz, S.; Ma, R.; Schmid, J.H.; Xu, D.X.; Delâge, A.; Lapointe, J.; Vachon, M.; Cheben, P. Compact and low power thermo-optic switch using folded silicon waveguides. *Opt. Express* **2009**, *17*, 10457–10465. [CrossRef] [PubMed]
28. Dong, P.; Qian, W.; Liang, H.; Shafiiha, R.; Feng, D.; Li, G.; Cunningham, J.E.; Krishnamoorthy, A.V.; Asghari, M. Thermally tunable silicon racetrack resonators with ultralow tuning power. *Opt. Express* **2010**, *18*, 20298–20304. [CrossRef] [PubMed]
29. Orlandi, P.; Morichetti, F.; Strain, M.J.; Sorel, M.; Melloni, A.; Bassi, P. Tunable silicon photonics directional coupler driven by a transverse temperature gradient. *Opt. Lett.* **2013**, *38*, 863–865. [CrossRef] [PubMed]
30. Yamada, M.; Sakuda, K. Analysis of almost-periodic distributed feedback slab waveguides via a fundamental matrix approach. *Appl. Opt.* **1987**, *26*, 3474–3478. [CrossRef]
31. Turner, A.C.; Manolatou, C.; Schmidt, B.S.; Lipson, M.; Foster, M.a.; Sharping, J.E.; Gaeta, A.L. Tailored anomalous group-velocity dispersion in silicon channel waveguides. *Opt. Express* **2006**, *14*, 4357–4362. [CrossRef] [PubMed]
32. Sahin, E.; Ooi, K.J.; Png, C.E.; Tan, D.T. Large, scalable dispersion engineering using cladding-modulated Bragg gratings on a silicon chip. *Appl. Phys. Lett.* **2017**, *110*, 161113. [CrossRef]
33. Samarelli, A.; Macintyre, D.S.; Strain, M.J.; De La Rue, R.M.; Sorel, M.; Thoms, S. Optical characterization of a hydrogen silsesquioxane lithography process. *J. Vac. Sci. Technol. B Microelectron. Nanometer Struct.* **2008**, *26*, 2290–2294. [CrossRef]

34. Roelkens, G.; Dumon, P.; Bogaerts, W.; Van Thourhout, D.; Baets, R. Efficient silicon-on-insulator fiber coupler fabricated using 248-nm-deep UV lithography. *IEEE Photonics Technol. Lett.* **2005**, *17*, 2613–2615. [CrossRef]
35. Skaar, J. Measuring the group delay of fiber Bragg gratings by use of end-reflection interference. *Opt. Lett.* **1999**, *24*, 1020–1022. [CrossRef] [PubMed]

micromachines

Article

Design of Ultra-Compact Optical Memristive Switches with GST as the Active Material

Ningning Wang [1], Hanyu Zhang [1], Linjie Zhou [1,*], Liangjun Lu [1], Jianping Chen [1] and B.M.A. Rahman [2]

[1] State Key Laboratory of Advanced Optical Communication Systems and Networks, Department of Electronic Engineering, Shanghai Jiao Tong University, Shanghai 200240, China

[2] Department of Electrical and Electronic Engineering, City, University of London, London EC1V 0HB, UK

* Correspondence: ljzhou@sjtu.edu.cn; Tel.: +86-186-2101-1480

Received: 21 May 2019; Accepted: 25 June 2019; Published: 5 July 2019

Abstract: In the following study, we propose optical memristive switches consisting of a silicon waveguide integrated with phase-change material $Ge_2Sb_2Te_5$ (GST). Thanks to its high refractive index contrast between the crystalline and amorphous states, a miniature-size GST material can offer a high switching extinction ratio. We optimize the device design by using finite-difference-time-domain (FDTD) simulations. A device with a length of 4.7 μm including silicon waveguide tapers exhibits a high extinction ratio of 33.1 dB and a low insertion loss of 0.48 dB around the 1550 nm wavelength. The operation bandwidth of the device is around 60 nm.

Keywords: optical switch; phase change material; integrated silicon photonic circuits; nanophononics

---

1. Introduction

Integrated chip-level photonic circuits play a pivotal role in optical communication systems and networks [1]. Active devices such as switches and modulators of high performances are highly demanded. The optical switch, as one of the most fundamental components, has got widely studied. Thermo-optic (TO) switches has been successfully demonstrated on the silicon-on-insulator (SOI) platform [1,2], which incorporates hundreds of passive optical components and active tuners. The TO effect is relatively slow, limiting the switching speed in the order of microsecond. On the other hand, the electro-optic (EO) switches based on the free-carrier plasma dispersion effect can have a much higher speed up to nanosecond. High-radix EO switches have also been realized [3,4]. Although the basic switching elements and the overall switching topology can be further optimized to improve the switching performances, there are fundamental limitations for the TO and EO switches. Firstly, the switching state can only be maintained when there is a sustainable external power supply. This leads to large static power consumption. Second, as the refractive index tuning efficiency in both TO and EO methods is relatively low, it requires a long active waveguide of at least 10's micrometers to make the switch, which is disadvantageous for high-density photonic integration.

Phase change materials (PCMs), which have an extra-high complex refractive index contrast between two phase states, can be used to mitigate the above issues [5–7]. In recent years, the combination of nanophotonic components and PCM has been studied intensively. For example, vanadium dioxide (VO_2) with semiconductor and metal phases [7] can be used to make optical switches. However, as VO_2 is a volatile PCM, it is difficult to maintain the metal phase without power consumption. In contrast, another common PCM, $Ge_2Sb_2Te_5$ (GST), has the "self-holding" feature, which means it does not need a continuous power supply to keep its state. Besides, the phase transition of GST occurs on a sub-nanosecond time scale [5]. Therefore, GST is a good material choice for implementing non-volatile miniature photonic devices [8,9].

Here, we propose ultra-compact optical memristive switches that exploit GST as the active material connecting two tapered silicon waveguides. The optical memristive switches can provide two or multiple distinct optical transmission states based on resistive switching with a memory effect, i.e., the switching is non-volatile [10–14]. Once the state of the switch is triggered, it can be kept with no power consumption. The silicon and GST waveguides are designed to have matched effective indices and modal profiles to minimize the mode transition loss. The switch with a 0.7-μm-long GST section exhibits an insertion loss of <0.5 dB and a high extinction ratio of >30 dB. The Si-GST hybrid platform enables dense integration of high-performance switching components, opening new avenues for future low-power non-volatile photonic integrated circuits.

2. Device Structure and Analysis

The phase change material, GST, has two phases: amorphous phase and crystalline phase. Figure 1a shows the refractive indices (both real and imaginary parts) between the two phases. The refractive index information is measured from a thin GST film by spectroscopic ellipsometry. The phase change can be induced thermally, optically, or electrically potentially with an ultra-high speed [15–18]. Additionally, both of two phases have the "self-holding" feature to maintain its state at room-temperature without continuous power supply. Figure 1b illustrates the device structure composed of a silicon waveguide inserted with a section of GST. When the GST is at the amorphous phase, light can be transmitted through with a high output power (ON-state). However, when GST becomes crystalline, light is highly absorbed by GST, so the output power is very low (OFF-state). In order to build a high-performance on-off switch, the device insertion loss (IL) of the ON-state should be low, and the output power extinction ratio (ER) between the ON-state and the OFF-state should be large.

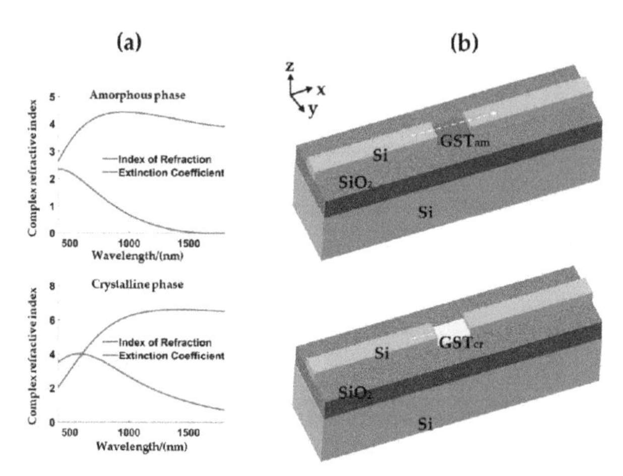

Figure 1. (**a**) Complex refractive index as a function of wavelength for amorphous and crystalline phases of GST material. (**b**) Schematic structure of the device showing ON-state and OFF-state optical transmissions.

The device insertion loss contains two parts, the GST absorption loss and the coupling loss between the silicon waveguide and the GST section. Both of these losses should be taken into consideration when we optimize the device. We first analyze the GST waveguide without considering its connection with the silicon waveguide. The effective complex refractive indices of GST waveguide are expressed as $n_{eff,am} = n_1 + ik_1$ and $n_{eff,cr} = n_2 + ik_2$ for its amorphous and crystalline phases, respectively. The optical transmission loss is determined by the attenuation coefficient, $k_{1,2}$. The transmissivity through the GST section, T, is given by

$$T = e^{\frac{-4\pi}{\lambda}k_{1,2}L} \tag{1}$$

the optical transmission loss at the two states can be written as,

$$\text{loss}_{\text{am}} = -10\log_{10}\cdot\text{T}_{\text{am}} = \frac{40\pi}{\lambda}\text{k}_1\text{L}\cdot\log_{10}e \tag{2}$$

$$\text{loss}_{\text{cr}} = -10\log_{10}\cdot\text{T}_{\text{cr}} = \frac{40\pi}{\lambda}\text{k}_2\text{L}\cdot\log_{10}e \tag{3}$$

the propagation extinction ratio (PER) of the GST waveguide between the ON-state and the OFF-state is written as:

$$\text{PER} = \frac{40\pi}{\lambda}(\text{k}_2 - \text{k}_1)\text{L}\cdot\log_{10}e \tag{4}$$

The coupling loss between the silicon waveguide and the GST waveguide is another contribution to the device loss. The power coupling rate (PCR) gives the total input coupling, taking into account both the modal overlap and the mismatch in effective indices between two modes. In simple cases, PCR is the product of the modal overlap integral (OI) and the Fresnel transmission rate (FR), written as [19,20]:

$$\text{PCR} = \text{OI}\cdot\text{FR} \tag{5}$$

the overlap integral gives the fractional power coupling from the silicon waveguide ($\vec{E_1}$, $\vec{H_1}$) into GST waveguide ($\vec{E_2}$, $\vec{H_2}$), given by [19].

$$\text{OI} = \text{Re}\cdot\left[\frac{\left(\int \vec{\text{E}}_1\times\vec{\text{H}}_2^{*}\cdot d\vec{\text{S}}\right)\left(\int \vec{\text{E}}_2\times\vec{\text{H}}_1^{*}\cdot d\vec{\text{S}}\right)}{\int \vec{\text{E}}_1\times\vec{\text{H}}_1\cdot d\vec{\text{S}}}\right]\cdot\frac{1}{\text{Re}\cdot\left(\int \vec{\text{E}}_2\times\vec{\text{H}}_2\cdot d\vec{\text{S}}\right)} \tag{6}$$

the Fresnel transmission rate is given by

$$\text{FR} = \frac{2n_{Si}\cdot\cos\,\text{i}_1}{n_{Si}\cdot\cos\,\text{i}_1 + n_{GST}\cdot\cos\text{i}_2} \tag{7}$$

where i_1 and i_2 represent the incident and refraction angles, respectively.

3. Design and Simulations

We first studied the GST waveguide using the finite-difference-time-domain (FDTD) simulations. The under-cladding of the device is a 2-μm-thick SiO_2 layer on the silicon substrate and the upper-cladding is air. The complex refractive indices of GST around the 1550 nm wavelength are taken as 3.98 + 0.024i and 6.49 + 1.05i for amorphous and crystalline states, respectively [9]. Figure 2a,c show the forward optical power transmission along the x-axis (P_x) in the amorphous GST waveguide when the GST waveguide width is 0.4 μm and 0.6 μm, respectively. Figure 2b,d show the P_x power distributions in their corresponding silicon waveguides, when their effective indices are matched with their amorphous GST waveguides. The heights of the GST and silicon waveguides are chosen as 0.05 μm and 0.22 μm. It can be seen that the optical power is mainly distributed outside the GST region for the 0.4-μm-wide GST waveguide, which has a high mode mismatch with the silicon waveguide. Therefore, we avoid the low-confinement GST waveguide in device optimization.

We next scanned the GST waveguide width and height in order to get an optimal design. Figure 3a,b show the propagation loss of the GST waveguide at the amorphous and crystalline states, respectively. The amorphous GST has a lower propagation loss. For a thin GST layer (less than ~75 nm), the loss decreases for a narrower GST waveguide. This is because the light cannot be well confined in the GST layer when the size is small, leading to an expanded mode. When the GST layer is thick enough (larger than ~75 nm), the loss is first reduced and then almost unchanged with an increasing waveguide width. On the other hand, the crystalline GST has good confinement of light even with a small waveguide size. It has a much larger propagation loss than the amorphous state. For a fixed width, the loss increases with the height and then slightly decreases; for a fixed height, the loss always decreases with a wider waveguide. Figure 3c shows the ER as a function of GST height with GST width as a variable. It generally follows the same trend as the waveguide propagation loss at the crystalline state.

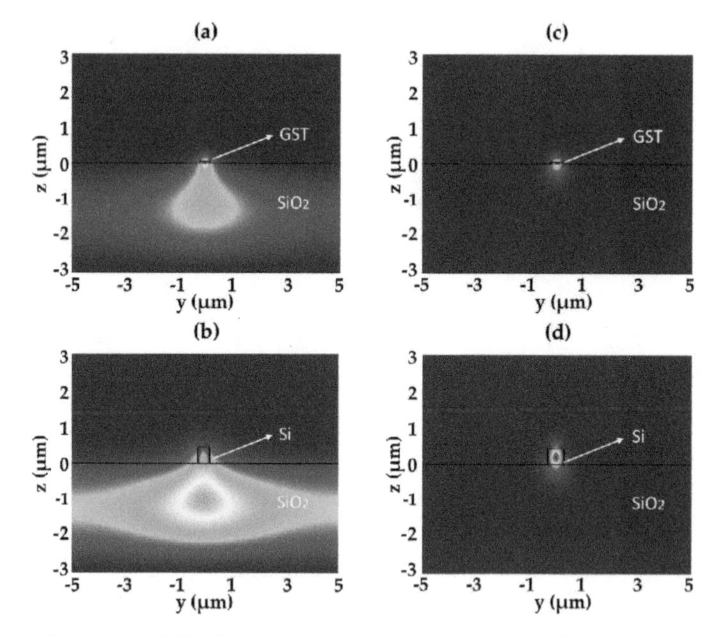

Figure 2. Cross-sectional P_x distributions in the (**a**) 0.4-μm-wide GST waveguide and (**b**) its corresponding silicon waveguide. Cross-sectional P_x distributions in the (**c**) 0.6-μm-wide GST waveguide and (**d**) its corresponding silicon waveguide. The GST layer thickness is 0.05 μm and the silicon waveguide layer thickness is 0.22 μm.

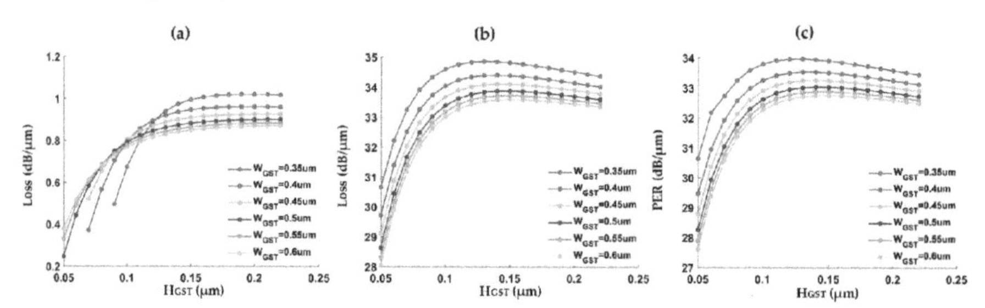

Figure 3. (**a**) Propagation loss of GST waveguide at the amorphous state. (**b**) Propagation loss of GST waveguide at the crystalline state. (**c**) Extinction ratio between the two states.

To get a balanced performance between the insertion loss and propagation extinction ratio, we chose $W_{gst} = 0.45$ μm and $H_{gst} = 0.12$ μm. The GST waveguide effective index is 2.292 + 0.0247i at the amorphous state. The silicon waveguide dimensions are chosen to be 0.45 μm (width) × 0.22 μm (height). The modal effective index is calculated to be 2.29, which is close to the effective index of the amorphous GST waveguide. Figure 4a,b show the mode profiles for the silicon and GST waveguides, respectively. Although their refractive indices are close, the mode profiles are still different, resulting in mode transition loss at the waveguide interface. The calculated PCR is 0.911. Figure 4c,d show the light propagation at the two GST states. The GST section length is L_{GST} = 0.6 μm. The transmission losses at the amorphous and crystalline states are 0.70 dB and 21.04 dB, respectively.

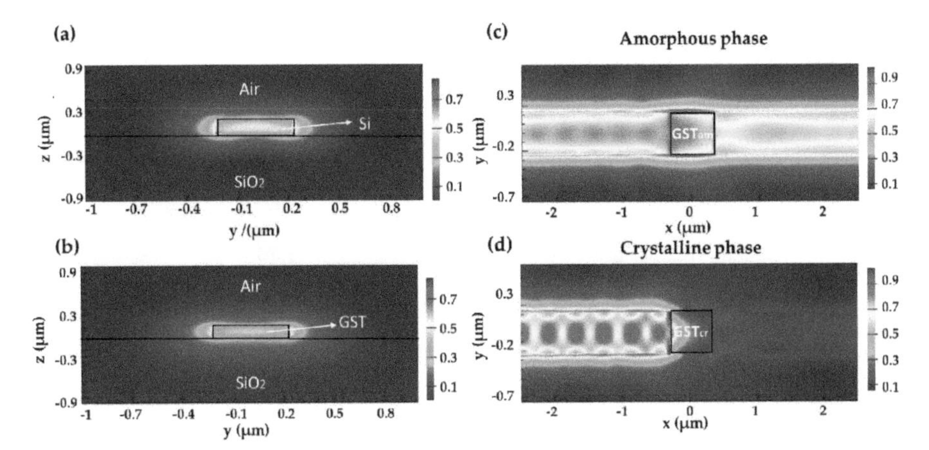

Figure 4. (**a**,**b**) Electric-field intensity mode profiles of (**a**) silicon waveguide and (**b**) amorphous GST waveguide. (**c**,**d**) Electric-field intensity distribution in the horizontal plane along the waveguide when GST is at (**c**) the amorphous state and (**d**) the crystalline state.

There is a compromise between the device IL and the ON-OFF switching ER. The IL is defined as the device optical transmission loss at the amorphous state. The ER is defined as the transmission loss difference between the amorphous state and the crystalline state. Figure 5 shows the optical transmission losses at the two states and the ER as a function of wavelength and GST length. The loss of ON-state (amorphous GST) increases when the wavelength deviates from the targeted 1550 nm wavelength because of the effective index mismatch. The loss of OFF-state (crystalline GST) does not monotonically increase with the GST length due to the Fabry-Perot resonant effect, given that the waveguide interface causes certain back reflection. It can be seen that when L_{GST} is 0.6 μm, the device IL is less than 1 dB, and the ON-OFF switching ER is more than 20 dB in an optical bandwidth of 70 nm. When the GST section is longer than 1 μm, the ER can be further improved but at the cost of a higher IL.

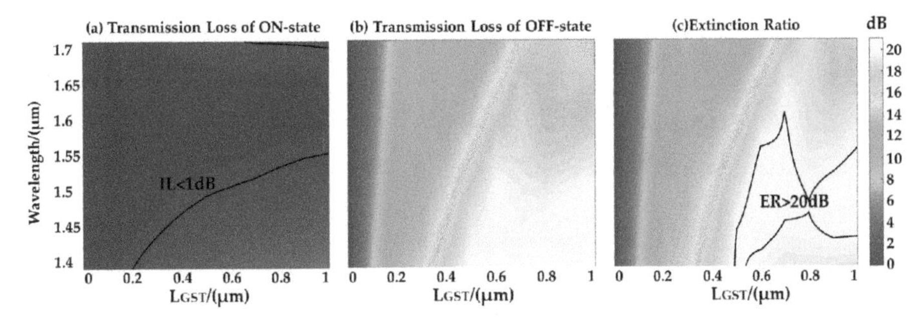

Figure 5. (**a**) Transmission loss of ON-state (amorphous phase). (**b**) Transmission loss of OFF-state (crystalline phase). (**c**) ON-OFF switching extinction ratio after phase change of GST.

In the above device, the slight difference in modal profiles results in extra scattering and reflection losses in the waveguide junction. To mitigate this issue, we revised the design, as shown in Figure 6a. The GST waveguide dimensions are the same as the previous design. The silicon waveguide has two layers of tapers with heights of 70 nm and 150 nm. The height values comply with the silicon etch depths in silicon photonics foundries [10] so that it can be conveniently integrated with other optical devices. The width of the top taper is reduced from 500 nm to 80 nm, causing the light field to be gradually squeezed to the bottom taper. To match the mode profile of the GST waveguide, the bottom taper end width is chosen to be 700 nm. The length of upper taper (L_1) and lower taper (L_2) are determined to be L_1 = 1.9 µm and L_2 = 2 µm. The transition loss through the two tapers is less than 0.1 dB.

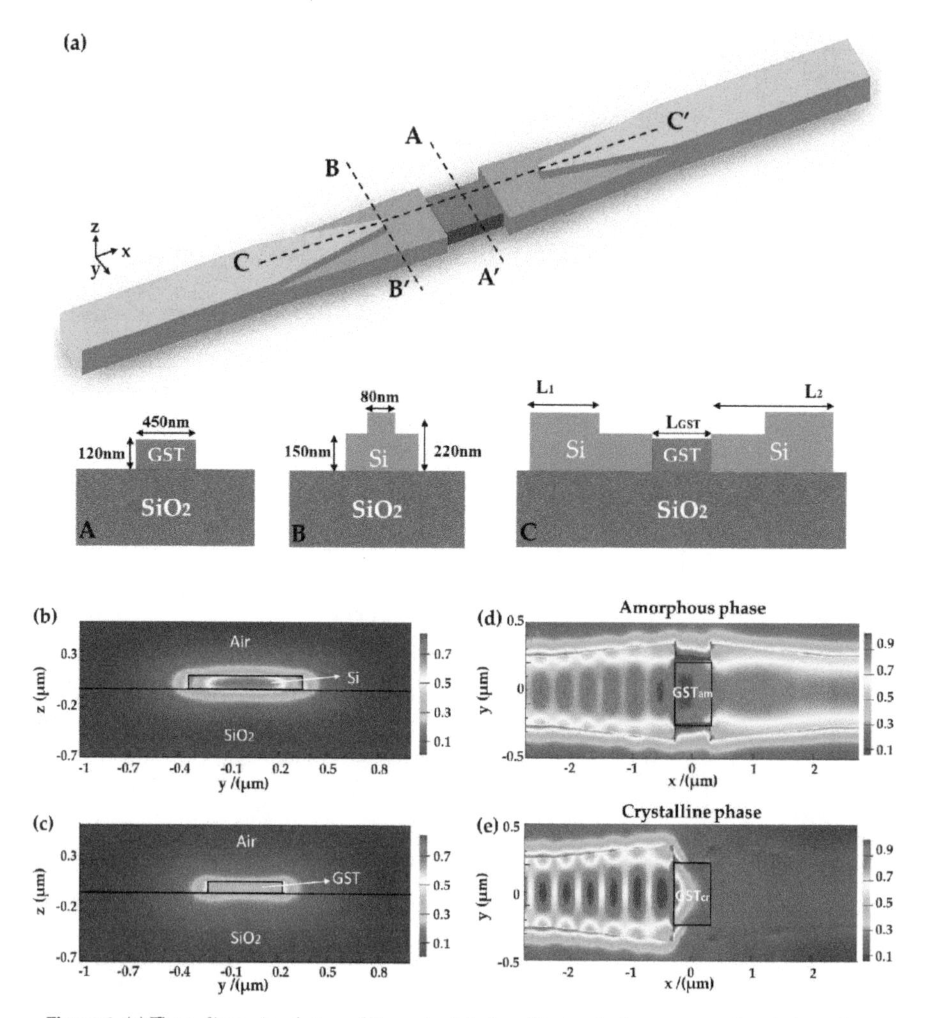

Figure 6. (**a**) Three-dimensional view of the revised design. The insets show cross-sectional views along the segments AA', BB', and CC'. (**b**,**c**) Electric-field intensity profiles of (**b**) the silicon waveguide and (**c**) the amorphous GST waveguide across the interface. (**d**,**e**) Electric-field intensity distributions in the horizontal plane along the waveguide when GST is at (**d**) the amorphous and (**e**) the crystalline states.

Figure 6b,c, show that the mode profiles of the silicon and GST waveguide match better than the previous design. The PCR is close to 0.968. Light propagates through the GST section with higher transitivity in the amorphous state but highly blocked in the crystalline state, as illustrated in Figure 6d,e.

Figure 7 shows the loss at the amorphous and crystalline states and the ER of the revised design. When L_{GST} = 0.7 μm, the device IL is 0.48 dB and the ER is as high as 33.1 dB at the wavelength of 1550 nm. In an optical bandwidth of 60 nm, the IL is less than 0.5 dB and the ER is larger than 30 dB.

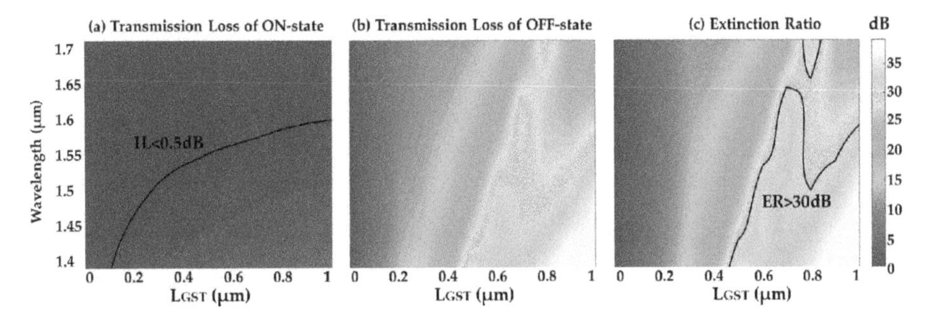

Figure 7. (**a**) Transmission loss of ON-state (amorphous phase). (**b**) Transmission loss of OFF-state (crystalline phase). (**c**) ON-OFF switching extinction ratio when GST phase change occurs.

4. Fabrication Tolerances

The proposed structures can be readily fabricated using complementary metal-oxide-semiconductor (CMOS) compatible fabrication processes. To investigate the robustness of our memristive switch to the alignment errors between the GST and Si waveguide layers, we simulated device performances (IL and ER) when the lateral alignment error increases. Table 1 shows the results. The operation wavelength is 1550 nm. The GST length is 0.7 μm. It can be seen that the IL increases and ER decreases when the misalignment rises from 0 to 20 nm. Nonetheless, the IL is still within 1 dB and the ER is larger than 30 dB for an alignment error up to 20 nm. It indicates that our device has a high tolerance to fabrication uncertainties.

Table 1. ON-state insertion loss and ON-OFF switching extinction ratio for different lateral alignment errors between the GST and Si waveguides.

Misalignment	IL (Insertion Loss)	ER (Extinction Ratio)
0 nm	0.48 dB	33.1 dB
5 nm	0.79 dB	32.6 dB
10 nm	0.80 dB	32.1 dB
15 nm	0.81 dB	31.7 dB
20 nm	0.82 dB	31.2 dB

5. Conclusions

GST is a promising material for optical switches due to its non-volatile high refractive index contrast between its amorphous and crystalline states. We have proposed an ultra-compact optical switch based on the phase change of GST to control the transmittance. We optimized the device to obtain a high ER and a low IL. The device with a 0.7-μm-long GST section can offer a high ER of more than 30 dB and a low IL of less than 0.5 dB with a broad optical bandwidth of 60 nm. The memristive switch shows high tolerance to misalignment between the GST and silicon waveguides. When the lateral alignment drift is up to 20 nm, the IL remains less than 1dB and the ER is larger than 30 dB. This ultra-small memristive switch can be applied in inter- and intra-chip optical communications.

Author Contributions: N.W. performed the simulations; L.Z., N.W., and H.Z. analyzed the data; L.Z., L.L., J.C., and B.M.A.R. supervised the project; N.W. wrote the paper.

Funding: This research was funded by the National Natural Science Foundation of China(NSFC) [61705129, 61535006] and the Municipal Science and Technology Major Project [2017SHZDZX03].

Conflicts of Interest: The authors declare no conflict of interest.

References

1. Dumais, P.; Goodwill, D.J.; Celo, D.; Jiang, J.; Zhang, C.; Zhao, F.; Tu, X.; Zhang, C.; Yan, S.; He, J.; et al. Silicon photonic switch subsystem with 900 monolithically integrated calibration photodiodes and 64-fiber package. *J. Lightwave Technol.* **2017**, *36*, 233–238. [CrossRef]
2. Suzuki, K.; Konoike, R.; Hasegawa, J.; Suda, S.; Matsuura, H.; Ikeda, K.; Namiki, S.; Kawashima, H. Low insertion loss and power efficient 32 × 32 silicon photonics switch with extremely-high-delta PLC connector. In *Optical Fiber Communication Conference*; Optical Society of America: Washington, DC, USA, 2018; p. Th4B.5. [CrossRef]
3. Lu, L.; Zhao, S.; Zhou, L.; Li, D.; Li, Z.; Wang, M.; Li, X.; Chen, J. 16 × 16 non-blocking silicon optical switch based on electro-optic Mach-Zehnder interferometers. *Opt. Express* **2016**, *24*, 9295–9307. [CrossRef] [PubMed]
4. Qiao, L.; Tang, W.; Chu, T. 32 × 32 silicon electro-optic switch with built-in monitors and balanced-status units. *Sci. Rep.* **2017**, *7*, 42306. [CrossRef] [PubMed]
5. Waldecker, L.; Miller, T.A.; Rudé, M.; Bertoni, R.; Osmond, J.; Pruneri, V.; Simpson, R.E.; Ernstorfer, R.; Wall, S. Time-domain separation of optical properties from structural transitions in resonantly bonded materials. *Nat. Mater.* **2015**, *14*, 991. [CrossRef] [PubMed]
6. Wuttig, M.; Bhaskaran, H.; Taubner, T. Phase-change materials for non-volatile photonic applications. *Nat. Photonics* **2017**, *11*, 465. [CrossRef]
7. Liu, M.; Hwang, H.Y.; Tao, H.; Strikwerda, A.C.; Fan, K.; Keiser, G.R.; Sternbach, A.J.; West, K.G.; Kittiwatanakul, S.; Lu, J.; et al. Terahertz-field-induced insulator-to-metal transition in vanadium dioxide metamaterial. *Nature* **2012**, *487*, 345. [CrossRef] [PubMed]
8. Zhang, H.; Zhou, L.; Rahman, B.M.A.; Wu, X.; Lu, L.; Xu, Y.; Xu, J.; Song, J.; Hu, Z.; Xu, L.; et al. Ultracompact Si-GST hybrid waveguides for nonvolatile light wave manipulation. *IEEE Photonics J.* **2017**, *10*, 1–10. [CrossRef]
9. Liang, H.; Soref, R.; Mu, J.; Majumdar, A.; Li, X.; Huang, W.P. Simulations of silicon-on-insulator channel-waveguide electrooptical 2 × 2 switches and 1 × 1 modulators using a $Ge_2Sb_2Te_5$ self-holding layer. *J. Lightwave Technol.* **2015**, *33*, 1805–1813. [CrossRef]
10. Emboras, A.; Goykhman, I.; Desiatov, B.; Mazurski, N.; Stern, L.; Shappir, J.; Levy, U. Nanoscale plasmonic memristor with optical readout functionality. *Nano Lett.* **2013**, *13*, 6151–6155. [CrossRef] [PubMed]
11. Hoessbacher, C.; Fedoryshyn, Y.; Emboras, A.; Hillerkuss, D.; Melikyan, A.; Kohl, M.; Sommer, M.; Hafner, C.; Leuthold, J. Latching Plasmonic Switch with High Extinction Ratio. In *2014 Conference on Lasers and Electro-Optics (CLEO)-Laser Science to Photonic Applications*; IEEE: Piscataway, NJ, USA, 2014; pp. 1–2.
12. Battal, E.; Ozcan, A.; Okyay, A.K. Resistive switching-based electro-optical modulation. *Adv. Opt. Mater.* **2014**, *2*, 1149–1154. [CrossRef]
13. Hoessbacher, C.; Fedoryshyn, Y.; Emboras, A.; Melikyan, A.; Kohl, M.; Hillerkuss, D.; Hafner, C.; Leuthold, J. The plasmonic memristor: A latching optical switch. *Optica* **2014**, *1*, 198–202. [CrossRef]
14. Ríos, C.; Stegmaier, M.; Hosseini, P.; Wang, D.; Scherer, T.; Wright, C.D.; Bhaskaran, H.; Pernice, W.H. Integrated all-photonic nonvolatile multi-level memory. *Nat. Photonics* **2015**, *9*, 725. [CrossRef]
15. Stegmaier, M.; Ríos, C.; Bhaskaran, H.; Pernice, W.H. Thermo-optical Effect in Phase-Change Nanophotonics. *ACS Photonics* **2016**, *3*, 828–835. [CrossRef]
16. Moriyama, T.; Kawashima, H.; Kuwahara, M.; Wang, X.; Asakura, H.; Tsuda, H. Small-sized Mach-Zehnder interferometer optical switch using thin film $Ge_2Sb_2Te_5$ phase-change material. In *Optical Fiber Communication Conference*; Optical Society of America: Washington, DC, USA, 2014; p. Tu3E. 4.
17. Kato, K.; Kuwahara, M.; Kawashima, H.; Tsuruoka, T.; Tsuda, H. Current-driven phase-change optical gate switch using indium-tin-oxide heater. *Appl. Phys. Exp.* **2017**, *10*, 072201. [CrossRef]
18. Lim, A.E.J.; Song, J.; Fang, Q.; Li, C.; Tu, X.; Duan, N.; Chen, K.K.; Tern, R.P.-C.; Liow, T.-Y. Review of silicon photonics foundry efforts. *IEEE J. Sel. Top. Quant.* **2013**, *20*, 405–416. [CrossRef]
19. Lumerical Knowledge Base. Available online: https://kb.lumerical.com (accessed on 3 July 2019).
20. Snyder, A.W.; Love, J. *Optical Waveguide Theory*; Chapman & Hall: London, UK, 1983.

 *micromachines*

Article

Translational MEMS Platform for Planar Optical Switching Fabrics

Suraj Sharma [1,*], Niharika Kohli [1], Jonathan Brière [2], Michaël Ménard [3] and Frederic Nabki [1]

1 Department of Electrical Engineering, Ecole de Technologie Supérieure, Montréal, QC H3C 1K3, Canada
2 AEPONYX Inc., Montréal, QC H3C 4J9, Canada
3 Department of Computer Science, Université du Québec à Montreal, Montréal, QC H2X 3Y7, Canada
* Correspondence: suraj.sharma.1@ens.etsmtl.ca; Tel.: +1-514-239-8995

Received: 7 June 2019; Accepted: 29 June 2019; Published: 30 June 2019

Abstract: While 3-D microelectromechanical systems (MEMS) allow switching between a large number of ports in optical telecommunication networks, the development of such systems often suffers from design, fabrication and packaging constraints due to the complex structures, the wafer bonding processes involved, and the tight alignment tolerances between different components. In this work, we present a 2-D translational MEMS platform capable of highly efficient planar optical switching through integration with silicon nitride (SiN) based optical waveguides. The discrete lateral displacement provided by simple parallel plate actuators on opposite sides of the central platform enables switching between different input and output waveguides. The proposed structure can displace the central platform by 3.37 μm in two directions at an actuation voltage of 65 V. Additionally, the parallel plate actuator designed for closing completely the 4.26 μm air gap between the fixed and moving waveguides operates at just 50 V. Eigenmode expansion analysis shows over 99% butt-coupling efficiency the between the SiN waveguides when the gap is closed. Also, 2.5 finite-difference time-domain analysis demonstrates zero cross talk between two parallel SiN waveguides across the length of the platform for a 3.5 μm separation between adjacent waveguides enabling multiple waveguide configuration onto the platform. Different MEMS designs were simulated using static structural analysis in ANSYS. These designs were fabricated with a custom process by AEPONYX Inc. (Montreal, QC, Canada) and through the PiezoMUMPs process of MEMSCAP (Durham, NC, USA).

Keywords: microelectromechanical systems (MEMS); electrostatic actuator; parallel plate actuation; optical switch; silicon-on-insulator (SOI); micro-platform; optical waveguide; silicon nitride photonics; integrated optics

---

1. Introduction

Over the years, micro devices with optical and microelectromechanical systems (MEMS) components known as micro-opto-electro-mechanical systems (MOEMS) have been developed for use in digital micro mirror displays [1] and laser scanners [2]. Development of such optical MEMS devices subsided due to immaturity of the technology and market penetration challenges [3]. However, with the world moving towards higher bandwidth optical fiber-based communication, MOEMS can help meet the ever-growing demand for power and transmission efficient integrated silicon photonics solutions. Conventional electronic data centers are often associated with high cost, and high energy and space consumption [4]. These technological, environmental and monetary constraints have paved the way for MEMS integration towards the development of hybrid optical data center designs such as Helios [5] and novel Scaled Out Optically Switched Network Architecture [6]. MEMS integration into data centers can reduce power consumption from 12.5 Watts per port for electronic switches to just 0.24 Watts per port for optical switches but with a re-configurability that is restricted to a few milliseconds [4]. Such data centers often rely upon 3-D MEMS with out-of-plane rotating micro-mirrors for beam steering inside an optical cross connect switch [7] using piezoelectric actuation [8] or electrostatic actuation [9,10]. Although 3-D MEMS allow the implementation of optical switches with a large number of ports, the development of such systems often suffers from fabrication and packaging constraints due to the complex structures and wafer bonding processes involved [11,12]. Thus, paving the way for simpler more affordable 2-D MEMS based integrated photonics solutions for switching applications.

Piezoelectric [13], electrothermal [14] and electrostatic [15] actuators provide precise mechanical motion at the micron scale. Piezoelectric actuators, although fast and suitable for applications with resonators, involve the use of complex piezoelectric materials such as AlN and PZT for actuation [16,17]. These materials can be integrated with silicon-on-insulator (SOI) technology to make optical switches [18] but the use of lead in PZT raises environmental concerns [19]. Alternative piezo materials, such as AlN, are difficult to reproduce with the same piezo properties because of their dependency on the film texture and dipole orientation, and that they can have large residual stress with even a slight change in the deposition parameters [20,21]. Electrothermal actuators produce large displacements at low actuation voltages but are slow, consume high power and produce heat during actuation [22,23]. This makes them undesirable for the green optical data centers envisioned for the future. Thus, low voltage electrostatic actuators based upon widely used comb drives and parallel plate designs become the right choice for planar optical switching applications [24,25]. These can also be fabricated with ease through commercial SOI microfabrication processes to validate MEMS designs before integration with optical waveguides [26,27].

Electrostatic actuators connected to a central platform have been demonstrated in the past for 2-D and 3-D MEMS based solutions such as optical scanners and cold atom detectors [28,29]. These designs have largely relied upon out-of-plane rotational motion of the central platform due to torsional beams [30–32]. A few translational MEMS structures exist, but they are largely designed for out-of-plane optics applications [33–35]. The MEMS for planar switching applications reported in the literature rely upon bringing movable waveguides closer to fixed waveguides in ON/OFF state [36,37] or as a 1×3 optical switch [38]. Through complex MEMS integration of soft polymer waveguides, a 2×2 optical switch has also been demonstrated [39]. Recent developments include planar switching done by adiabatic coupling between waveguides through vertical actuation at very low voltage [40,41], and butt coupling through in-plane rotational actuation [42].

Accordingly, in this work, we present a translational MEMS platform capable of motion along two axes using multiple electrostatic actuators. A detailed overview of a translational MEMS platform compatible with different planar optical switching configurations is presented in Section 2 along with optical design considerations. Results of EigenMode Expansion (EME) and Finite Difference Time Domain (FDTD) simulations for the butt coupling of SiN waveguides and the cross-talk between parallel SiN waveguides are also presented in this section. Alignment tolerance simulations show the potential for efficient optical switching with the proposed MEMS platform. The evolution of the actuator and spring designs along with the critical design choices made for a simple switching approach are also discussed. Design of the translational MEMS platform and the critical design parameters and dimensions are presented in Section 3. The fabrication process used, and the analysis of the fabricated devices are discussed in Section 4. In this section, the test setup used for the actuation experiments and the results obtained are also reported. A discussion of these results is presented in Section 5 and is followed by the envisioned future work and concluding remarks in Section 6.

2. Design Considerations

2.1. Translational MEMS Platform for Optical Switching

Previous devices developed by our research group relied upon a rotational MEMS platform for planar optical switching using SiN waveguides surrounded by a SiO_2 cladding [42]. The device uses 5° of its total 9.5° of rotation on each side to form a crossbar switch requiring 113 V to actuate and having a 1.3 mm by 1 mm footprint. Also, the air gap closing actuator designed operates at 118 V with a minimal gap of 250 nm upon actuation.

In this work, a unique translational MEMS platform capable of bi-axial motion is proposed and demonstrated. The lateral motion of the platform is bi-directional and can provide optical switching in 1 × 3 or 2 × 2 crossbar switch configurations. The longitudinal motion of the central platform is unidirectional and designed to completely close the air gap between waveguides on the substrate and the platform to achieve highly efficient butt-coupling. The design operates at a reduced actuation voltage for both switching and gap closing motions compared to that reported in [42], and the device footprint is smaller. Figure 1 shows illustrations for the translational MEMS platform as a 1 × 3 optical switch (Figure 1a) and as a 2 × 2 crossbar optical switch (Figure 1b). These structures are meant to include integrated SiN waveguides which are not the focus on this work but that have been demonstrated in [42,43]. A cross sectional representation of the entire optical MEMS stack envisioned is also shown in Figure 1c.

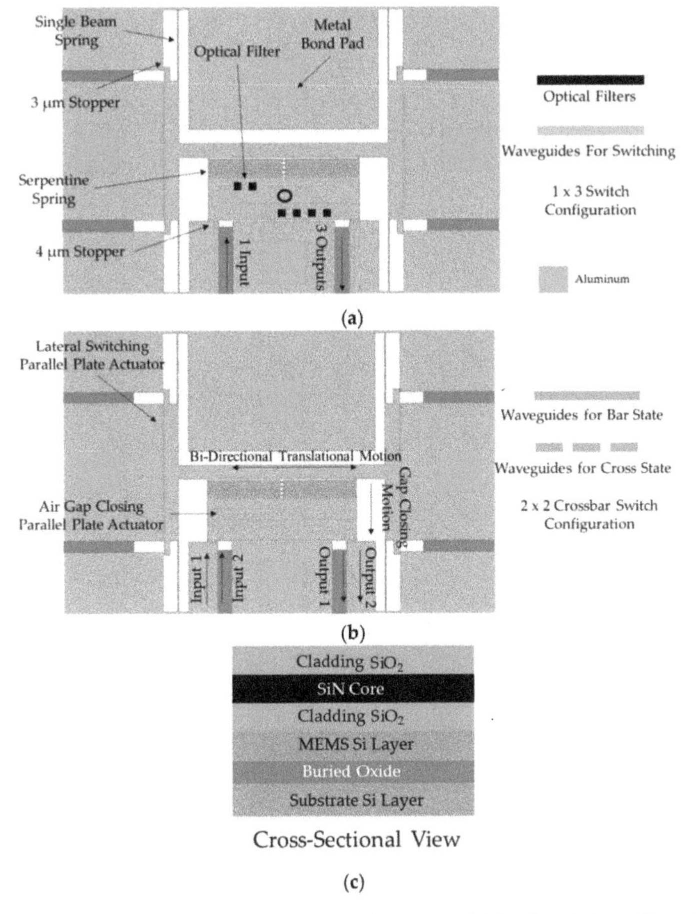

Figure 1. Illustrations of the proposed translational actuator: (**a**) 1 × 3 switch configuration with integrated optical filters; (**b**) 2 × 2 crossbar switch configuration; (**c**) cross sectional view of the optical MEMS stack proposed. The color scheme to represent the different materials is consistent throughout the figure.

In the 1 × 3 optical switch configuration, the central platform accommodates three separate SiN waveguides. The input side on the left of the platform has one waveguide on the fixed substrate. The output side on the right of the platform is designed to have three separate waveguides. Four symmetrical single silicon beam springs support the entire MEMS structure. These beams provide the necessary spring action to allow lateral displacement through parallel plate actuators on the opposite side of the platform. The central platform is connected to the lateral actuators through a central beam and a serpentine spring structure. This design choice decouples the lateral and longitudinal motion of the platform. The serpentine spring enables longitudinal displacement through a parallel plate actuator at the bottom of the platform. Whereas the parallel plate actuators on the opposite sides of the platform are designed to provide discrete lateral displacement of 3 µm on each side, the bottom parallel plate actuator is designed to close the 4 µm air gap between the platform and the substrate. The discrete lateral displacement of 3 µm on either side along with the neutral position of zero lateral displacement provides 3 switching possibilities to form a 1 × 3 switch. The platform is large enough to integrate optical filters, such as Bragg gratings and ring resonators [44,45], and it can be used to select among a bank of filters to implement discretely tunable devices. Examples of optical filters are also illustrated in Figure 1a.

Figure 1b shows the MEMS platform envisioned as a 2 × 2 crossbar switch. There are two input and output waveguides on the substrate. The actuation mechanism and operational parameters remain the same as the 1 × 3 switch configuration, but the platform accommodates four SiN waveguides. When the platform is actuated to the right, the waveguides on the platform shown as solid lines in Figure 1b are aligned with the input and output waveguides. In this position light travels from input 1 to output 1 and from input 2 to output 2, creating the 'bar' configuration. When the platform is actuated to the left, waveguides on the platform shown as dotted lines in Figure 1b are aligned to the input and output waveguides. The optical signal then propagates from input 1 to output 2 and from input 2 to output 1, creating the 'cross' configuration. In both configurations, the gap closing mechanism provides highly efficient butt-coupling between the waveguides. The discrete motion of the platform also eliminates optical losses due to displacement / voltage fluctuations in the system, as the MEMS platform is designed to operate in the pull-in state for both the lateral and longitudinal actuators.

2.2. Optical Design Considerations

Our work involves the validation of the MEMS [27] structures, prior to employing the commercial process enabling the addition of optical waveguides which has been developed by our group in collaboration with AEPONYX. The commercial process used to validate the MEMS requires a minimal gap of ~3 µm. This constraint leads to a significant air gap between the fixed and moving waveguides envisioned in a planar optical MEMS device. In the previous rotational MEMS developed by our group, the input and output waveguides were located on top of the gap closing actuator due to design constraints [42]. This enables the air gap to be reduced to only 250 nm as the rotational platform that is grounded cannot come in contact with the gap closing actuator that is kept at a high DC voltage. If the two come in contact, shorting during actuation would damage the MEMS device. This phenomenon can be prevented by dimpled structures but leads to a residual air gap even after gap closing. However, on the translational MEMS platform shown in Figure 1, the input and output waveguides are separated from the gap closing actuator. The air gap between the platform and the fixed section of the switch (with input / output waveguides) is designed to be 4 µm whereas the gap for the bottom parallel plate actuator is designed to be 6 µm. As a result, the platform and the fixed section of the switch can both be grounded to eliminate shorting during gap closing actuation. This provides complete gap closing between waveguides eliminating any significant residual air gap.

EME analysis using MODE Solutions from Lumerical® (Vancouver, Canada) was performed to study the effect of an air gap on optical signal transmission between two butt-coupled SiN channel waveguides with a core of 435 nm × 435 nm and with a top and bottom SiO_2 cladding thickness of 3.4 µm for both the TE and TM modes. All of the optical simulations shown in this section were performed at a wavelength of 1550 nm. EME results show a transmission efficiency of over 99% for direct butt-coupling between these waveguides with an air gap of 50 nm or less, which is reduced to 33% when the gap is 3 μm. To reduce the expansion of the light beams in the gap and increase the coupling between the two waveguides, we introduced inverted tapers where the core width narrows down to 250 nm at both waveguide edges. The optimal length of the tapers was found to be 20 μm. The transmission efficiency is almost 100% with a 50 nm air gap and even on increasing the gap to 3 μm, the coupling efficiency dropped to 83% for the transverse electric (TE) mode and 74% for transverse magnetic (TM) mode in waveguides with inverted tapers. This result is shown in Figure 2a and demonstrates that a high coupling efficiency can be obtained even if fabrication imperfections limit the minimum size of the gap. Furthermore, the ability to reduce the gap to dimensions significantly smaller than the wavelength of light (which is typically around 1.3 μm or 1.5 μm in telecommunication applications) remove the need for an antireflection coating at the interface of the waveguides. When the gap size is larger than approximately half a wavelength, multiple beam interference phenomena can occur because of reflections at the waveguide interfaces, which explains the undulations that are visible in Figure 2a.

The dimensions of the central platform in an optical MEMS device as shown in Figure 1 are highly critical. The platform must accommodate at least three waveguides to operate as a 1 × 3 switch and four waveguides to operate as a 2 × 2 crossbar switch. Another important design consideration is the width of the gap closing interface between the platform and the substrate. The platform must be able to accommodate the number of waveguides envisioned with minimal optical cross-talk and optimal bending radius for low propagation losses [46]. Therefore, to have an estimate of the number of waveguides that can be implemented on the platform, we studied the cross-talk between two parallel SiN waveguides as function of the gap between them. 2.5D FDTD analysis were performed for 435 nm × 435 nm waveguides with a 3.4 μm thick top and bottom SiO_2 cladding where the total length of the inner waveguide is 565.5 μm with a bending radius of 75 μm. It was found that for the TE mode the field remains confined in the input waveguide and does not couple to the adjacent outer waveguide when the gap between them is 3.5 μm or greater. The simulated propagation loss in the input waveguide of length 565.5 μm is only 0.01 dB for a 3.5 μm gap between adjacent waveguides. Results of the 2.5D FDTD cross-talk simulations are shown in Figure 2b. It can also be observed that the cross-talk for the TM mode is smaller than the TE mode and becomes negligible at a gap of 3.0 μm.

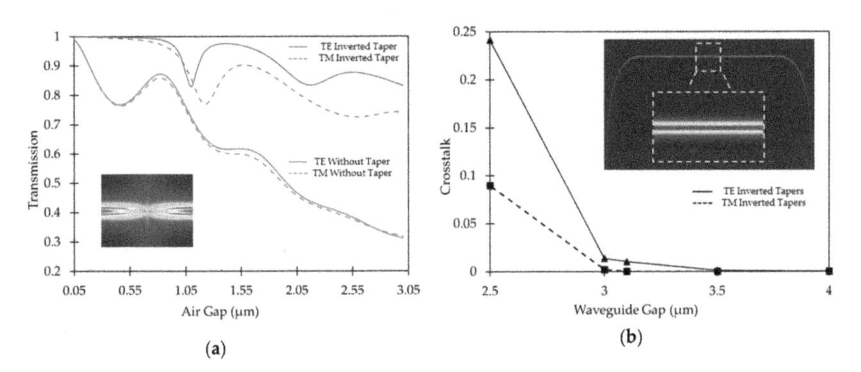

Figure 2. Optical simulation results for: (**a**) EME analysis showing transmission efficiency for TE and TM modes between two butt-coupled SiN waveguides as a function of air gap with and without inverted tapers. The inset shows the top-view of the magnitude of the electric field of TE mode for butt-coupling with inverted tapers at a gap of 500 nm; (**b**) 2.5D FDTD analysis showing cross-talk for TE and TM modes as a function of the gap between two SiN waveguides with 90° bends and 75 μm bending radius. Inset shows top-view of the magnitude of electric field of TE mode for the complete optical path with two parallel SiN waveguides at a gap of 3.1 μm. Image shows that the field remains completely confined in the input waveguide.

The optical simulation results show that a rectangular platform of 150 µm by 520 µm is large enough to accommodate four separate SiN waveguides with a 75 µm bending radius. Also, the gap closing interface between the platform and the fixed section of the switch is 35 µm wide and can easily accommodate three separate 435 nm wide SiN waveguides with a 3.5 µm gap between them. These can be fabricated with inverted tapers having tip-width of 250 nm and 20 µm length in the coupling region at the edges for minimal optical loss. The transverse horizontal and vertical alignment tolerance between the butt-coupled waveguides with and without tapers were also analyzed as shown in Figure 3. The inverted tapered structures have a high alignment tolerance providing a transmission of more than 80% in case of the TE mode and of more than 70% in case of the TM mode even when one waveguide is displaced by 700 nm relative to the other. These transmission coefficients were obtained with an air-gap of 250 nm between the waveguides. Therefore, the proposed switch has high fabrication tolerances in comparison to typical silicon photonic devices implemented with SOI wafers that have a 220 nm thick device layer.

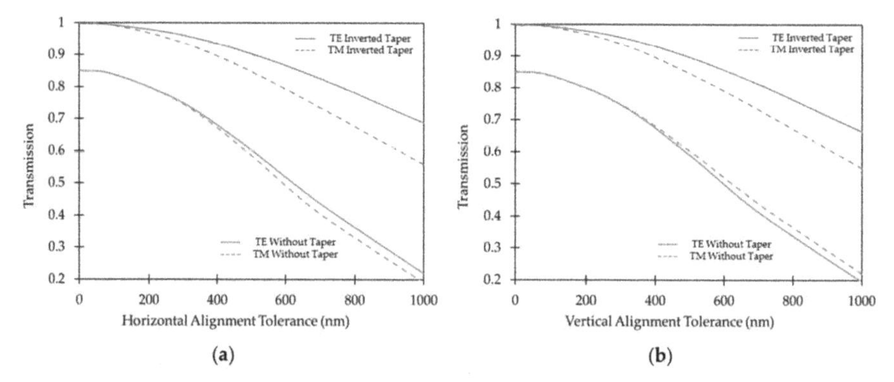

Figure 3. Optical simulation results for EME simulations showing the transmission efficiency of TE and TM modes between two butt-coupled SiN waveguides with and without inverted tapers as a function of: (**a**) horizontal and (**b**) vertical alignment tolerance.

2.3. MEMS Design Considerations

The initial MEMS actuator choice for the translational platform was to use of a unidirectional comb drive for lateral switching whereas the gap closing actuator was the same parallel plate actuator discussed above. This first MEMS design incorporated serpentine spring structures for both lateral and longitudinal motions. Since comb drives enable large controlled displacements, the MEMS was designed in ANSYS using static structural analysis to provide up to 6 µm of displacement at ~220 V. This design was fabricated by AEPONYX with an in-house microfabrication process for MEMS based on SOI technology. However, the fabricated devices showed rotational effects in the comb drive after a displacement of 2.39 µm at 100 V during testing. Figure 4 shows micrographs of a device during actuation tests along with the actuation curves for simulation and experimental results. The experimental measurements appear to follow a linear relationship in comparison to the simulation results because the displacements recorded during the experiment are limited to the beginning of the polynomial actuation curve where the slope is increasing slowly. Before we could observe the non-linear behaviour of the actuator, the comb drive-based actuator rotated inhibiting further actuation. SEM analysis of the MEMS device showed some fabrication discrepancies. Fabricated dimensions varied from 2.35 µm to 2.58 µm in the comb drive compared to the design dimensions of 3 µm. This varies the gap between the drive fingers in different regions of the comb. These observations are shown in Figure 5.

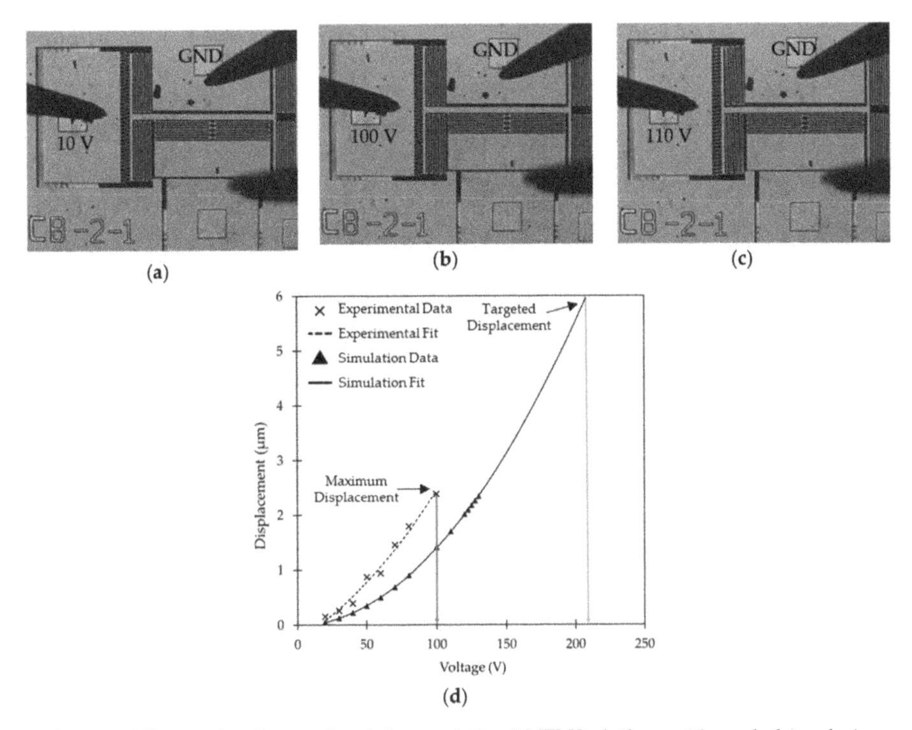

Figure 4. Microscopic micrographs of the translational MEMS platform with comb drive during actuation: (**a**) at 10 V; (**b**) maximum displacement at 100 V; (**c**) rotation at 110 V; (**d**) experimental and simulation based lateral switching displacement v/s actuation voltage curves for the translational MEMS design with comb drive.

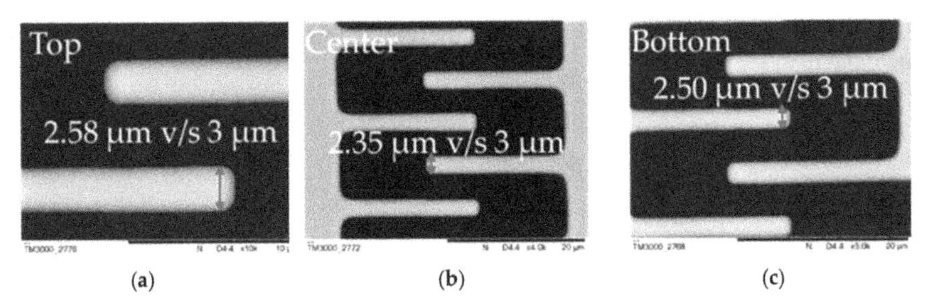

Figure 5. Fabricated v/s design dimensions through SEM micrograph analysis of the drive fingers for comb drive-based translational MEMS platform in various parts of the actuator: (**a**) top; (**b**) center; (**c**) bottom.

The vertical in-plane stiffness of the main horizontal beam was further analyzed following the actuation results obtained. A static structural analysis of the structure was performed using ANSYS. A load force of 10 μN was simulated on the top left corner of the device model to verify the vertical stiffness of the main horizontal beam. These simulation results are shown in Figure 6a. The MEMS design has a low vertical stiffness as the 10 μN force applied led to a total maximum deformation of 322 nm. The fabrication discrepancies described earlier are assumed to make the electrostatic field generated by the fabricated comb drive slightly asymmetrical compared to the simulated model with an ideal comb drive. This is due to the varying gap between the fabricated fingers in different regions of the actuator. These fabrication geometry discrepancies combined with the low vertical stiffness of the system make the structure highly susceptible to the rotational effect observed. Designs for the lateral switching actuator and serpentine spring were modified to the final iteration shown in Figure 6b. Parallel plate actuators were chosen for lateral switching to simplify fabrication. The vertical stiffness was increased through a single beam spring for lateral actuation that is anchored on two ends unlike the previous serpentine spring design. A static structural simulation for a 10 μN force on the top left corner of the new design yielded only 1.18 nm of total maximum deformation, more than 300× reduction over the prior design shown in Figure 6a.

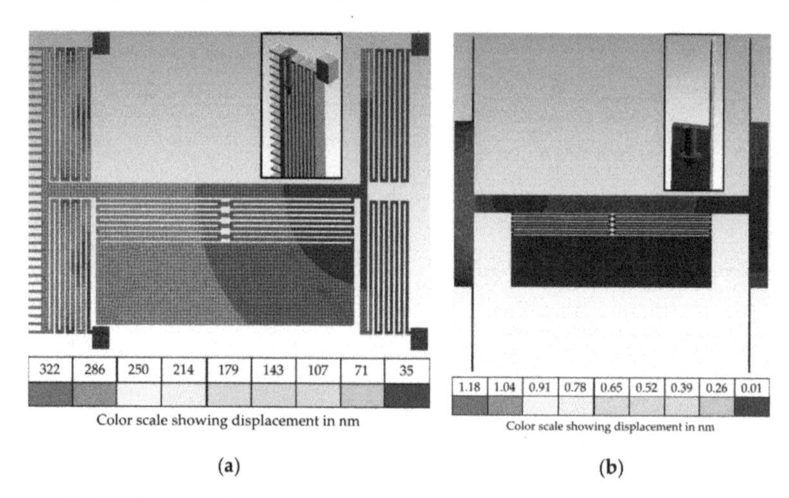

Figure 6. Total deformation heatmap when 10 μN of force is applied on the top left corner of the structure (force location shown in image insets): (**a**) comb drive and serpentine spring design; (**b**) parallel plate and single beam spring design.

The completely parallel plate actuation-based design with a single beam spring was successful in eliminating any rotation due to the fabrication discrepancies caused by the complex comb drive structure, and to increase the vertical in-plane spring constant. In order to achieve the same targeted 6 μm of displacement as the comb drive, two lateral parallel plate actuators were designed on opposite sides of the platform.

3. Final Translational MEMS Platform

The final iteration of the translational MEMS platform was designed on the basis of the optical design considerations and comb drive-based MEMS results discussed in the previous sections. Two parallel plate actuators were implemented on the opposite sides of the central rectangular platform for lateral displacement. Another parallel plate actuator was created on the bottom of the actuator for air gap closing. All actuators were designed as parallel-plate and operated under the pull-in effect [3]. In case of a parallel plate actuator with initial gap (d) and total overlap area (A) between the plates, the actuation voltage (V_p) where the pull-in effect occurs is given by:

$$V_p = \sqrt{\left(\frac{8kd^3}{27\varepsilon A}\right)} \quad (1)$$

where k is the spring constant of the system in the direction of actuation and ε is permittivity of the dielectric medium. Also, the maximum controlled displacement (X_p) for these actuators before pull-in is given by:

$$X_p = \frac{d}{3} \quad (2)$$

These lateral actuators were designed with an initial gap (d) of 4 μm making the maximum displacement (X_p) before pull-in to be 1.3 μm. Similarly, for the longitudinal actuator with an initial gap (d) of 6 μm, the maximum displacement (X_p) before pull-in is 2 μm. Since the pull-in effect enables a quick and large displacement, the two parallel plates in the actuator tend to snap together. In order to prevent shorting during pull-in, 10 μm long stoppers at a 3 μm gap (less than the actuator initial gap of 4 μm) were added at the two ends of both the lateral actuators. These stoppers also provide the necessary 3 μm of maximum displacement to translate the waveguides on the platform and form a 2 × 2 crossbar optical switch as discussed earlier. Similar 35 μm wide stoppers forming a 4 μm gap (less than the actuator initial gap of 6 μm) were built for the longitudinal actuator. These stoppers are larger than the lateral stoppers in order to accommodate multiple SiN waveguides. Images of the fabricated MEMS device along with the critical stopper and actuator dimensions are shown in Figure 7.

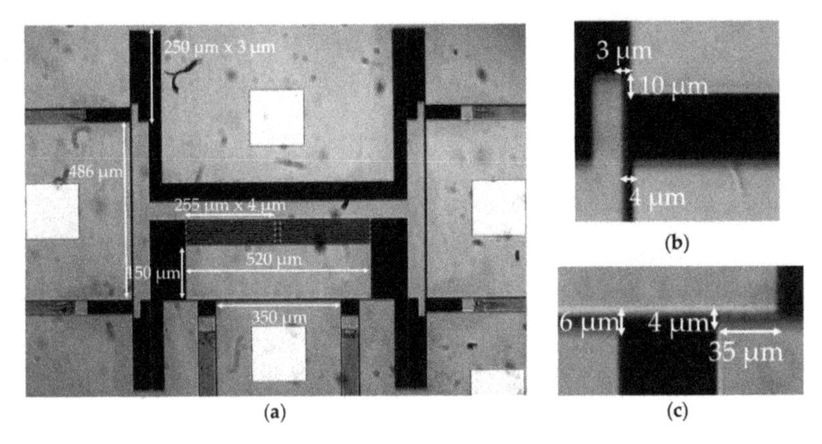

Figure 7. Micrographs of the final translational MEMS platform with critical design dimensions for: (**a**) platform, springs and actuators; (**b**) lateral switching actuator and stopper; (**c**) longitudinal gap closing actuator and stopper.

A static structural analysis for the device model in ANSYS was performed to estimate the lateral and longitudinal spring constants of the MEMS. The spring constant of the lateral actuators' springs was found to be 15.37 N/m, whereas the spring constant of the platform's serpentine spring is 1.81 N/m. The lower stiffness of the serpentine spring reduces the downward electrostatic force needed to close the gap. This helps limit the impact of gap closing actuation upon the lateral actuators and prevent any rotation of the platform. Since the stiffness of the serpentine spring is considerably lower, the gap closing actuator dimensions are different from the lateral actuators. The initial gap and length of the gap closing actuator was kept at 6 μm and 350 μm, respectively, whereas the initial gap and length of the lateral actuator was kept at 4 μm and 486 μm, respectively. These choices were made to enable the operation of all the actuators within a small voltage range. Theoretical calculations using the spring constant simulation results presented in this section predict a pull-in voltage of ~82 V and ~61 V for lateral switching and longitudinal gap closing, respectively. Modal analysis was also performed with the device model in ANSYS to obtain the resonance frequencies of the structure. The resonance frequency of the gap closing actuator was found to be 4.6 kHz whereas that of the switching actuator structure was 9.2 kHz. This limits the operational frequency for the switch at ~4.6 kHz. The MEMS cell as demonstrated can be used in both a 2 × 2 crossbar switch configuration as well as a 1 × 3 switch configuration.

4. Experimental Results

4.1. Microfabrication Results

The MEMS devices were fabricated with a commercial process (PiezoMUMPs by MEMSCAP) [27]. The process uses SOI technology with a 10 μm device layer. SEM micrographs of the fabricated structures with measured critical dimensions are presented in Figure 8.

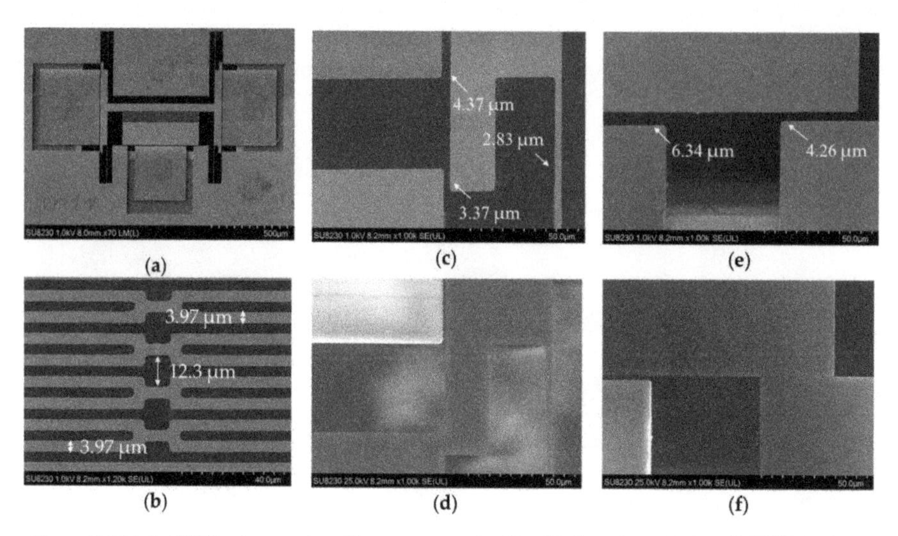

Figure 8. Detailed SEM micrographs with measurements of the fabricated translational MEMS platform: (**a**) final translational MEMS platform; (**b**) serpentine spring; (**c**) lateral actuator and spring's fabricated dimensions; (**d**) lateral actuation during high power imaging; (**e**) gap closing actuator's fabricated dimensions; (**f**) gap closing during high power imaging.

An analysis of the SEM micrographs showed that the fabricated dimensions were slightly different from the design dimensions. The lateral actuator gap increased from 4 μm to 4.37 μm and the stopper gap from 3 μm to 3.37 μm. The gap closing actuator gap increased from 6 μm to 6.34 μm and the stopper gap increased from 4 μm to 4.26 μm. These slight variations should increase the actuation voltage due to increased gap between the actuator plates. However, the spring beam dimensions were also smaller by a margin of ~0.17 μm for the lateral spring beams and by a margin of ~0.03 μm for the serpentine spring beams. This lowers the spring stiffness thereby negating the effect of the increase of the actuator gaps to some extent. A video showing both lateral switching and gap closing actuators in motion during SEM imaging is provided in the Supplementary Materials section. Actuation tests were performed to study the impact of these fabrication variations upon the actuation voltage. The test setup used, and the results obtained are presented in Section 4.2.

4.2. Actuation Test Results

Different fabricated devices were tested using a Wentworth probe station with a Bausch & Lomb microscopic system. Four DC probes were used during these tests. High voltage DC sources were used to provide the necessary voltage for actuation. A high resolution camera from Omax was used to image the devices during these tests. The actuator was grounded through a 100 kΩ resistor to prevent any device damage due to high current during actuation. Detailed image of the test setup used along with a schematic of the test circuit for the actuation experiments is given in Figure 9.

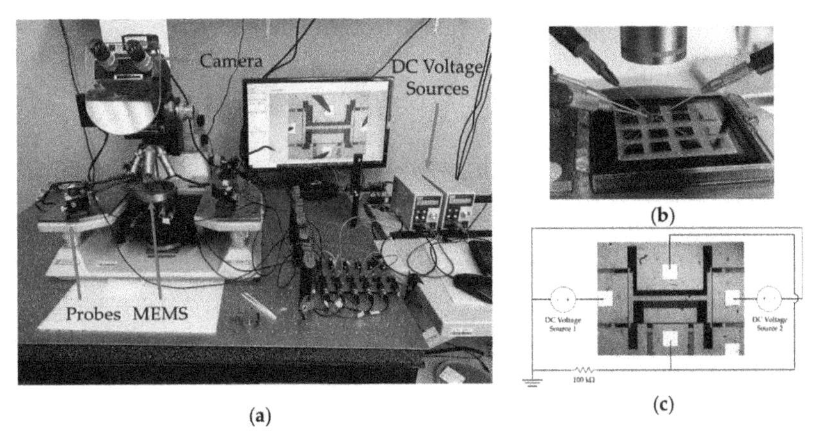

Figure 9. (**a**) Experimental actuation setup used; (**b**) zoomed in image of the probes on the MEMS device during tests; (**c**) schematic of the test circuit used for lateral actuation experiments.

The lateral actuator on both the sides showed pull-in at an actuation voltage of 65 V. Since after pull-in the movable actuator plate snaps towards the fixed plate, the total displacement obtained should be equivalent to the gap between the actuator plates. However, 3 μm stoppers were included specifically in the design to prevent any shorting through contact between the two actuator plates. The fabricated dimensions for the devices discussed in Section 4.1 showed that a 3.37 μm stopper gap was fabricated instead. Our lateral actuator shows 3.37 μm of displacement for the central platform at just 65 V of pull-in voltage. Similarly, the results for the longitudinal gap closing actuator show pull-in at 50 V. The fabricated dimensions for the gap closing stoppers was 4.26 μm instead of 4 μm as per the design. The gap closing actuator provides a 4.26 μm displacement to the central platform at a pull-in voltage of just 50 V. No stiction issues or damage to the fabricated stoppers were observed after repeated actuation. The measured displacement for different actuation voltages of the lateral switching actuator follows a linear trend over the range of voltages used in the experimental characterization in comparison to the non-linear behavior of the gap-closing actuator before electrostatic pull-in. This can be explained by the difference in spring stiffness between the lateral switching and gap-closing actuators. The spring constant values for the single beam spring design of the lateral switching actuator in the simulation model was found to be 15.37 N/m, whereas that for the multi beam serpentine spring design of the gap-closing actuator is only 1.18 N/m. Therefore, the non-linear response of the gap-closing can be observed by applying a much smaller force or equivalently, a smaller actuation voltage. Also, the spring for the lateral switching actuator is similar to a clamped-clamped beam system which follows linear displacement as per small beam deflection theory up to a quarter of the beam thickness following which non-linear displacement can be observed for larger displacements [47]. Since the thickness of the SOI device layer used is 10 μm, the maximum displacement observed before pull-in is much lower than a quarter of the silicon beam thickness (i.e., 2.5 μm resulting in the linear behavior of the fabricated actuator). The simulation model for the lateral switching actuator shows an initial linear behavior which becomes non-linear with a larger displacement than seen in the measurements, which could be due to the higher stiffness of the spring by the actuator in the simulation model, due to the geometry variations resulting from the fabrication process, as seen in Figure 8. The experimental actuation voltage to reach pull-in was lower in comparison to the simulated model for both the lateral switching actuator and the gap-closing actuator. This can be explained by the difference in the fabricated and simulated silicon beam dimensions. As discussed earlier in Section 4.1, the width of the fabricated silicon beams in the spring structure was slightly less than in the simulation model. This reduces the stiffness of the spring leading to lower experimental actuation voltages for the simulated displacements compared to simulated actuation voltages. The experimental displacement before pull-in was also observed to be larger in comparison to the simulated model. This is because the fabricated gap between the parallel plates of the actuator was larger, as discussed earlier in Section 4.1.

Simulation v/s experimental results are presented in Figure 10 and they show the pull-in voltage for both lateral switching and longitudinal gap closing actuators. Micrographs of the bi-directional lateral switching action at 65 V combined with vertical gap closing at 50 V along with the neutral position of the actuator are shown in Figure 11.

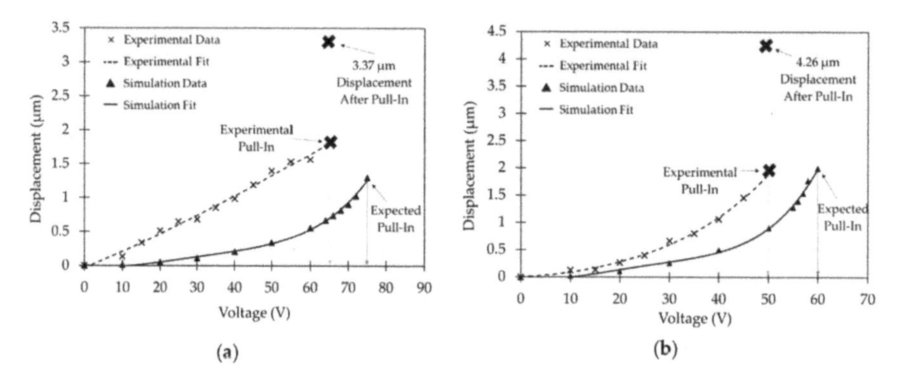

Figure 10. Experimental and simulation based displacement v/s actuation voltage results along with relevant pull-in voltages and maximum displacement obtained for (**a**) the lateral switching actuator; (**b**) the gap closing actuator.

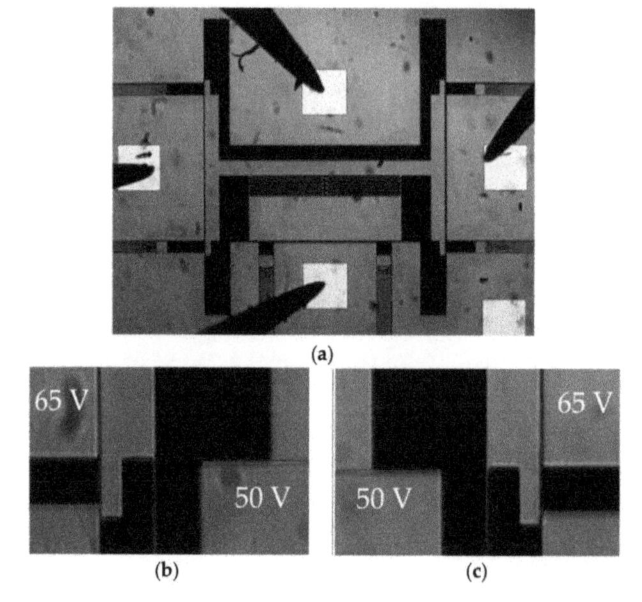

Figure 11. (**a**) Microscopic micrographs of the device at 0 V during actuation tests with probes on MEMS; (**b**) zoomed in image of the left switching actuator at 65 V and gap closing actuator at 50 V; (**c**) zoomed in image of the right switching actuator at 65 V and gap closing actuator at 50 V.

5. Discussion

Different translational MEMS actuator designs were implemented and tested. The comb drive-based actuator design showed a maximum displacement of 2.38 μm before a rotational effect occurred at 110 V precluding further motion. This maximum displacement of 2.38 μm achieved for the central platform could not provide the displacement of at least 6 μm needed for successful operation as a 2 × 2 crossbar switch. The MEMS design was improved through the incorporation of single beam spring for lateral switching actuation. Silicon beams anchored at two ends provide the necessary higher vertical stiffness and eliminated any rotational effect due to the serpentine spring structure with low vertical stiffness. Incorporation of simple parallel plate actuator design for lateral switching actuators eliminates any rotational effect caused due to fabrication discrepancies in the comb drive geometry. Two parallel plate actuators on the opposite side of the central platform provide the necessary discrete displacement of at least 6 μm (i.e., 3 μm on each side) after pull-in for lateral switching. Parallel plate actuator designed for air gap closing motion of the central platform provides zero gap between platform and the substrate upon actuation. Soft spring design for the central platform using serpentine spring system also ensures zero impact of the air gap closing motion upon the lateral switching motion of the platform.

The state-of-the-art planar optical switch developed in [41] demonstrates low loss switches which rely upon polarization sensitive vertical adiabatic coupling between polysilicon ridge waveguides. Although polarization insensitive switches based upon polysilicon waveguides have also been realized, these involve a complex 20 masks fabrication process with 3 waveguide layers [48]. State-of-the-art 2 × 2 MEMS switches with zero gap butt-coupling between suspended and fixed waveguides has also been demonstrated in the past through the incorporation of soft polymer waveguides over the MEMS actuator and had a low switching speed of < 0.5 ms. This approach involves a complex bonding process between the polymer waveguides and MEMS structures [39]. Also, the actuator springs need to be precompressed into latching position using probes under a microscope to provide the zero gap coupling between waveguides and optical fibers.

The translational platform presented in this work is designed to be integrated with polarization insensitive square SiN waveguides in a single optics layer with SiO_2 cladding for less stringent packaging requirements. The actuator springs designed do not require any complex assembly procedure before switching operation for zero gap closing either. This is due to the independent spring design for lateral switching and air gap closing motions which provides bi-axial motion to the central platform necessary for its operation in 2 × 2 crossbar switch configuration. Recently, a rotational MEMS platform demonstrated crossbar switching capability at 118 V with the ability to reduce the air gap between fixed and movable waveguides down to 250 nm at an actuation voltage of 113 V [42]. The translational platform presented in this work operates at a much lower voltage of 65 V for 2 × 2 crossbar switching and 50 V for air gap closing. The rectangular platform design also provides the unique possibility to integrate SiN based optical filters on the platform itself in 1 × 3 switch configuration.

Although the design choice of parallel plate-based actuation for lateral switching makes the alignment of the optical waveguide mask with the MEMS mask during microfabrication critical, previously SiN waveguides have been successfully integrated with high precision [42]. Also, optical simulations of the alignment tolerances showed more than 80% efficiency for 700 nm of misalignment and more than 96% efficiency for less than 300 nm of misalignment with a 250 nm gap between the waveguides. Stepper tools for lithography can be used to precisely align the MEMS layer with the waveguides during microfabrication process. Optical simulations show that the gap closing motion of the platform can provide over 99% transmission efficiency for butt-coupling waveguides. Waveguides with inverted tapers can provide more than 83% efficiency even when there is a separation of 3 μm between them. The design is capable of minimizing the optical losses due to the air gap. The effect of surface roughness of the fabricated devices upon the minimal gap achievable should not cause significant optical losses either.

6. Conclusions

In this work, a translational MEMS platform was presented for planar optical switching applications. The lateral switching actuator designed for this translational MEMS device operates at an actuation of voltage of 65 V while closing the air gap completely at just 50 V. The ability to integrate up to four SiN waveguides with minimal crosstalk on the large 150 μm by 520 μm platform provides 2 × 2 crossbar switching capability. A 2 × 2 crossbar switch can be realized with just one core switch cell in a smaller device footprint of 1 mm by 1 mm when compared to [42]. This switch also operates at a much lower voltage when compared to the rotational MEMS platform designed for planar crossbar switching. It can also be used to demonstrate a wavelength channel selection system through integration with SiN based optical filters on the central platform for Reconfigurable Optical Add–Drop Multiplexer (ROADM) applications [49,50]. In the future, we aim to integrate SiN waveguides and optical filters with the fabrication process demonstrated previously [42]. Spring stiffness and actuator dimensions will be further optimized so that both lateral and gap closing actuators operate at the same voltage. Actuator stopper dimensions will also be optimized for minimal stiction and high reliability.

Supplementary Materials: The following are available online at http://www.mdpi.com/2072-666X/10/7/435/s1, Video S1: Translational MEMS platform in motion.

Author Contributions: S.S. designed the MEMS devices, performed experiments and analyzed the data; N.K. performed the optical simulations; J.B. contributed MEMS expertise; F.N. and M.M. contributed expertise, direction, materials and analysis tools.

Funding: This research was funded by the Natural Sciences and Engineering Research Council of Canada (NSERC) and the Regroupement Stratégique en Microsystèmes du Québec (ReSMiQ).

Acknowledgments: The authors would like to thank CMC Microsystems for providing the software tools and enabling device fabrication. The authors also thank AEPONYX Inc. for access to their test facilities, device fabrication and financial support.

Conflicts of Interest: The authors declare no conflict of interest.

References

1. Monk, D.W.; Gale, R.O. The Digital Micromirror Device for Projection Display Deformable Membrane Displays. *Microelectron. Eng.* **1995**, *27*, 489–493. [CrossRef]
2. Holmström, S.T.S.; Baran, U.; Urey, H. MEMS laser scanners: A review. *J. Microelectromech. Syst.* **2014**, *23*, 259–275. [CrossRef]
3. Kaajakari, V. *Practical MEMS: Design of Microsystems, Accelerometers, Gyroscopes, RF MEMS, Optical MEMS, and Microfluidic Systems*; Small Gear Publishing: Las Vegas, NV, USA, 2009; ISBN 0982299109.
4. Hammadi, A.; Mhamdi, L. A survey on architectures and energy efficiency in Data Center Networks. *Comput. Commun.* **2014**, *40*, 1–21. [CrossRef]
5. Farrington, N.; Porter, G.; Radhakrishnan, S.; Bazzaz, H.H.; Subramanya, V.; Fainman, Y.; Papen, G.; Vahdat, A. Helios: A Hybrid Electrical/Optical Switch Architecture for Modular Data Centers. *ACM SIGCOMM Comput. Commun. Rev.* **2010**, *40*, 339–350. [CrossRef]
6. Ménard, F.D.; Bérard, M.; Prescott, R. Scaled Out Optically Switched (SOOS) network architecture for Web Scale Data Centers. *2015 IEEE Opt. Interconnects Conf. (OI)* **2015**, *15*, 90–91.
7. Kim, J.; Nuzman, C.J.; Kumar, B.; Lieuwen, D.F.; Kraus, J.S.; Weiss, A.; Lichtenwalner, C.P.; Papazian, A.R.; Frahm, R.E.; Basavanhally, N.R.; et al. 1100 × 1100 Port MEMS-Based Optical Crossconnect With 4-dB Maximum Loss. *IEEE Photonics Technol. Lett.* **2003**, *15*, 1537–1539. [CrossRef]
8. Truex, T.A.; Bent, A.A.; Hagood, N.W. *Beam-Steering Optical Switch Fabric Utilizing Piezoelectric Actuation Technology*; Continuum Photonics Inc.: Billerica, MA, USA, 2003.
9. Fernandez, A. Modular MEMS design and fabrication for an 80 × 80 transparent optical cross-connect switch. *Proc. SPIE* **2004**, *5604*, 208–217.
10. Aksyuk, V.A.; Pardo, F.; Bolle, C.A.; Arney, S.; Giles, C.R.; Bishop, D.J. Lucent Microstar micromirror array technology for large optical crossconnects. *MOEMS Miniaturized Syst.* **2003**, *4178*, 320–324.
11. Li, S.; Xu, J.; Zhong, S.; Wu, Y. Design, fabrication and characterization of a high fill-factor micromirror array for wavelength selective switch applications. *Sens. Actuators A Phys.* **2011**, *171*, 274–282. [CrossRef]
12. Liu, Y.; Xu, J.; Zhong, S.; Wu, Y. Large size MEMS scanning mirror with vertical comb drive for tunable optical filter. *Opt. Lasers Eng.* **2013**, *51*, 54–60. [CrossRef]

13. Koh, K.H.; Lee, C.; Kobayashi, T. A piezoelectric-driven three-dimensional MEMS VOA using attenuation mechanism with combination of rotational and translational effects. *J. Microelectromech. Syst.* **2010**, *19*, 1370–1379. [CrossRef]
14. Microactuator, I.; Man, A.; Kwan, H.; Member, S.; Song, S.; Member, S.; Lu, X.; Lu, L.; Teh, Y.; Member, S.; et al. Improved Designs for an Electrothermal In-Plane Microactuator. *J. Microelectromech. Syst.* **2012**, *21*, 586–595.
15. Member, S.; Lohmann, A. MOEMS Tuning Element for a Littrow External Cavity Laser. *J. Microelectromech. Syst.* **2003**, *12*, 921–928.
16. Puder, J.M.; Bedair, S.S.; Pulskamp, J.S.; Rudy, R.Q.; Polcawich, R.G.; Bhave, S.A. Higher dimensional flexure mode for enhanced effective electromechanical coupling in PZT-on-silicon MEMS resonators. In Proceedings of the 2015 Transducers—2015 18th International Conference on Solid-State Sensors, Actuators and Microsystems (TRANSDUCERS), Anchorage, AK, USA, 21–25 June 2015; pp. 2017–2020.
17. Cassella, C.; Chen, G.; Qian, Z.; Hummel, G.; Rinaldi, M. RF passive components based on aluminum nitride cross-sectional lamé-mode MEMS resonators. *IEEE Trans. Electron Devices* **2017**, *64*, 237–243. [CrossRef]
18. Kim, S.J.; Cho, Y.H.; Nam, H.J.; Bu, J.U. Piezoelectrically pushed rotational micromirrors using detached PZT actuators for wide-angle optical switch applications. *J. Micromech. Microeng.* **2008**, *18*, 125022. [CrossRef]
19. Panda, P.K.; Sahoo, B. PZT to lead free piezo ceramics: A review. *Ferroelectrics* **2015**, *474*, 128–143. [CrossRef]
20. Tonisch, K.; Cimalla, V.; Foerster, C.; Romanus, H.; Ambacher, O.; Dontsov, D. Piezoelectric properties of polycrystalline AlN thin films for MEMS application. *Sens. Actuators A Phys.* **2006**, *132*, 658–663. [CrossRef]
21. Iborra, E.; Olivares, J.; Clement, M.; Vergara, L.; Sanz-Hervás, A.; Sangrador, J. Piezoelectric properties and residual stress of sputtered AlN thin films for MEMS applications. *Sens. Actuators A Phys.* **2004**, *115*, 501–507. [CrossRef]
22. Liu, L.; Pal, S.; Xie, H. Sensors and Actuators A: Physical MEMS mirrors based on a curved concentric electrothermal actuator. *Sens. Actuators A. Phys.* **2012**, *188*, 349–358. [CrossRef]
23. Peters, T.J.; Tichem, M. Electrothermal actuators for SiO_2 photonic MEMS. *Micromachines* **2016**, *7*, 200. [CrossRef]
24. Li, J.; Zhang, Q.X.; Liu, A.Q. Advanced fiber optical switches using deep RIE (DRIE) fabrication. *Sens. Actuators A Phys.* **2003**, *102*, 286–295. [CrossRef]
25. Sabry, Y.M.; Eltagoury, Y.M.; Shebl, A.; Soliman, M.; Sadek, M.; Khalil, D. In-plane deeply-etched optical MEMS notch filter with high-speed tunability. *J. Opt.* **2015**, *17*, 125703. [CrossRef]
26. Brière, J.; Beaulieu, P.-O.; Saidani, M.; Nabki, F.; Menard, M. Rotational MEMS mirror with latching arm for silicon photonics. *Proc. SPIE* **2015**, *9375*, 937507.
27. Cowen, A.; Hames, G.; Glukh, K.; Hardy, B. *PiezoMUMPs Design Handbook*; rev. 1.3; MEMSCAP Inc.: Bernin, France, 2014.
28. Chu, H.M.; Hane, K. Design, fabrication and vacuum operation characteristics of two-dimensional comb-drive micro-scanner. *Sens. Actuators A Phys.* **2011**, *165*, 422–430. [CrossRef]
29. Srinivasan, P.; Gollasch, C.O.; Kraft, M. Three dimensional electrostatic actuators for tunable optical micro cavities. *Sens. Actuators A Phys.* **2010**, *161*, 191–198. [CrossRef]
30. Lee, L.P.; Kwon, S.; Member, S.; Milanovic, V. Vertical Combdrive Based 2-D Gimbaled Micromirrors With Large Static Rotation by Backside Island Isolation. *IEEE J. Sel. Top. Quantum Electron.* **2004**, *10*, 498–504.
31. Zhao, R.; Qiao, D.; Song, X.; You, Q. The exploration for an appropriate vacuum level for performance enhancement of a comb-drive microscanner. *Micromachines* **2017**, *8*, 126. [CrossRef]
32. Hung, A.C.L.; Lai, H.Y.H.; Lin, T.W.; Fu, S.G.; Lu, M.S.C. An electrostatically driven 2D micro-scanning mirror with capacitive sensing for projection display. *Sens. Actuators A Phys.* **2015**, *222*, 122–129. [CrossRef]
33. Zhang, W.; Li, P.; Zhang, X.; Wang, Y.; Hu, F. InGaN/GaN micro mirror with electrostatic comb drive actuation integrated on a patterned silicon-on-insulator wafer. *Opt. Express* **2018**, *26*, 7672. [CrossRef]
34. Dziuban, P.; Laszczyk, K.; Bargiel, S.; Gorecki, C.; Kr, J.; Callet, D.; Frank, S. A two directional electrostatic comb-drive X–Y micro-stage for MOEMS applications. *Sens. Actuators A Phys.* **2010**, *163*, 255–265.
35. Xue, G.; Toda, M.; Ono, T. Assembled comb-drive XYZ-microstage with large displacements for low temperature measurement systems. In Proceedings of the 2015 28th IEEE International Conference on Micro Electro Mechanical Systems (MEMS), Estoril, Portugal, 18–22 January 2015.
36. Bulgan, E.; Kanamori, Y.; Hane, K. Submicron silicon waveguide optical switch driven by microelectromechanical actuator. *Appl. Phys. Lett.* **2008**, *92*, 1–4. [CrossRef]

37. Abe, S.; Hane, K. Variable-gap silicon photonic waveguide coupler switch with a nanolatch mechanism. *IEEE Photonics Technol. Lett.* **2013**, *25*, 675–677. [CrossRef]
38. Munemasa, Y.; Hane, K. Compact 1 × 3 silicon photonic waveguide switch based on precise investigation of coupling characteristics of variable-gap coupler. *Jpn. J. Appl. Phys.* **2013**, *52*, 06GL15. [CrossRef]
39. Liu, H.B.; Chollet, F. Moving polymer waveguides and latching actuator for 2 × 2 MEMS optical switch. *J. Microelectromech. Syst.* **2009**, *18*, 715–724.
40. Han, S.; Seok, T.J.; Quack, N.; Yoo, B.-W.; Wu, M.C. Large-scale silicon photonic switches with movable directional couplers. *Optica* **2015**, *2*, 1–6. [CrossRef]
41. Seok, T.J.; Quack, N.; Han, S.; Muller, R.S.; Wu, M.C. Large-scale broadband digital silicon photonic switches with vertical adiabatic couplers. *Optica* **2016**, *3*, 64. [CrossRef]
42. Brière, J.; Elsayed, M.Y.; Saidani, M.; Bérard, M.; Beaulieu, P.O.; Rabbani-Haghighi, H.; Nabki, F.; Ménard, M. Rotating circular micro-platform with integrated waveguides and latching arm for reconfigurable integrated optics. *Micromachines* **2017**, *8*, 354. [CrossRef] [PubMed]
43. Ménard, M.; Elsayed, M.Y.; Brière, J.; Rabbani-Haghighi, H.; Saidani, M.; Bérard, M.; Ménard, F.; Nabki, F. Integrated optical switch controlled with a MEMS rotational electrostatic actuator. In Proceedings of the Photonics in Switching, New Orleans, LA, USA, 24–27 July 2017.
44. Tabti, B.; Nabki, F.; Ménard, M. Polarization insensitive Bragg gratings in silicon nitride waveguides. In Proceedings of the Integrated Photonics Research, Silicon, and NanoPhotonics, New Orleans, LA, USA, 24–27 July 2017.
45. Gondarenko, A.; Levy, J.S.; Lipson, M. High confinement micron-scale silicon nitride high Q ring resonator. *Opt. Express* **2009**, *17*, 11366. [CrossRef] [PubMed]
46. Jones, A.M.; DeRose, C.T.; Lentine, A.L.; Trotter, D.C.; Starbuck, A.L.; Norwood, R.A. Ultra-low crosstalk, CMOS compatible waveguide crossings for densely integrated photonic interconnection networks. *Opt. Express* **2013**, *21*, 12002. [CrossRef]
47. Legtenberg, R.; Groeneveld, A.W.; Elwenspoek, M. Comb-drive actuators for large displacements. *J. Micromech. Microeng.* **1996**, *6*, 320–329. [CrossRef]
48. Han, S.; Seok, T.J.; Yu, K.; Quack, N.; Muller, R.S.; Wu, M.C. Large-Scale Polarization-Insensitive Silicon Photonic MEMS Switches. *J. Light. Technol.* **2018**, *36*, 1824–1830. [CrossRef]
49. Nakamura, S.; Yanagimachi, S.; Takeshita, H.; Tajima, A.; Hino, T.; Fukuchi, K. Optical Switches Based on Silicon Photonics for ROADM Application. *IEEE J. Sel. Top. Quantum Electron.* **2016**, *22*, 185–193. [CrossRef]
50. Strasser, T.A.; Wagener, J.L. Wavelength-selective switches for ROADM applications. *IEEE J. Sel. Top. Quantum Electron.* **2010**, *16*, 1150–1157. [CrossRef]

Article

An Integrated Germanium-Based THz Impulse Radiator with an Optical Waveguide Coupled Photoconductive Switch in Silicon

Peiyu Chen [1,*], Mostafa Hosseini [2] and Aydin Babakhani [2]

[1] Department of Electrical and Computer Engineering, Rice University, Houston, TX 77005, USA
[2] Department of Electrical and Computer Engineering, University of California at Los Angeles, Los Angeles, CA 90095, USA; mostafahosseini@ucla.edu (M.H.); aydinbabakhani@ucla.edu (A.B.)
* Correspondence: opt.cpy@gmail.com

Received: 5 May 2019; Accepted: 16 May 2019; Published: 31 May 2019

Abstract: This paper presents an integrated germanium (Ge)-based THz impulse radiator with an optical waveguide coupled photoconductive switch in a low-cost silicon-on-insulator (SOI) process. This process provides a Ge thin film, which is used as photoconductive material. To generate short THz impulses, N++ implant is added to the Ge thin film to reduce its photo-carrier lifetime to sub-picosecond for faster transient response. A bow-tie antenna is designed and connected to the photoconductive switch for radiation. To improve radiation efficiency, a silicon lens is attached to the substrate-side of the chip. This design features an optical-waveguide-enabled "horizontal" coupling mechanism between the optical excitation signal and the photoconductive switch. The THz emitter prototype works with 1550 nm femtosecond lasers. The radiated THz impulses achieve a full-width at half maximum (FWHM) of 1.14 ps and a bandwidth of 1.5 THz. The average radiated power is 0.337 μW. Compared with conventional THz photoconductive antennas (PCAs), this design exhibits several advantages: First, it uses silicon-based technology, which reduces the fabrication cost; second, the excitation wavelength is 1550 nm, at which various low-cost laser sources operate; and third, in this design, the monolithic excitation mechanism between the excitation laser and the photoconductive switch enables on-chip programmable control of excitation signals for THz beam-steering.

Keywords: germanium; integrated optics; optoelectronics; photoconductivity; silicon photonics; terahertz

---

1. Introduction

Sandwiched between traditional microwave and optical spectrums, terahertz (THz) technology has attained great scientific interest in recent decades. Compared to a THz continous-wave (CW) signal, THz impulses feature ultra-wide bandwidth, usually larger than 1 THz. This wide frequency band allows THz impulse to be used for various applications, such as, biology and medicine sciences [1], environmental monitoring [2,3], chemical sensing [4,5], high-resolution three-dimensional imaging [6–9], nondestructive evaluation [10], and high-speed wireless communication link [11].

Researchers have been investigating various technologies that can produce high power and wideband THz impulses. There are two technical solutions. One solution is to use fully-electronics technology; the other is to rely on optoelectronics methods. In recent years, silicon-based fully-electronics THz impulse radiators have been reported using CMOS or BiCMOS process technologies [12–16]. These fully-electronics devices produce picosecond impulses that cover the lower end of THz spectrum (less than 1.1 THz). Additionally, these designs feature the benefits of low cost, high scalability, and low power consumption.

The more widely used and traditional solution of THz impulse generation is photoconductive antennas (PCAs) based on optoelectronics technology [17]. As shown in Figure 1, a conventional THz PCA has two parts of metal contacts fabricated on a photoconductive semiconductor substrate, with a gap between the two metal contacts. The two metal contacts also operate as on-chip THz antennas. Conventional THz PCAs are usually triggered by free-space femtosecond laser, which is incident onto the photoconductive semiconductor substrate through the gap between the metal contacts. The semiconductor substrate has appropriate bandgap energy so that the incident femtosecond laser pulses are absorbed and photocarriers are generated. A DC bias voltage is applied across the gap through the metal contacts, building up electric fields in the substrate below the gap. Consequently, photocarriers drift and are eventually collected by the metal contacts before photocarrier recombinations occur. The induced ultrafast photocurrent drives the on-chip THz antennas that produce THz impulse radiations. A silicon lens is usually attached to the backside of the substrate to increase radiation efficiency.

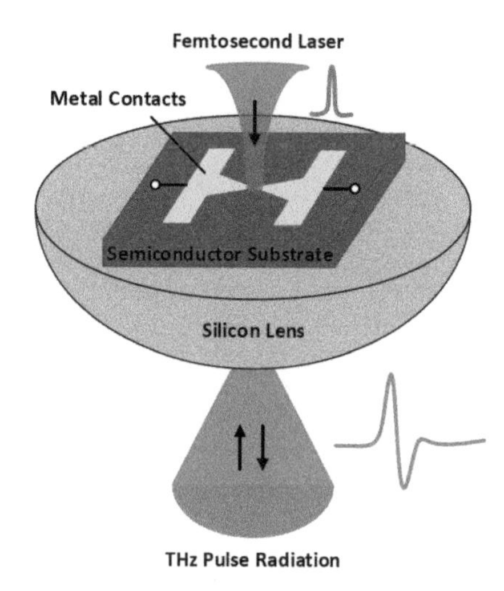

Figure 1. A conventional THz photoconductive antenna (PCA) emitter.

Admittedly, researchers have attempted to exploit the unique properties of various nanostructures, such as plasmonic structures [18,19] and optical nano-antennas [20,21], to improve radiation bandwidth, radiated power, and efficiency [22,23], but they all share the same fundamental limitations on the photoconductive semiconductor substrate and free-space optical excitation scheme:

First, to ensure large radiation bandwidth, the semiconductor substrate must be an ultrafast photoconductive material that should exhibit short photocarrier lifetime and high carrier mobility. Radiation-damaged silicon-on-sapphire (RDSOS) and low-temperature-grown GaAs (LT-GaAs) are widely used for THz PCAs [24]. Due to the complicated fabrication procedures of these materials, the cost of conventional THz PCAs is high. Another limitation on substrate material is that these materials are usually not compatible with 1550 nm laser excitations. There are various low-cost laser sources at 1550 nm regime for fiber-optic communications. Therefore, designing a THz PCA that operates with a 1550 nm laser source is a cost-effective strategy to reduce the cost further. Due to the aforementioned limitations, the first technical challenge of this work is to use low-cost and ultrafast photoconductive semiconductor materials that operate at 1550 nm region.

Conventional PCAs require free-space optical excitation scheme, which also exhibits two limitations. First, this excitation scheme requires accurate optical alignments to ensure that the femtosecond laser is aligned onto the tiny gap between the metal contacts. Optical alignment has stringent requirements on system stability, and consequently, is not suitable for portable applications. Second, because the optical excitation signal propagates in the free space before it is absorbed by the substrate, programmable control on the excitation signal is usually performed in the free space by using a mechanical translation stage with retroreflector mirrors [2,10]. Therefore, the free-space optical excitation scheme in conventional THz PCAs prevents the implementation of fully integrated THz PCA phased arrays, resulting in the second technical challenge of this work: design an on-chip optical excitation scheme for THz impulse radiator chips.

In this work, we present a Germanium (Ge)-based THz impulse radiator in silicon that resolves the aforementioned limitations. Following the operation principle of THz photoconductive antennas [17], our proposed design features three advantages: first, due to the bandgap energy of Ge, which is used as the photoconductive substrate, the prototype THz impulse radiator can be excited by 1550 nm femtosecond laser sources. As a result, various low-cost laser sources in this wavelength regime can be used to reduce cost. Second, it incorporates a waveguide-coupled photoconductive switch that provides a monolithic interaction between the optical excitation signal and the photoconductive substrate. This novel coupling mechanism eliminates the existing obstacle of achieving on-chip programmable control of excitation laser. Third, this optoelectronics device was batch fabricated by a silicon photonics process foundry, similar to microelectronic chip tape-out: we completed device design using the computer-aided design (CAD) tool and sent the design layout file to the foundry for batch fabrication. In this work, there is no need for post-processing in clean room, which can significantly reduce the cost for mass production.

This paper is an extension of [25] with extensive details of device design and analysis of the simulated and measured results. The remaining context of this paper is organized as follows. Section 2 presents the design techniques and system architecture of the prototype THz impulse emitter. Section 3 describes the measurement results, followed by conclusions in Section 4.

2. Design Techniques and System Architecture

The silicon photonics process technology used in this work implements photonics devices using CMOS-compatible technologies. As a result, this technology can significantly reduce the cost of integrated photonics devices in mass production. The silicon photonics process technology is provided by the Institute of Microelectronics (IME), Agency for Science Technology and Research (A*STAR), Singapore [26]. This process has a silicon-on-insulator substrate, and it provides a module library with various pre-designed integrated passive devices, such as single-mode optical waveguides and optical grating couplers. Apart from the passive devices, Ge-based active devices, such as ring modulators and photodetectors, are available for use in this process. Both passive and active devices are optimized for the 1550 nm regime.

2.1. Ultrafast Germanium Thin Film

The silicon photonics process technology used in the work provides Ge as the photoconductive material at 1550 nm wavelength. As discussed in Section 1, there are two requirements for the photoconductive materials of THz impulse radiators. The first requirement is that the photocarrier lifetime of the material should be on the order of sub-picoseconds; the second requirement is that the photocarrier mobility should be high enough to produce large photocurrent. A previous work [27] has demonstrated that photocarrier lifetime of Ge thin films can be reduced to the order of sub-picoseconds by implanting O^+ ions without sacrificing photocarrier mobility, which can remain at about 100 $cm^2/(Vs)$. These results can be applied to the present investigation. The process technology can fabricate Ge thin films with 500 nm thickness, and it provides phosphorus implant with a dose of 4×10^{15} cm^{-2}. Therefore, the first technical challenge of this work, low-cost and ultrafast photoconductive semiconductor material that operates with 1550 nm excitation laser source, is resolved.

2.2. *Waveguide-Coupling THz Photoconductive Switch*

The second technical challenge is to design an on-chip optical excitation scheme for THz impulse radiators. Figure 2 demonstrates a conceptual illustration of the proposed waveguide-coupling excitation mechanism. In this scheme, the free-space 1550 nm femtosecond laser is firstly coupled into the on-chip optical waveguides through an integrated optical grating coupler. Then, the femtosecond laser propagates in the optical waveguide and reaches at silicon photonics devices that can modulate the excitation optical signal by performing amplitude modulation or phase modulation. The modulated femtosecond laser continues to propagate in the optical waveguide until it is absorbed by the Ge thin film, which excites the photoconductive switch that drives the on-chip THz antennas to radiate THz impulses.

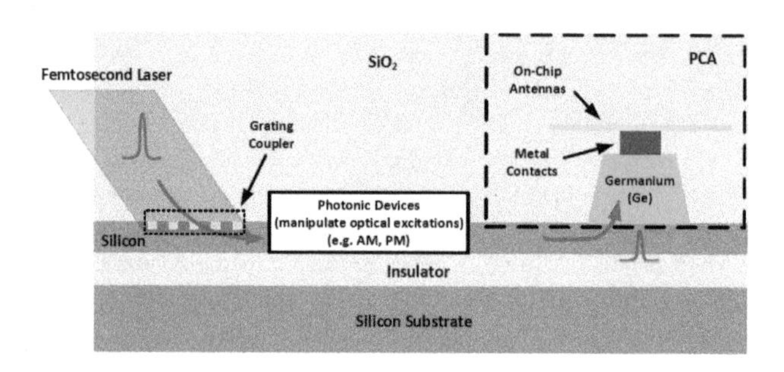

Figure 2. A conceptual illustration of the proposed optical waveguide coupling excitation scheme.

Figure 3 presents the structure of the implemented Ge-based waveguide-coupling photoconductive switch. The femtosecond excitation laser travels to the photoconductive switch through an integrated optical waveguide. A tapered transition is designed between the optical waveguide and the switch to reduce undesired optical reflections. A Phosphorus-doped Ge thin film is grown on the silicon layer. The N++ (Phosphorus) implant layer is split into two parts to prevent a large dc current produced under dc biasing. To increase the transient response speed of the photoconductive switch, the spacing between the two metal electrodes is set to the minimum value allowed by the DRC rule of the process technology. Compared with conventional THz photoconductive switches, the proposed waveguide-coupling solution exhibits a significant advantage: In this design, the femtosecond laser arrives at the photoconductive switch from the substrate side rather than through the gap between the metal electrodes as in conventional designs. Therefore, when a small spacing between the metal electrodes is required for enhancing the transient response speed, the optical excitation signal will not be blocked by the small gap, and consequently, complicated plasmonics-related simulation and design can be avoided. As shown in Figure 3, on-chip THz antennas are connected to the two metal electrodes.

The Ge thin film is optimized to increase both absorption efficiency and conversion efficiency at 1550 nm regime. The length of Ge thin film is designed to be 20 µm to ensure high absorption efficiency. Figure 4a shows that the 20 µm Ge thin film can absorb almost all the incident optical excitation signal at 1550 nm. The conversion efficency boost can be explained by investigating the optical mode distributions at different propagation distances within the photoconductive switch at 1550 nm. As shown in Figure 4b, the optical mode size is expanded transversly along propagation in the Ge thin film. As a result, more photocarriers are generated at closer locations to the metal electrodes, followed by being converted to photocurrent before recombination occurs. As a result, the conversion efficiency from photocarriers to photocurrent in the Ge thin film is increased.

The efficiency of THz photoconductive switches can be potentially enhanced by introducing artificial 3D architectures that exhibit tailored optoelectronic properties. Artificial 3D structures have been proposed and demonstrated for various THz applications, such as THz lasers, THz photodetectors, and THz polarizers [28,29]. Additionally, antenna performance can also be improved by utilizing noval nanomaterials [30–33].

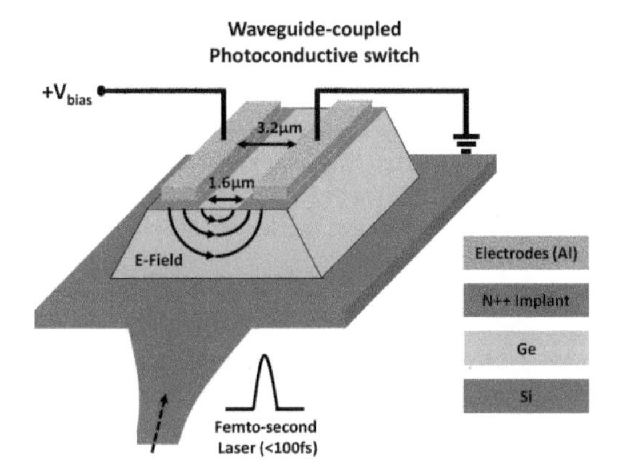

Figure 3. Structure of the proposed waveguide-coupling THz photoconductive switch.

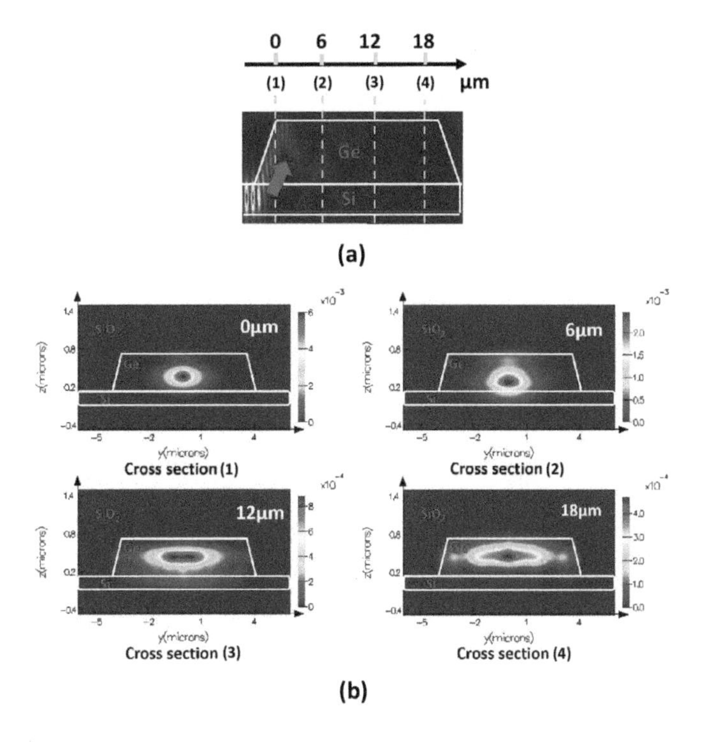

Figure 4. (**a**) Simulated optical absorption at 1550 nm within the proposed photoconductive switch. (**b**) Simulated 1550 nm optical mode distributions at different propagation distances within the proposed photoconductive switch.

2.3. System Architecture

Figure 5 demonstrates the chip micrograph of the proposed Ge-based optical waveguide coupled THz impulse radiator using an SOI-based silicon photonics process technology. It occupies a small die area of 440 µm × 680 µm. A bowtie antenna is designed on the top metal layer to reduce conductive loss. Metal vias connect the photoconductive switch and the antenna. An integrated optical grating coupler with a typical insertion loss of 4.37 dB [26] is used to couple the free-space femtosecond laser into the on-chip waveguide, where the femtosecond laser propagates and is eventually absorbed by the photoconductive switch. Given that the length of the integrated waveguide is smaller than 300 µm in the prototype chip, its loss and dispersion effects are negligible, i.e., 0.05 dB loss and 1.3 fs pulse-width broadening. The main focus of this work is to design the Ge-based waveguide-coupled photoconductive switch. Therefore, pre-designed optical grating couplers and optical waveguides in the process development kit (PDK) are used in the design phase. These passive components can be further improved through custom design (loss and dispersion effects are challenging in the large-scale system integration, where much longer waveguide routing is required. The loss and dispersion performance of integrated grating coupler and waveguide can be further improved by custom design, which, however, is not the focus of this paper.). The chip package is also shown in Figure 5. A highly resistive silicon lens is attached to the backside of the chip to increase radiation efficiency.

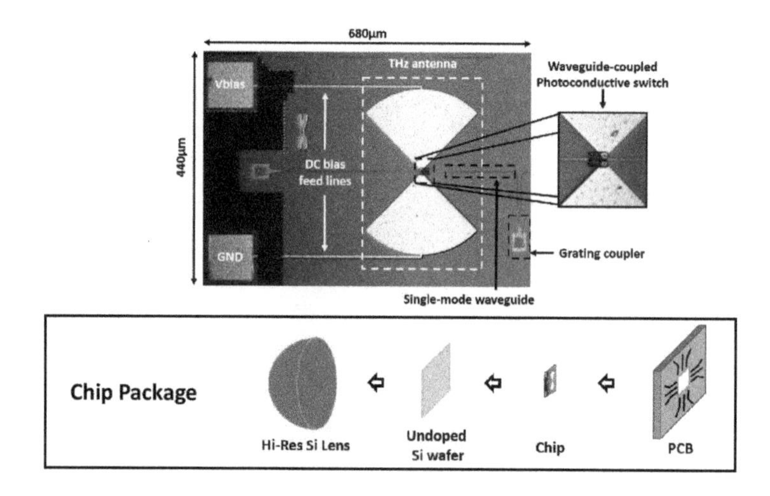

Figure 5. Micrograph of the prototype THz impulse radiator chip and chip package.

3. Measurement Results

The prototype THz impulse radiator chip was characterized in both time domain and frequency domain using an Advantest THz-TDS system (TAS7500TS), which is based on asynchronous optical sampling mechanism [34,35]. The characterization setup is demonstrated in Figure 6. A 50 fs pump laser beam from the Advantest THz-TDS system was coupled to the free space, and then focused onto the integrated grating coupler in the prototype chip. The focusing lens has a NA of 0.5. The chip was mounted on a rotation stage, which can be adjusted to achieve maximum coupling efficiency for the integrated grating coupler. A THz polyethylene lens focused the THz radiation to the THz detector. To measure average radiated power of the prototype chip, a calibrated pyroelectric detector, which is sensitive from 20 GHz to 1.5 THz, was utilized with a mechanical chopper modulating the pump femtosecond laser beam in the free space.

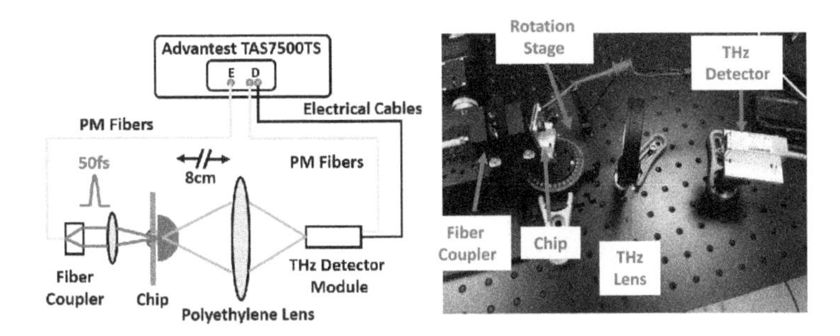

Figure 6. Measurement setup for the prototype THz impulse radiator chip.

Figure 7 presents the measured THz impulse radiated by the prototype chip. With a maximum bias voltage of 3.5 V, the prototype chip radiates THz impulses with an FWHM of 1.14 ps. Its frequency spectrum, shown in Figure 7b, is obtained by performing DFT on the measured time-domain waveform. The radiated THz impulse has a peak frequency component at 176 GHz, and it has an SNR > 1 bandwidth of 1.5 THz. The measured average radiated power is 0.337 μW, and its DC-to-RF conversion efficiency is 3.6×10^{-5}.

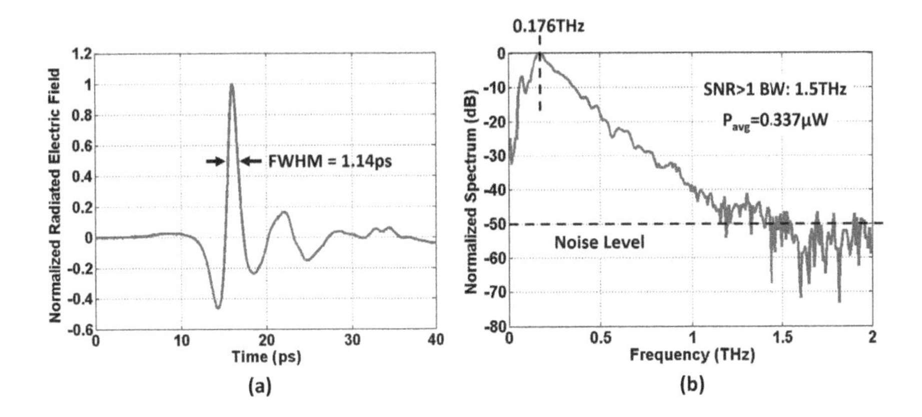

Figure 7. (**a**) Measured time-domain waveform of the radiated THz impulse. (**b**) Measured frequency-domain spectrum of the radiated THz impulse.

Figure 8 demonstrates the measured effects of bias voltage on the radiated THz impulses. Theoretically, by increasing the bias voltage, the generated photocarriers in the Ge thin film have a higher drift velocity (before saturation happens), inducing a stronger transient current, and consequently, producing stronger THz impulse radiation. Measured results confirmed this theoretical prediction. When the bias voltage increased from 2 V to 3.5 V, the measured THz impulse had a larger peak amplitude (Figure 8a), a greater SNR > 1 bandwidth (Figure 8b), and a stronger average radiated power (Figure 8c) (we did not detect THz emission based on surface optical rectification [36,37] when the dc bias voltage is 0 V. One possible reason is that the waveguide-coupling mechanism, which forms a horizontal coupling interaction (with respect to the semiconductor substrate surface) between the optical excitation and the photoconductive material, may weaken the nonlinear optical rectification induced THz emission. Further investigation is needed). In this design, the maximum bias voltage is 3.5 V, and is limited by the current capacity of the electrical vias in the prototype chip.

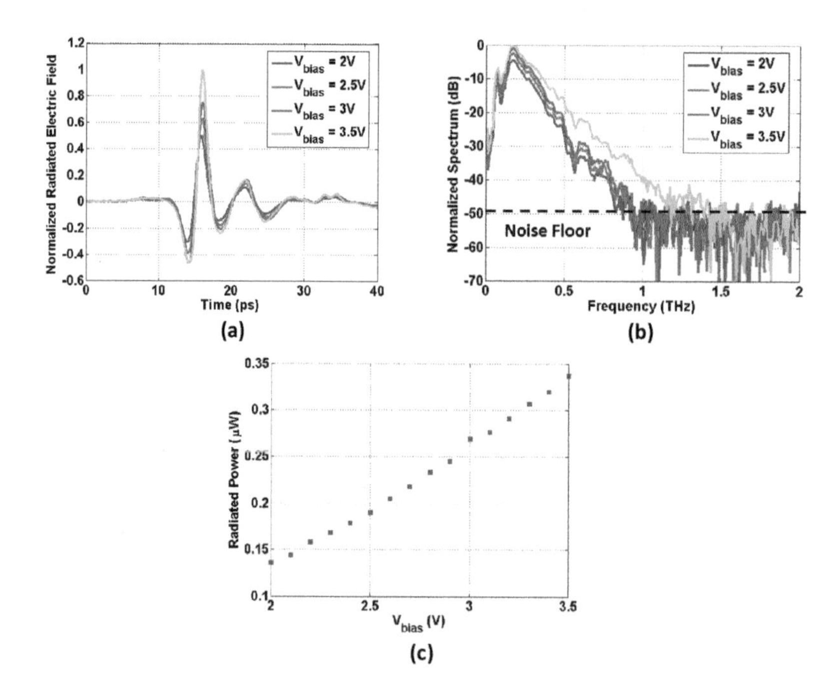

Figure 8. Measured effects of bias voltage on the radiated THz impulse. (**a**) Effects on the peak amplitude of the radiated THz impulse. (**b**) Effects on the SNR > 1 bandwidth of the radiated THz impulse. (**c**) Effects on the average radiated power.

4. Conclusions

In this work, an integrated Ge-based THz impulse radiator with an optical waveguide coupling scheme was implemented using an SOI-based silicon photonics process technology provided by IME A*STAR, Singapore. In the prototype chip, a phosphorus-doped Ge thin film is designed to reduce the photocarrier's lifetime for faster transient response speed. Additionally, it enables the THz impulse radiator to operate with 1550 nm femtosecond lasers. The proposed optical waveguide coupling photoconductive switch provides a monolithic on-chip excitation scheme between the incident femtosecond laser and the photoconductive material. This monolithic configuration facilitates full-system integration of THz impulse radiators and other silicon photonics modules. The prototype CMOS-compatible THz impulse radiator chip can radiate 1.14 ps THz impulses with an SNR > 1 bandwidth of 1.5 THz and an average radiated power of 0.337 μW. The DC-to-RF conversion efficiency of the prototype chip is 3.6×10^{-5}.

Author Contributions: Conceptualization, P.C. and A.B.; methodology, P.C.; measurement and analysis, P.C. and M.H.; writing—original draft preparation and revision, P.C.

Funding: This research was funded by NSF Career Award.

Conflicts of Interest: The authors declare no conflict of interest.

References

1. Siegel, P.H. Terahertz technology in biology and medicine. *IEEE Trans. Microw. Theory Tech.* **2004**, *52*, 2438–2447. [CrossRef]
2. Siegel, P.H. Terahertz technology. *IEEE Trans. Microw. Theory Tech.* **2002**, *50*, 910–928. [CrossRef]
3. Aggrawal, H.; Chen, P.; Assefzadeh, M.M.; Jamali, B.; Babakhani, A. Gone in a Picosecond: Techniques for the Generation and Detection of Picosecond Pulses and Their Applications. *IEEE Microw. Mag.* **2016**, *17*, 24–38. [CrossRef]
4. Liu, L.; Jiang, Z.; Rahman, S.; Shams, M.I.B.; Jing, B.; Kannegulla, A.; Cheng, L.J. Quasi-Optical Terahertz Microfluidic Devices for Chemical Sensing and Imaging. *Micromachines* **2016**, *7*, 75. [CrossRef]
5. Alfihed, S.; Bergen, M.H.; Ciocoiu, A.; Holzman, J.F.; Foulds, I.G. Characterization and Integration of Terahertz Technology within Microfluidic Platforms. *Micromachines* **2018**, *9*, 453. [CrossRef]
6. Friederich, F.; von Spiegel, W.; Bauer, M.; Meng, F.; Thomson, M.D.; Boppel, S.; Lisauskas, A.; Hils, B.; Krozer, V.; Keil, A.; et al. THz Active Imaging Systems With Real-Time Capabilities. *IEEE Trans. Terahertz Sci. Technol.* **2011**, *1*, 183–200. [CrossRef]
7. Chen, P.; Babakhani, A. 3-D Radar Imaging Based on a Synthetic Array of 30-GHz Impulse Radiators With On-Chip Antennas in 130-nm SiGe BiCMOS. *IEEE Trans. Microw. Theory Tech.* **2017**, *65*, 4373–4384. [CrossRef]
8. Chen, P.; Babakhani, A. A 30GHz impulse radiator with on-chip antennas for high-resolution 3D imaging. In Proceedings of the 2015 IEEE Radio and Wireless Symposium (RWS), San Diego, CA, USA, 25–28 January 2015; pp. 32–34.
9. Malhotra, I.; Jha, K.R.; Singh, G. Terahertz antenna technology for imaging applications: A technical review. *Int. J. Microw. Wirel. Technol.* **2018**, *10*, 271–290. [CrossRef]
10. Tonouchi, M. Cutting-edge terahertz technology. *Nat. Photonics* **2007**, *1*, 97–105. [CrossRef]
11. Nagatsuma, T.; Ducournau, G.; Renaud, C.C. Advances in terahertz communications accelerated by photonics. *Nat. Photonics* **2016**, *10*, 371–379. [CrossRef]
12. Chen, P.; Wang, Y.; Babakhani, A. A 4ps amplitude reconfigurable impulse radiator with THz-TDS characterization method in 0.13 µm SiGe BiCMOS. In Proceedings of the 2016 IEEE MTT-S International Microwave Symposium (IMS), San Francisco, CA, USA, 22–27 May 2016; pp. 1–4.
13. Chen, P.; Assefzadeh, M.M.; Babakhani, A. A Nonlinear Q-Switching Impedance Technique for Picosecond Pulse Radiation in Silicon. *IEEE Trans. Microw. Theory Tech.* **2016**, *64*, 4685–4700. [CrossRef]
14. Assefzadeh, M.M.; Babakhani, A. Broadband THz spectroscopic imaging based on a fully-integrated 4 × 2 Digital-to-Impulse radiating array with a full-spectrum of 0.03–1.03THz in silicon. In Proceedings of the 2016 IEEE Symposium on VLSI Technology, Honolulu, HI, USA, 14–16 June 2016; pp. 1–2.
15. Assefzadeh, M.M.; Chen, P.; Babakhani, A. High-power THz pulse radiation with GHz repetition rate in silicon. In Proceedings of the 2016 41st International Conference on Infrared, Millimeter, and Terahertz Waves (IRMMW-THz), Copenhagen, Denmark, 25–30 September 2016; pp. 1–3.
16. Assefzadeh, M.M.; Babakhani, A. Broadband Oscillator-Free THz Pulse Generation and Radiation Based on Direct Digital-to-Impulse Architecture. *IEEE J. Solid-State Circuits* **2017**, *52*, 2905–2919. [CrossRef]
17. Auston, D.H. Picosecond optoelectronic switching and gating in silicon. *Appl. Phys. Lett.* **1975**, *26*, 101–103. [CrossRef]
18. Berry, C.W.; Jarrahi, M. Terahertz generation using plasmonic photoconductive gratings. *New J. Phys.* **2012**, *14*, 105029. [CrossRef]
19. Jooshesh, A.; Bahrami-Yekta, V.; Zhang, J.; Tiedje, T.; Darcie, T.E.; Gordon, R. Plasmon-Enhanced below Bandgap Photoconductive Terahertz Generation and Detection. *Nano Lett.* **2015**, *15*, 8306–8310. [CrossRef] [PubMed]
20. Lepeshov, S.; Gorodetsky, A.; Krasnok, A.; Toropov, N.; Vartanyan, T.A.; Belov, P.; Alú, A.; Rafailov, E.U. Boosting Terahertz Photoconductive Antenna Performance with Optimised Plasmonic Nanostructures. *Sci. Rep.* **2018**, *8*, 6624. [CrossRef]
21. Yardimci, N.T.; Jarrahi, M. Nanostructure-Enhanced Photoconductive Terahertz Emission and Detection. *Small* **2018**, *14*, 1802437. [CrossRef] [PubMed]
22. Burford, N.M.; Evans, M.J.; El-Shenawee, M.O. Plasmonic Nanodisk Thin-Film Terahertz Photoconductive Antenna. *IEEE Trans. Terahertz Sci. Technol.* **2018**, *8*, 237–247. [CrossRef]

23. Bashirpour, M.; Forouzmehr, M.; Hosseininejad, S.E.; Kolahdouz, M.; Neshat, M. Improvement of Terahertz Photoconductive Antenna using Optical Antenna Array of ZnO Nanorods. *Sci. Rep.* **2019**, *9*, 1414. [CrossRef]
24. Ding, R.; Baehr-Jones, T.; Pinguet, T.; Li, J.; Harris, N.C.; Streshinsky, M.; He, L.; Novack, A.; Lim, A.E.J.; Liow, T.Y.; et al. A silicon platform for high-speed photonics systems. In Proceedings of the OFC/NFOEC, Los Angeles, CA, USA, 4–8 March 2012; pp. 1–3.
25. Chen, P.; Hosseini, M.; Babakhani, A. An integrated germanium-based optical waveguide coupled THz photoconductive antenna in silicon. In Proceedings of the 2016 Conference on Lasers and Electro-Optics (CLEO), San Jose, CA, USA, 5–10 June 2016; pp. 1–2.
26. Novack, A.; Liu, Y.; Ding, R.; Gould, M.; Baehr-Jones, T.; Li, Q.; Yang, Y.; Ma, Y.; Zhang, Y.; Padmaraju, K.; et al. A 30 GHz silicon photonic platform. In Proceedings of the 10th International Conference on Group IV Photonics, Seoul, Korea, 28–30 August 2013; pp. 7–8.
27. Sekine, N.; Hirakawa, K.; Sogawa, F.; Arakawa, Y.; Usami, N.; Shiraki, Y.; Katoda, T. Ultrashort lifetime photocarriers in Ge thin films. *Appl. Phys. Lett.* **1996**, *68*, 3419–3421. [CrossRef]
28. Ferguson, B.; Zhang, X.C. Materials for terahertz science and technology. *Nat. Mater.* **2002**, *1*, 26–33. [CrossRef]
29. Komatsu, N.; Gao, W.; Chen, P.; Guo, C.; Babakhani, A.; Kono, J. Modulation-Doped Multiple Quantum Wells of Aligned Single-Wall Carbon Nanotubes. *Adv. Funct. Mater.* **2017**, *27*, 1606022. [CrossRef]
30. Burke, P.J. Carbon Nanotube Devices for GHz to THz Applications. In Proceedings of the Optics East, Philadelphia, PA, USA, 25–28 October 2004; pp. 52–62.
31. Bengio, E.A.; Senic, D.; Taylor, L.W.; Tsentalovich, D.E.; Chen, P.; Holloway, C.L.; Babakhani, A.; Long, C.J.; Novotny, D.R.; Booth, J.C.; et al. High efficiency carbon nanotube thread antennas. *Appl. Phys. Lett.* **2017**, *111*, 163109. [CrossRef]
32. Bandaru, P.R. Electrical Properties and Applications of Carbon Nanotube Structures. *J. Nanosci. Nanotechnol.* **2007**, *7*, 1239–1267. [CrossRef]
33. Bengio, E.A.; Senic, D.; Taylor, L.W.; Headrick, R.J.; King, M.; Chen, P.; Little, C.A.; Ladbury, J.; Long, C.J.; Holloway, C.L.; et al. Carbon Nanotube Thin Film Patch Antennas for Wireless Communications. *Appl. Phys. Lett.* **2019**, to be online soon.
34. Yasui, T.; Saneyoshi, E.; Araki, T. Asynchronous optical sampling terahertz time-domain spectroscopy for ultrahigh spectral resolution and rapid data acquisition. *Appl. Phys. Lett.* **2005**, *87*, 061101. [CrossRef]
35. Chen, P.; Assefzadeh, M.M.; Babakhani, A. Time-Domain Characterization of Silicon-Based Integrated Picosecond Impulse Radiators. *IEEE Trans. Terahertz Sci. Technol.* **2017**, *7*, 599–608. [CrossRef]
36. Chuang, S.L.; Schmitt-Rink, S.; Greene, B.I.; Saeta, P.N.; Levi, A.F.J. Optical rectification at semiconductor surfaces. *Phys. Rev. Lett.* **1992**, *68*, 102–105. [CrossRef] [PubMed]
37. Peters, L.; Tunesi, J.; Pasquazi, A.; Peccianti, M. High-energy terahertz surface optical rectification. *Nano Energy* **2018**, *46*, 128–132. [CrossRef]

Article

Geometrical Representation of a Polarisation Management Component on a SOI Platform

Massimo Valerio Preite [1], Vito Sorianello [2], Gabriele De Angelis [2] and Marco Romagnoli [2] and Philippe Velha [1,*]

[1] Scuola Superiore Sant'Anna—TeCIP Institute, Via Moruzzi 1, 56124 Pisa, Italy; valeriopreite@gmail.com
[2] CNIT—Laboratory of Photonic Networks, Via Moruzzi 1, 56124 Pisa, Italy; vito.sorianello@cnit.it (V.S.); gabriele.deangelis@cnit.it (G.D.A.); marco.romagnoli@cnit.it (M.R.)
* Correspondence: p.velha@santannapisa.it; Tel.: +39-050-88-2187

Received: 19 April 2019; Accepted: 26 May 2019; Published: 30 May 2019

Abstract: Grating couplers, widely used in Silicon Photonics (SiPho) for fibre-chip coupling are polarisation sensitive components, consequently any polarisation fluctuation from the fibre optical link results in spurious intensity swings. A polarisation management componentis analytically considered, coupled with a geometrical representation based on phasors and Poincaré sphere, generalising and simplifying the treatment and understanding of its functionalities. A specific implementation in SOI is shown both as polarisation compensator and polarisation controller, focusing on the operative principle. Finally, it is demonstrated experimentally that this component can be used as an integrated polarimeter.

Keywords: Silicon Photonics; off-chip coupling; polarisation controller; integrated polarimeter; polarisation multiplexing; polarisation shift keying

1. Introduction

Silicon photonics, thanks to its compatibility with CMOS technology, is imposing itself for large scale fabrication of low cost and small footprint photonic integrated circuits (PICs). One of Silicon Photonics main challenges is coupling light in and out from the chip in an efficient and practical way. The high index contrast between silicon and silica enables the use of grating couplers (GC), that, to our knowledge, are the most widespread off chip coupling solution, mainly thanks to the design flexibility deriving from the fact that they can be placed nearly everywhere on the chip and are not constrained to the chip edge. An important limitation of GCs is their polarisation sensitivity; consequently, the input polarisation fluctuations that regularly occur in optical fibres as a result of deformations or temperature changes translate into random spurious amplitude modulations. Other coupling schemes, such as butt or end fire coupling, are possible, but they are not of interest for this paper.

The problem of assuring polarisation tracking was broadly addressed decades ago, as it is critical for the working of coherent optical systems, given the need to match the time varying State of Polarisation (SOP) of the input signal to the local oscillator's one. The proposed solutions were based on fibre squeezers [1], lithium-niobate integrated devices [2,3] and Planar Lightwave Circuit (PLC) technology [4].

In [5], the proof of concept of Caspers et al. [6] was further developed into a fully functional building block with two independent phase shifters and was integrated at the two ends of the bus waveguide in a subsystem presented in [7] , thus making it transparent to polarisation fluctuations. Until now, the strategy to cope with polarisation fluctuations has been to separate the signal in two orthogonal polarisations and duplicate the circuits , in what is called polarisation diversity scheme. In this example, the compensator is able to convert any SOP into the standard TE mode of Silicon Photonics waveguide eluding a duplicated circuit.

The current article extends the results in [5,6], and aims at developing a simple but precise and exhaustive geometrical picture based on both phasor and Bloch sphere representation. The use of this pictorial representation is illustrated by solving the problem of the frequency response together with its effect on the point representing the SOP on the Poincaré sphere.

In polarisation tracking, sometimes it is not enough to have a device that can compensate all possible SOPs, but it may be desirable to have an endless system, i.e., one where resets—when the physical quantity producing the phase shift reaches any end of its range and a phase jump of an integer number of 2π must be applied—are collision-free and do not provoke transmission disruptions nor information losses. Even better is a reset-free system, that is, one in which potentially unlimited phase shifts can be achieved with the control quantity limited in a finite interval. Usual wave plate transformers possess this last property [2] and have been implemented in lithium-niobate [8,9]. Quarter- and half-wave plates can be implemented in PLC [10], thus even in SiPho. Hence, the same functionality of the device in [2] can be achieved in SiPho, at the price of a greater complexity.

This paper is organised in the following way: after a short description of the device structure in Section 1.1, Section 2 analyses single wavelength operation, while Section 3 examines the frequency response. In both cases, both operating modes are analysed: here "compensator" refers to the case in which the light enters from the 2DGC and the circuit acts so that all the power is routed toward one of its two ends, while "controller" means the reciprocal case where the input is one of the two waveguides and the device settings determine the polarisation exiting from the 2DGC.

In Section 2.1, a phasor representation illustrates the particular case of the polarisation compensation. Two classes of periodical solutions are found, for any input SOP. As a corollary, the knowledge of the phase shifts needed to compensate the input of an unknown SOP can be used to measure without ambiguity this SOP. For the first time, it is demonstrated that a polarimeter can be made in a silicon photonics platform.

Next, several properties are derived thanks to the graphical representation as a function of two physical quantities: the applied phase shifts.

In Section 2.2, the operation as polarisation controller is examined, proving that all the SOPs can be generated and the effect of the two phase shifters on the corresponding point on Poincaré sphere is described.

Section 3 examines the frequency response for both the compensator and the generator operation. Using the expression provided in [11] for the frequency and temperature dependence of SOI effective index, an expression for the phase shift frequency dependence is derived. A transcendental implicit equation for the -3 dB bandwidth is found and it is shown that the frequency response is non trivial depending in both the considered input SOP and the chosen phase shift pair. Then, those results are transposed to Poincaré sphere.

Finally, Section 4 presents an experimental test of the derived results.

1.1. Device Schematic

The device exploits a 2D grating coupler (2DGC) to split the incoming field into two orthogonal components. These two components are then fed into two distinct integrated waveguides.

The first stage introduces a first phase shift, then the fields in the two branches are combined and again split by means of an MMI, and undergo another phase shift.

Eventually, a second MMI recombines the fields and sends its output to the two exits.

In [7], the top output is terminated on a monitoring photodiode, while the bottom one is connected to the rest of the PIC. The photodiode current is minimised during the module tuning to make sure that all the optical power goes to the bottom output and therefrom to the downhill PIC. For consistency with that article and the work in [5], in the rest of this paper, we consider that only port "B" in Figure 1 is used.

However, that is not the only possible arrangement. On the contrary, each port can be connected to a distinct photonic module, which processes the information encoded on the SOP orthogonal to that of the other output, for instance in a POLarisation Shift Keying (POLSK) scheme.

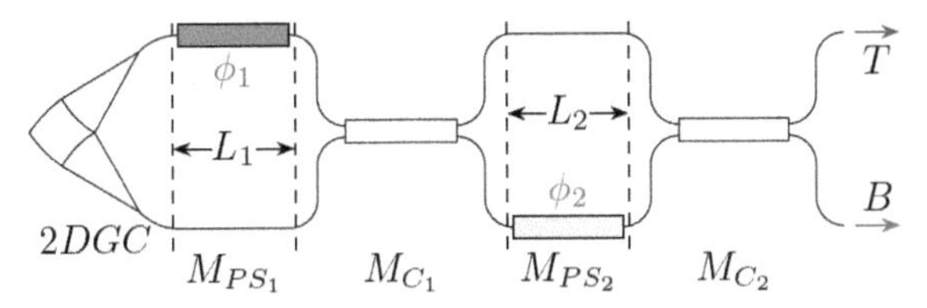

Figure 1. Schematic of the integrated polarisation controller. The labels M_i refer to the transfer matrix associated with a given circuit section. The block on the picture leftmost portion is a schematic of the 2DGC. The "T" and "B" labels denote the top and bottom outputs, respectively.

2. Single Wavelength Operation

2.1. Compensator

The overall device, excluding the grating couplers, can be viewed as the cascade of four blocks; correspondingly, its total transfer matrix is the product of the four individual blocks matrices, which are phase shifters (PS) and couplers (C), respectively, and read:

$$M_{PS} = \begin{bmatrix} e^{i\,\Delta\varphi} & 0 \\ 0 & 1 \end{bmatrix} \qquad M_C = \begin{bmatrix} \cos(\kappa) & -i\sin(\kappa) \\ -i\sin(\kappa) & \cos(\kappa) \end{bmatrix} = \frac{1}{\sqrt{2}}\begin{bmatrix} 1 & -i \\ -i & 1 \end{bmatrix} \tag{1}$$

where $\Delta\varphi = \Delta_{tb}\,(\beta L) = \Delta_{tb}(\beta)L$ and the function Δ_{tb} denotes the difference of the quantity between brackets for the top and bottom arm. It is applied just to the propagation constant $\boldsymbol{\beta}$ as the two arms are ideally of equal length. The coupling coefficient is assumed to be $\kappa = \pi/2$ (3 dB coupler).

Thus, the overall transfer matrix reads:

$$T = M_{C_2} M_{PS_2} M_{C_1} M_{PS_1} = \frac{e^{-i\,\psi}}{2}\begin{bmatrix} e^{i\,\phi_1}\left(1-e^{i\,\phi_2}\right) & -i\left(1+e^{i\,\phi_2}\right) \\ -i\,e^{i\,\phi_1}\left(1+e^{i\,\phi_2}\right) & -\left(1-e^{i\,\phi_2}\right) \end{bmatrix} \tag{2}$$

where ψ is an arbitrary absolute phase shift, and ϕ_1 and ϕ_2 are the applied phase shifts, as shown in Figure 1. As input, we consider a generic polarisation, described by the Jones vector in three equivalent forms:

$$A_{in} = e^{i\,\delta_1}\begin{pmatrix} a_1 \\ a_2\,e^{i\,\delta} \end{pmatrix} = e^{i\,\delta_1}\begin{pmatrix} \cos\alpha \\ \sin\alpha\,e^{i\,\delta} \end{pmatrix} = \begin{pmatrix} q_1 \\ q_2 \end{pmatrix} \tag{3}$$

where q_1 and q_2 are defined as two complex values representing the input vector polarisation. The main hypothesis of this works is that the 2DGC is supposed to split the TE and TM components without affecting their relative amplitude and phases (attenuating or phase shifting them in the same way, i.e., without crosstalk), so that the input vector to the rest of the circuit can be assumed to coincide with the above Jones vector.

Thus, the output complex amplitudes vector from the circuit is:

$$A_{out} = \begin{pmatrix} a_{up} \\ a_{bottom} \end{pmatrix} = T\,A_{in} = \frac{e^{-i\,\vartheta}}{2}\begin{pmatrix} a_1\left(1-e^{i\,\phi_2}\right) - i\,a_2\,e^{i\,(\delta-\phi_1)}\left(1+e^{i\,\phi_2}\right) \\ -i\,a_1\left(1+e^{i\,\phi_2}\right) - a_2\,e^{i\,(\delta-\phi_1)}\left(1-e^{i\,\phi_2}\right) \end{pmatrix} \tag{4}$$

To send all the power in the bottom waveguide, the condition to be fulfilled is:

$$A_{out} = \begin{pmatrix} a_{up} \\ a_{bottom} \end{pmatrix} = e^{i\theta} \begin{pmatrix} 0 \\ 1 \end{pmatrix} \tag{5}$$

Again, the absolute phase shift term in front of the matrix, $\vartheta = \psi + \delta_1 + \phi_1$, can be ignored, whereas the one in front of the desired output, $\boldsymbol{\theta}$, is kept for the sake of completeness.

2.1.1. Graphical Solutions in the Phasor Space

We introduce a graphical representation based on phasors, commonly used in quantum mechanics, to offer an intuitive view of the components' behaviour. The two terms of the vector of Equation (4) are represented using the convention that a phase shift $e^{i\varphi}$ corresponds to a counter clockwise rotation of angle φ. The first term of the top part $a_1(1 - e^{i\phi_2})$ brings us to point A and the second term $i\, a_2\, e^{i\,(\delta-\phi_1)}\,(1 + e^{i\,\phi_2})$ brings us to point B.
The fruitfulness of this pictorial approach lies in the immediate and simple way to find the solutions to Equation (4), that is, of solving the problem.

In order for the difference appearing in Equation (4) first row to be zero, its two terms must be equal, that is to say, the corresponding phasors tips A and B must coincide in Figure 2a (small red circle).

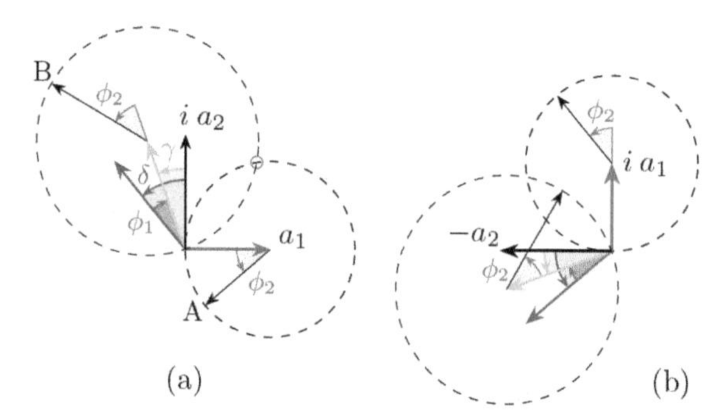

Figure 2. Phasor diagrams for the two components of A_{out}; actually, it shows the difference, not the sum, of the two terms of each component, as it is easier to visualise. (**a**) Top component of Equation (4). When the tips of the two vectors lie in the second intersection, circled in red, between the circumferences, then the first component of A_{out} is zero. (**b**) Bottom component.

This can happen only if both tips lie on the intersection between the dashed circumferences (the other intersection, where the phasors "nocks" are, is not a solution, as ϕ_2 should be simultaneously 0 for a_1 and π for a_2); this means that the four represented arrows form a closed quadrilateral (Figure 3a).

In such a circumstance, the quadrilateral opposite angles on the tips of a_1 and $a_2\, e^{i(\gamma+\pi/2)}$ are supplementary by construction, and, therefore, such must be the other two angles as well.

Furthermore, as shown in Figure 3a, those angles are congruent, because they are the sum of angles on the base of two isosceles triangles (because they are inscribed in the two circumferences). Thus, they must equal a right angle, in order to be supplementary.

In turn, that imposes the following relationships:

$$\delta - \phi_1 = n\pi \qquad \text{or} \qquad \phi_1 = \delta + n\pi \tag{6}$$

Thus, the phase shift introduced by the first stage must match that between the input Jones vector components, δ modulo π.

The requirement of cancelling the first component of Equation (4) translates into

$$e^{i\phi_2} = \frac{a_1 - i\,a_2\,e^{i\delta-\phi_1}}{a_1 + i\,a_2\,e^{i\delta-\phi_1}} = \frac{a_1 \mp i\,a_2}{a_1 \pm i\,a_2} = e^{\mp i\,2\arctan(a_2/a_1)}$$
$$\text{i.e.,} \qquad \phi_2 = \mp 2\arctan(a_2/a_1) = \mp 2\alpha \tag{7}$$

This condition can be deduced more easily from Figure 3a, considering the rectangular triangles with catheti of length a_1 and a_2. Note that the parity of n in Equation (6) determines the sign of the solution for ϕ_2, as well as which pair of diagrams is to be considered in Figure 3.

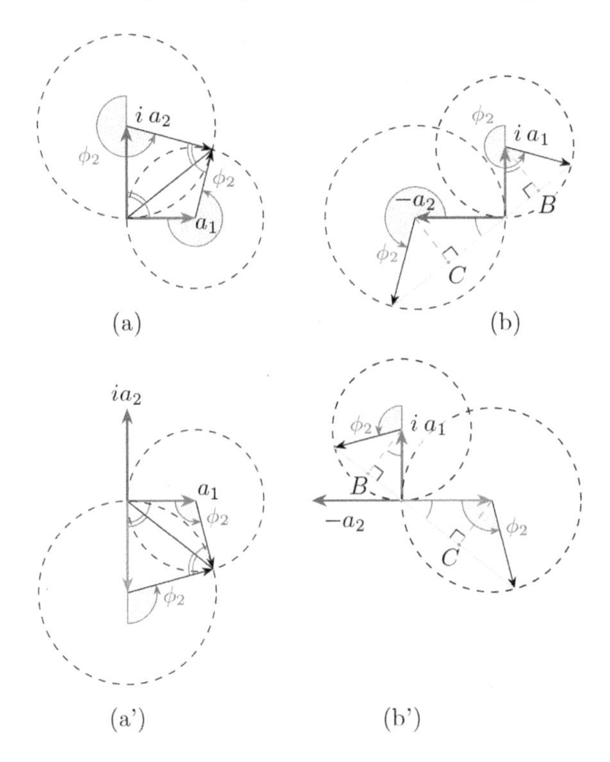

Figure 3. Phasor diagrams for the two components of $\boldsymbol{A}_{out}$, when both conditions of Equation (6) and (7) are fulfilled. (**a**) The four vectors form a quadrilateral. Note that Equation (7) can be deduced observing that the quadrilateral is the union of two congruent rectangle triangles with catheti a_1 and a_2 and thus the green angle equals twice $\arctan(a_2/a_1)$ (or the explementary angle, for a', when n is odd). (**b**) When Equation (6) is fulfilled, then the segments joining the origin with the points on the circumferences lie on the same line. The construction in figure shows that, when Equation (7) holds too, the segment between those two points is twice the hypotenuse of the rectangle triangle with a_1 and a_2 as catheti, which corresponds to the norm (power) of the input vector. Notice that, if one of the two angles (but not both) is changed by π, then the role in (a) and (b) gets reversed, i.e., the whole power goes in the top port.

Now, considering the bottom component, it can be shown to be

$$-\,i\,e^{-i\,(\psi+\delta_1+\phi_1)}\,(a_1 \mp i\,a_2) \tag{8}$$

Even though the results can be obtained through straightforward calculations, a geometrical resolution based on the phasor representation offers a quicker and more insightful method.

Starting from Figure 2b, the term can be set to zero rotating the phasors such that their tips meet at the intersection of the circles, as in Figure 3a(a').

When looking at the phasor of the bottom term of Equation (4) when the top term is equal to zero, we obtain the representation of Figure 3b(b'). Dropping the heights (points B and C), it is clear that the length of the bases' sum is twice the hypotenuse of the rectangle triangle with a_1 and a_2 as catheti, as shown in Figure 3b(b').

This shows that it is possible, at least at one single wavelength, to compensate the polarisation in such a way that the amplitude at one exit port is zero while the other is maximum.

Summarising, the two solutions for the phase shifts to be applied are:

$$\phi_1' = \delta + 2n\,\pi \qquad \phi_2' = -2\arctan(a_2/a_1) + 2m\,\pi \tag{9a}$$

$$\phi_1'' = \delta + (2n+1)\pi \qquad \phi_2'' = 2\arctan(a_2/a_1) + 2m\,\pi \tag{9b}$$

Until now, the Jones representation (J) has been used but a more complete approach can be derived using Stokes parameters. The Stokes parameters are defined in terms of *Pauli matrices* σ_i as (eq. 2.5.24 [12])

$$s_i \doteq J^\dagger \sigma_i J \tag{10}$$

with the notation of Equation (3) they become

$$\begin{aligned}
s_0 &= |q_1|^2 + |q_2|^2 && = a_1^2 + a_2^2 && = 1 \\
s_1 &= |q_1|^2 - |q_2|^2 && = a_1^2 - a_2^2 && = \cos(2\alpha) \\
s_2 &= q_1 q_2^* + q_2 q_1^* && = 2\,a_1 a_2 \cos\delta && = \sin(2\alpha)\ \cos\delta \\
s_3 &= i(q_1 q_2^* - q_2 q_1^*) && = 2\,a_1 a_2 \sin\delta && = \sin(2\alpha)\ \sin\delta
\end{aligned} \tag{11}$$

so the quantities appearing in Equation (9) can be expressed as:

$$2\alpha = 2\arctan(a_2/a_1) = \arccos(s_1) \qquad \delta = \arctan\left(\frac{s_3}{s_2}\right) \tag{12}$$

One must note that the signs depend on the particular placement of the phase shifters that has been considered. If one shifter is moved to the other arm, this will result in a sign change in the formula above. In addition, using a push–pull configuration would halve the phase shift to be applied on each individual heater. The practical advantage is that, if a negative shift in the $(-\pi, 0)$ interval is to be applied, actually a shift between π and 2π would have to be used in a single heater configuration, whereas in the symmetric case a positive shift between 0 and π would suffice.

2.1.2. Intensity Surface

It is interesting to consider, for a given input polarisation state, how the output intensity depends on the two phase shifts (that are not necessarily set to the values which yield perfect compensation). To do so, we expand the squared modulus of Equation (4) bottom component:

$$P_L = \left|\frac{1}{2}\left[-i\,a_1\left(1+e^{i\,\phi_2}\right) - \ a_2\,e^{i\,(\delta-\phi_1)}\left(1-e^{i\,\phi_2}\right)\right]\right|^2 \tag{13}$$

It is convenient to use the form of Equation (3) in terms of the angles α and δ.

After several passages and trigonometric identities, the above formula can be shown to become:

$$P_L = \cos^2(\alpha + \phi_2/2)\ \cos^2\left(\frac{\delta-\phi_1}{2}\right) + \cos^2(\alpha - \phi_2/2)\ \sin^2\left(\frac{\delta-\phi_1}{2}\right) = E(\phi_1,\phi_2) + O(\phi_1,\phi_2) \tag{14}$$

Note that P_L is a function of ϕ_1, ϕ_2, whereas α and δ are parameters, which identify the input SOP.

Thus, the normalised power exiting from the lower branch is given by the sum of two surfaces O and E (Figure 4a,b) consisting in the square product of trigonometric functions.

Consequently, each surface is bounded between 0 and 1, and vanishes on lines parallel to the coordinate axes, spaced by 2π.

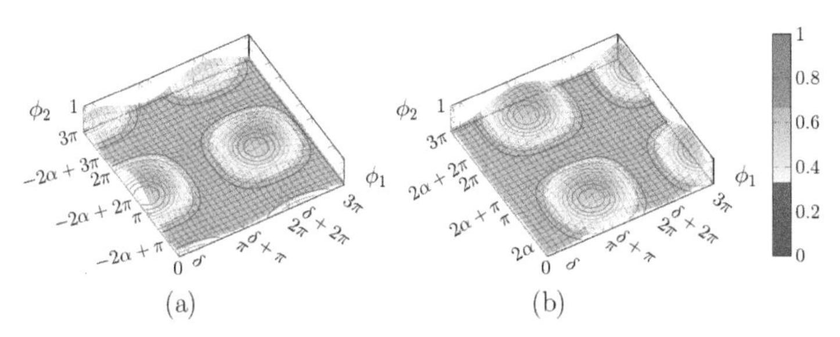

Figure 4. Plot of (**a**) $E(\phi_1, \phi_2)$ and (**b**) $O(\phi_1, \phi_2)$.

For the first surface, E, the maxima are located in (cf. Equation (9a)):

$$\phi_2' = -2\,\alpha + 2m\pi \qquad\qquad \phi_1' = \delta + 2n\,\pi \tag{15}$$

and adding an odd multiple of π to either angle would cancel the first term.

Instead, for the second surface, O, the maxima are located in (cf. Equation (9b)):

$$\phi_2'' = \;\; 2\,\alpha + 2m\pi \qquad\qquad \phi_1'' = \delta + (2n+1)\pi \tag{16}$$

Thanks to the fact that the two sets of solutions have the optimal value for ϕ_1 differing by odd multiples of π, the maxima of one surface lie on the null contour lines of the other and vice versa; this guarantees that the value of 1 will not be exceeded. The maxima position depends on the input polarisation state: the surfaces shift accordingly.

Both surfaces are shifted by the same amount and in the same direction along the ϕ_1 axis upon a change in δ, whereas a variation in α produces an equal and opposite shift along ϕ_2. Thus, changing δ results in a mere shift of the total surface (Figure 5b), but acting on α brings about a deformation of the surface (Figure 5c).

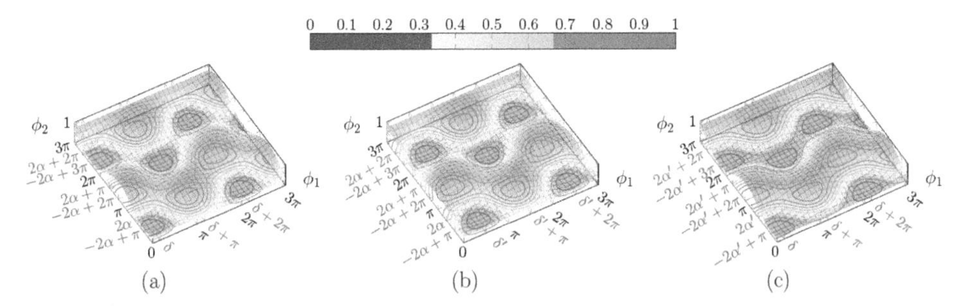

Figure 5. Plot of $P_L(\phi_1, \phi_2)$ for: (**a**) $\alpha = \pi/3$ and $\delta = \pi/6$; (**b**) same α but $\delta = \pi/2$, where the surface is shifted along ϕ_1, but retains the same shape as (**a**); and (**c**) $\alpha = 3/8\,\pi$ but same δ as in (**a**). Note that the surface has a different shape with respect to (**a**), even though the extrema lie in the same ϕ_1 values. Red and green ticks correspond to the first and second solution classes, respectively.

As shown in Equation (12), $\boldsymbol{\alpha}$ depends on $\mathbf{s_1}$ alone, so the surface shape depends on it only, i.e., is the same for all the points on "parallels" of Poincaré sphere (Section 2.2.1) with the same value of s_1, while the surfaces for different "longitudes" differ by a shift along ϕ_1.

In the limit case when the ϕ_2 values for the two solution sets maxima coincide, we have:

$$-2\,\alpha + 2m\pi = 2\,\alpha + 2n\pi \quad \Rightarrow \quad \alpha = n\,\frac{\pi}{2} \tag{17}$$

Distinguishing between even and odd multiples of $\pi/2$, and neglecting the factor $m\,\pi$ (as the sign change it could introduce is cancelled by the square), we find:

$$n = 2m \qquad P_L = \cos^2(\phi_2/2) \tag{18a}$$

$$n = 2m + 1 \qquad P_L = \sin^2(\phi_2/2) \tag{18b}$$

It is evident that the first phase shifter has no effect, as there is no dependence on ϕ_1, as displayed in Figure 6.

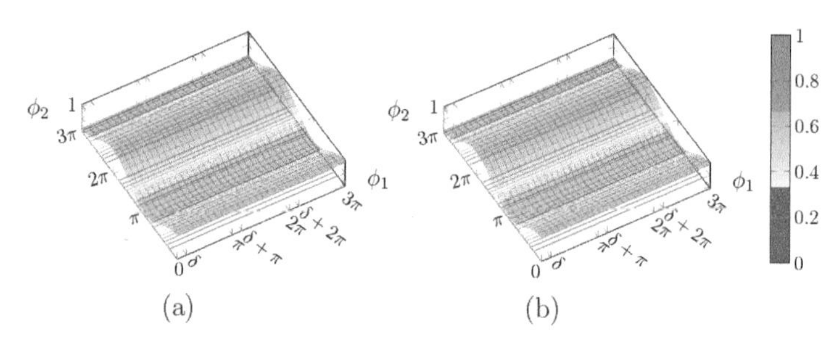

Figure 6. Plot of $P_L(\phi_1, \phi_2)$ for: **(a)** $\alpha = 0$; and **(b)** $\alpha = \pi/2$.

In fact, those two solutions correspond, respectively, to

$$\alpha = m\pi \qquad \to \tan\alpha = \frac{a_2}{a_1} = 0 \qquad \to a_2 = 0 \quad s_1 = 1 \tag{19a}$$

$$\alpha = \frac{\pi}{2} + m\pi \qquad \to \tan\alpha = \frac{a_2}{a_1} = \pm\infty \qquad \to a_1 = 0 \quad s_1 = -1 \tag{19b}$$

that is to say, the power flows completely either in the upper or lower coupler branch, respectively; in turn, this means that the input polarisation is one of the two principal SOPs of the 2DGC, which here are assumed by convention as horizontal or vertical.

As one of the two amplitudes is zero, its phase is not defined, thus there is no need for phase compensation.

Surfaces corresponding to input polarisations with opposite values of s_1, i.e., lying on opposite "parallels", possess the same shape, except for a reflection around the ϕ_1 axis.

In turn, this means that, for polarisations with opposite values of s_1, the values for α are complementary

$$\alpha^+ + \alpha^- = \pi/2 \tag{20}$$

The intensity surfaces for orthogonal polarisations are complementary, i.e., their sum equals to 1 (Figure 7). In fact, if (ϕ_1, ϕ_2) are set so that all the power exits from the bottom port, then, when the orthogonal polarisation is fed into the circuit, the result is to have no power at the bottom port, in that it exits all from the top one. This fact is useful for polarisation multiplexing.

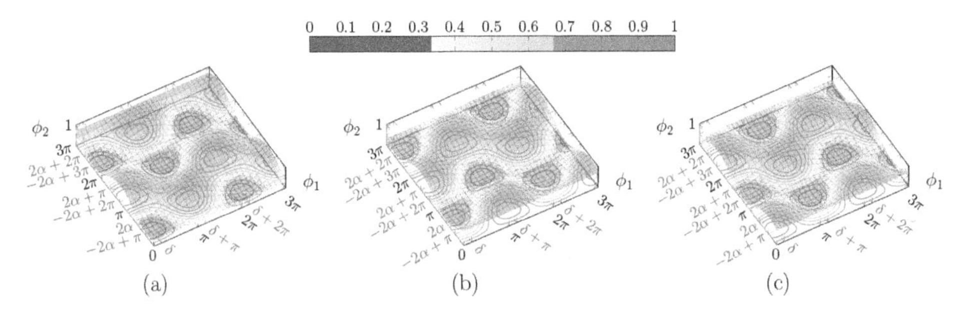

Figure 7. Plot of $P_L(\phi_1, \phi_2)$ for: (**a**) $\alpha = \pi/3$ and $\delta = \pi/6$; (**b**) $\alpha' = \pi/2 - \alpha$ and same δ; and (**c**) $\alpha' = \pi/2 - \alpha$ and $\delta' = \delta + \pi$. This is the complementary surface of (**a**). Note that the two solution classes are now inverted.

An important remark to be done is that there are always two maxima in the square with one vertex on the origin and side length of 2π in the positive axes direction (given that the thermo optic effect is exploited, it is possible to apply phase shifts of just one sign); i.e., phase shifts no greater than 2π are needed in order to recover any possible polarisation (although this does not guarantee that a drifting SOP can be tracked without interruptions, i.e., with endless operation). The blue lines connecting the different maxima in Figure 8 follow a path, which in the worst case scenario in Figure 9 has a minimum along the curve of 0.5. That means that it is possible to hop from one maxima to another suffering at most a 3 dB loss. This can be particularly useful to reduce power consumption and to avoid hitting the physical limits of the heaters. Finally, it must be highlighted that, contrarily to what Figure 5 may suggest, generally maxima and minima do not lie on lines parallel to the bisectors, as shown in Figure 8. This only occurs if $\alpha = \pi/4$ (Figure 9).

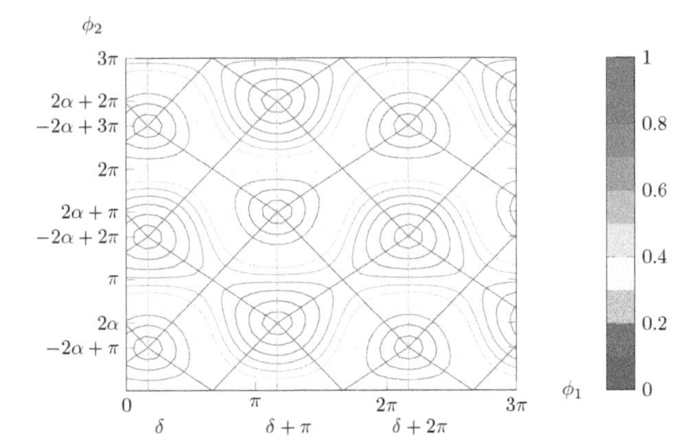

Figure 8. Contour plot for $\alpha = 55°$ and $\delta = 30°$. In general, maxima and minima are not placed on a square grid. Blue lines join maxima lying on the same "ridge" or minima in the same "valley", whereas red lines connect maxima separated by a "valley" or minima with a "ridge" in between, respectively.

Another important feature that is clearly apparent in Figure 8 is that for $\phi_2 = m\pi$ the intensity is constant, independently from the value of ϕ_1. In particular, Equation (14) tells that

$$P_L(\phi_1, 2n\pi) = \cos^2 \alpha \qquad\qquad P_L(\phi_1, (2n+1)\pi) = \sin^2 \alpha \tag{21}$$

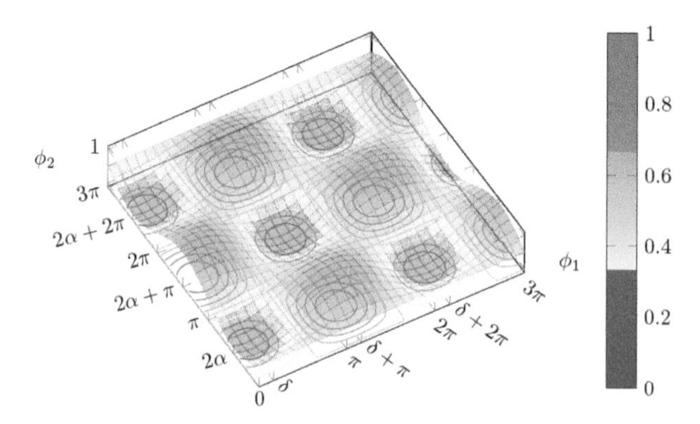

Figure 9. Plot of $P_L(\phi_1, \phi_2)$ for $\alpha = \pi/4$ and $\delta = \pi/6$. Note that now the surface displays a check pattern.

2.1.3. Use as Polarimeter

As shown in Section 2.1.1, we found the phase shifts ϕ_1 and ϕ_2 needed to convert the known input polarisation into the standard "TE" mode or, in other words, to inject all the power in one single output waveguide. Actually, two pairs were found for each SOP and one may wonder if it is possible to reverse the problem, i.e., to determine the SOP from the knowledge of the phase shifts that achieve optimal conversion. We show that said problem is solvable and that the existence of two solution classes generates no ambiguity, within a π shift range.

Hence, let us suppose that there exists a pair of polarisations $\mathfrak{p}$ and $\widetilde{\mathfrak{p}}$ for which there is ambiguity, that is to say, the second class of solutions of $\widetilde{\mathfrak{p}}$ coincides with the first one of $\mathfrak{p}$ or vice versa:

$$\begin{cases} \phi_1' &= \delta + 2n\pi &= \widetilde{\phi_1}'' &= \widetilde{\delta} + (2m+1)\pi \\ \phi_2' &= -2\alpha + 2l\pi &= \widetilde{\phi_2}'' &= 2\widetilde{\alpha} + 2q\pi \end{cases}$$
$$\begin{cases} \phi_1'' &= \delta + (2n+1)\pi &= \widetilde{\phi_1}' &= \widetilde{\delta} + 2m\pi \\ \phi_2'' &= 2\alpha + 2l\pi &= \widetilde{\phi_2}' &= -2\widetilde{\alpha} + 2q\pi \end{cases} \tag{22}$$

In both cases, the solution is:

$$\widetilde{\delta} = \delta + (2m+1)\pi \qquad \widetilde{\alpha} = -\alpha + q\pi \tag{23}$$

If we substitute back in the input Jones vector, we find the same vector, except for a change of sign:

$$\widetilde{A_{in}} = \begin{pmatrix} \cos(\widetilde{\alpha}) \\ \sin(\widetilde{\alpha})\, e^{i\,\widetilde{\delta}} \end{pmatrix} = (-1)^q \begin{pmatrix} \cos(\alpha) \\ -\sin(\alpha)(-e^{i\,\delta}) \end{pmatrix} = (-1)^q\, A_{in} \tag{24}$$

This means that there is no ambiguity.

Once the offsets and the power needed to produce a 2π phase shift are known, the applied phase shifts (φ_1, φ_2) can be computed.

Next, exploiting the device periodicity, we restrict to the first period (which on the (ϕ_1, ϕ_2) plane is a square of side 2π with one vertex on the origin), and consider the reminders $\widehat{\varphi}_1, \widehat{\varphi}_2$ modulo 2π:

$$\widehat{\varphi}_1 \equiv \varphi_1 \mod 2\pi \qquad \widehat{\varphi}_2 \equiv \varphi_2 \mod 2\pi \tag{25}$$

It is possible to figure out to which class belongs the solution at hand; in fact, considering that 2α ranges from 0 to π, we immediately conclude that:

$$\begin{aligned} \widehat{\varphi}_2 &= 2\pi - 2\alpha \quad \in (\pi, 2\pi) \quad \text{if} \quad \varphi_2 = \phi_2' = -2\alpha + 2n\pi \\ \widehat{\varphi}_2 &= \quad\; 2\alpha \quad \in (0, \pi) \quad \text{if} \quad \varphi_2 = \phi_2'' = \;\; 2\alpha + 2n\pi \end{aligned} \tag{26}$$

In the first case, the remainder of the first phase equals the longitude δ:

$$\widehat{\varphi}_1 = \delta \tag{27}$$

while, in the second, we need to further distinguish between the lower and upper half. In fact, it is:

$$\phi_1'' = \delta + (2m+1)\pi \quad \rightarrow \quad \widehat{\varphi}_1 = \begin{cases} \delta - \pi & \text{if } \in (0, \pi) \\ \delta + \pi & \text{if } \in (\pi, 2\pi) \end{cases} \tag{28}$$

The situation is shown in Figure 10.

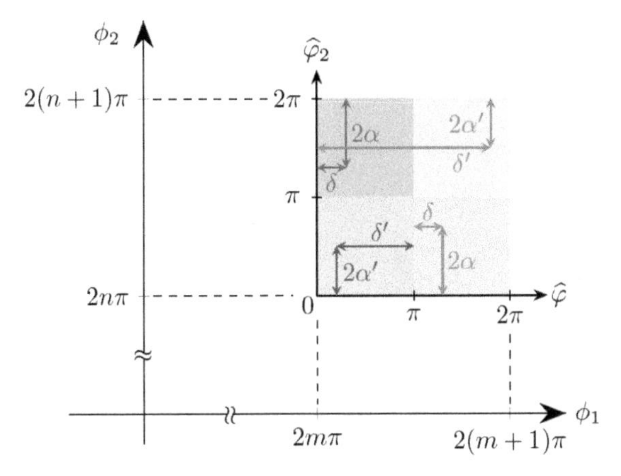

Figure 10. Diagram for the determination of the input SOP: ϕ_1 and ϕ_2 are the actual applied phase shifts, whereas $\widehat{\varphi_1}$ and $\widehat{\varphi_2}$ are the remainders; the new origin is placed in $(2m\pi, 2n\pi)$. The top quadrants (shaded in red and yellow) correspond to the first solution, (ϕ_1', ϕ_2'), while the lower ones (blue and green) to the second solution. Polarisations with the same value of δ lie in diametrically opposed quadrants: the red and green concern the case of $\delta \in (0, \pi)$, whilst the yellow and blue ones to $\delta \in (\pi, 2\pi)$.

Now that univocal behaviour has been proven, we can take advantage of Equation (21) to determine the SOP. It tells that

$$P_L(\phi_1, 2n\pi) - P_L(\phi_1, (2n+1)\pi) = \cos^2\alpha - \sin^2\alpha = \cos 2\alpha \qquad \frac{P_L(\phi_1, (2n+1)\pi)}{P_L(\phi_1, 2n\pi)} = \tan^2\alpha \tag{29}$$

i.e., α can be extrapolated from two measurements, e.g., at $\phi_2 = 0$ and π, up to its sign

$$2\alpha = \pm \arccos\left(P_L(\phi_1, 0) - P_L(\phi_1, \pi)\right) \qquad \alpha \approx \pm\sqrt{\frac{P_L(\phi_1, \pi)}{P_L(\phi_1, 0)}} \tag{30}$$

The first expression is more accurate when α is around 0 or $\pi/2$ ($P_L(\phi_1, 0) \gg P_L(\phi_1, \pi)$ or vice versa) but becomes sensitive to measurement errors for $\alpha \approx \pi/4$ ($P_L(\phi_1, 0) \approx P_L(\phi_1, \pi)$), while the second expression has a complimentary behaviour.

Next step is to set ϕ_2 to this value and repeat the previous procedure on ϕ_1, getting:

$$
\begin{array}{lll}
\phi_2 = 2\alpha & \phi_1 = 0 & P_L(0,2\alpha) = \cos^2 2\alpha + \sin^2 2\alpha \, \sin^2(\delta/2) \\
 & \phi_1 = \pi & P_L(\pi,2\alpha) = \cos^2 2\alpha + \sin^2 2\alpha \, \cos^2(\delta/2) \\
\phi_2 = -2\alpha & \phi_1 = 0 & P_L(0,-2\alpha) = \cos^2 2\alpha + \sin^2 2\alpha \, \cos^2(\delta/2) \\
 & \phi_1 = \pi & P_L(\pi,-2\alpha) = \cos^2 2\alpha + \sin^2 2\alpha \, \sin^2(\delta/2)
\end{array} \quad (31)
$$

In conclusion,

$$
P_L(0,2\alpha) - P_L(\pi,2\alpha) = -\sin^2 2\alpha \, \cos\delta \qquad P_L(0,-2\alpha) - P_L(\pi,-2\alpha) = \sin^2 2\alpha \, \cos\delta \quad (32)
$$

This allows finding δ. As with Equation (30), this formula is accurate for $P_L(0,\pm 2\alpha) \gg P_L(\pi,\pm 2\alpha)$ or vice versa, corresponding to δ around 0 or π and $2\alpha \approx \pi/2$; however, it is vulnerable when $P_L(0,\pm 2\alpha) \approx P_L(\pi,\pm 2\alpha)$.

Noticing that Equation (31) can be expressed as

$$
\begin{array}{ll}
P_L(0,2\alpha) = 1 - [\sin 2\alpha \, \cos(\delta/2)]^2 & P_L(0,-2\alpha) = 1 - [\sin 2\alpha \, \sin(\delta/2)]^2 \\
P_L(\pi,2\alpha) = 1 - [\sin 2\alpha \, \sin(\delta/2)]^2 & P_L(\pi,-2\alpha) = 1 - [\sin 2\alpha \, \cos(\delta/2)]^2
\end{array} \quad (33)
$$

one can use

$$
\tan^2(\delta/2) = \frac{1 - P_L(\pi,2\alpha)}{1 - P_L(0,2\alpha)} \quad \text{or} \quad \tan^2(\delta/2) = \frac{1 - P_L(0,-2\alpha)}{1 - P_L(\pi,-2\alpha)} \quad (34)
$$

that are better when $P_L(0,\pm 2\alpha) \approx P_L(\pi,\pm 2\alpha)$.

Both Equations (32) and (34) have issues for $\alpha = 0$ or $\pi/2$, but this is not a problem, in that δ is not well defined around those values (Figure 6).

2.2. Controller

If the device is used in the other direction, i.e., sending a lightwave in the lower arm and letting it exit from the 2D grating coupler, as depicted in Figure 11, then it can be used to control the polarisation state of output wave, that is to say, as a polarisation controller.

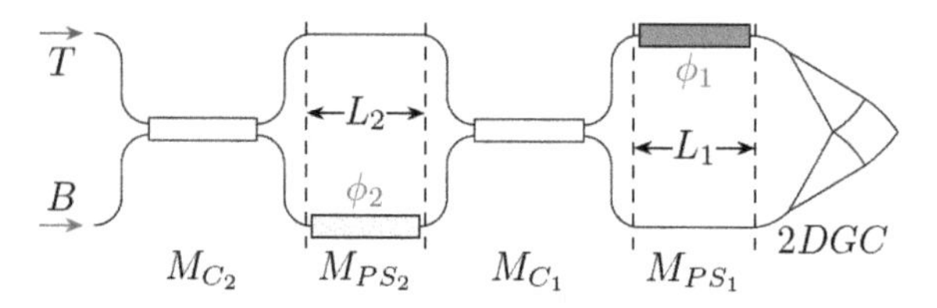

Figure 11. Schematic of the integrated polarisation controller. The labels are as in Figure 1.

The overall transfer matrix is the adjoint of the one in Equation (2):

$$
\bar{T} = M_{PS_1}^{\dagger} \, M_{C_1}^{\dagger} \, M_{PS_2}^{\dagger} \, M_{C_2}^{\dagger} = \left(M_{C_2} M_{PS_2} M_{C_1} M_{PS_1}\right)^{\dagger} = T^{\dagger} = \frac{e^{i\psi}}{2} \begin{bmatrix} e^{-i\phi_1}\left(1 - e^{-i\phi_2}\right) & i\,e^{-i\phi_1}\left(1 + e^{-i\phi_2}\right) \\ i\left(1 + e^{-i\phi_2}\right) & -\left(1 - e^{-i\phi_2}\right) \end{bmatrix} \quad (35)
$$

The input vector will always be of the form:

$$
A_{in} = a \begin{pmatrix} 0 \\ 1 \end{pmatrix} \quad (36)
$$

Thus, the output polarisation is:

$$A_{out} = \tilde{T} A_{in} = a\,\frac{e^{i\,\psi}}{2}\begin{pmatrix} i\,e^{-i\,\phi_1}\,(1+e^{-i\,\phi_2}) \\ -\quad (1-e^{-i\,\phi_2}) \end{pmatrix} = \begin{pmatrix} a_1\,e^{i\,\delta_1} \\ a_2\,e^{i\,\delta_2} \end{pmatrix} \tag{37}$$

Since the device is assumed to be lossless, power is conserved:

$$||A_{out}||^2 = s_0 \doteq a_1^2 + a_2^2 = a^2 \tag{38}$$

The components amplitude and phase equal:

$$\begin{aligned} a_1 &= a\sqrt{\frac{1+\cos\phi_2}{2}} & \delta_1 &= \frac{\pi}{2} - \phi_1 + \arctan\left(\frac{-\sin\phi_2}{1+\cos\phi_2}\right) \\ a_2 &= a\sqrt{\frac{1-\cos\phi_2}{2}} & \delta_2 &= \pi + \arctan\left(\frac{\sin\phi_2}{1-\cos\phi_2}\right) \end{aligned} \tag{39}$$

Consequently, the phase difference is

$$\delta = \delta_2 - \delta_1 = \phi_1 + \frac{\pi}{2}\,[1 + \mathrm{sgn}\,(\sin\phi_2)] \tag{40}$$

The *Stokes parameters* (Equation (11)) can be explicitly derived as:

$$\begin{aligned} s_1 &= s_0\,\cos(-\phi_2) \\ s_2 &= s_0\,\sin(-\phi_2)\,\cos\phi_1 \\ s_3 &= s_0\,\sin(-\phi_2)\,\sin\phi_1 \end{aligned} \tag{41}$$

It is convenient to take the minus sign in front of ϕ_2, as the phase shifter is assumed to be placed in the lower arm. However, when compared with their usual form (1.4.2 [13]),

$$\begin{cases} s_1 = s_0\,\cos 2\chi\,\cos 2\psi \\ s_2 = s_0\,\cos 2\chi\,\sin 2\psi \\ s_3 = s_0\,\sin 2\chi \end{cases} \tag{42}$$

it is evident (cf. Figure 12) that the axes undergo the cyclic permutation and that the new and old angles are connected by the relation (which is not the only possible solution):

$$\begin{aligned} s_1 &\to s_2 & \\ s_2 &\to s_3 & 2\chi &\to \pi/2 - (-\phi_2) \\ & & 2\psi &\to \phi_1 \\ s_3 &\to s_1 & \end{aligned} \tag{43}$$

In the case at hand, it is more convenient to refer the polar and azimuthal angles not to the $s_1 - s_2$ plane, as usual, but to the s_1 axis and the $s_2 - s_3$ plane, respectively. The situation is depicted in Figure 12a.

Another important remark is that there are again, as it is expected thanks to reciprocity, two solution sets, which result in the same point, as in Equation (9):

$$\begin{aligned} s_1 &= \cos(-\varphi_2') & &= \cos(-\varphi_2'') & &= \cos 2\alpha \\ s_2 &= \sin(-\varphi_2')\,\cos\varphi_1' & &= \sin(-\varphi_2'')\,\cos\varphi_1'' & &= \sin 2\alpha\,\cos\delta \\ s_3 &= \sin(-\varphi_2')\,\sin\varphi_1' & &= \sin(-\varphi_2'')\,\sin\varphi_1'' & &= \sin 2\alpha\,\sin\delta \end{aligned} \tag{44}$$

In practice, when a given SOP corresponding to a point on Poincaré sphere is to be generated, the phases to be applied are, considering both solution sets:

$$\phi_1 = \arctan\left(\frac{s_3}{s_2}\right) - m\pi \qquad\qquad \phi_2 = -(-1)^m \arccos\left(\frac{s_1}{s_0}\right) + 2n\pi \tag{45}$$

(to be compared with Equation (9)).

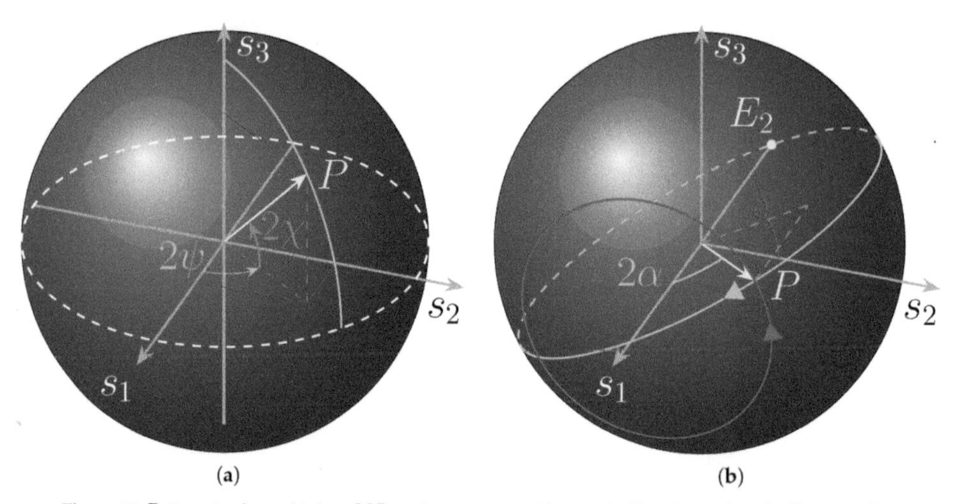

Figure 12. Poincaré sphere. (**a**) Any SOP can be represented by a point P on its surface, by Equation (42). The usual spherical coordinates are used. (**b**) Poincaré sphere with the angles as in Equation (41). The point P corresponds to a generic polarisation state. The red and green circles display the trajectory on the sphere when a complete sweep is performed on the angle δ (ϕ_1) and α (ϕ_2), respectively, while the other is held constant. Notice the direction of the arrows on the circles: scanning on δ results in a counterclockwise rotation, whereas the opposite happens for α, as the phase shifter is assumed to be in the lower arm.

2.2.1. Properties of the Poincaré Sphere

At this point, it must be recalled that the two forms of Stokes parameters in Equations (41) and (42) refer to the observer (on our case, the 2DGC) and the polarisation ellipse frames, respectively.

The flexibility offered by this method of representation provides a physical and insightful point of view particularly adapted to solve graphically otherwise cumbersome algebraic systems. In addition, let us point out that, as in [14,15], this representation using a Bloch sphere can be conveniently used for a quantum treatment of the device as there is a direct correspondence from Stokes parameters to the density operator. In the observer frame, the polarisation ellipse with semi axes a and b is tilted by an angle ψ with respect to the x axis and is inscribed inside a rectangle of sides a_1 and a_2 (Figure 13). The vertical component of the Jones vector A_{out} is phase shifted by an angle δ (Equation (40)).

The angles α and χ are defined as:

$$\tan\alpha = \frac{a_2}{a_1} \qquad\qquad \tan\chi = \mp\frac{b}{a} \tag{46}$$

They are connected to the aspect ratio of the black and green rectangles in Figure 13. The angle χ is usually called ellipticity and α is termed auxiliary angle.

The Jones vector in the observer frame can be expressed as

$$A = \begin{pmatrix} \cos\alpha \\ \sin\alpha \, e^{i\delta} \end{pmatrix} \tag{47}$$

with respect to the basis given by the horizontal and vertical linear polarisations (TE and TM or H and V) usually used in quantum optics.

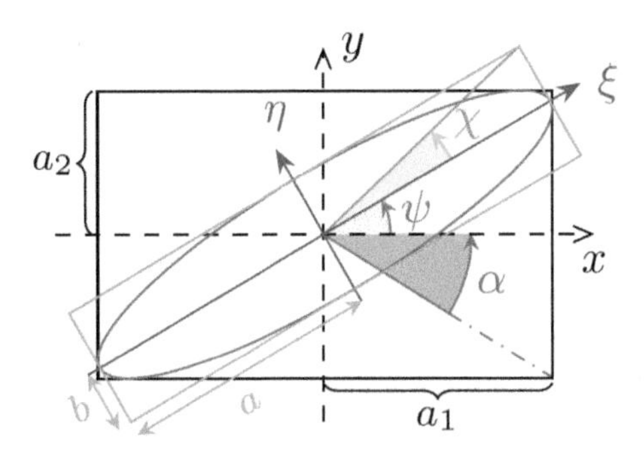

Figure 13. Elliptically polarised wave seen in the observer (xOy) and in the ellipse ($\xi O\eta$) frame. The vibrational ellipse is for the electric field. The semi-major axis ξ is tilted from x by the angle ψ. The semi- axes length are a and b, not to be confused with the amplitudes a_1 and a_2 of the oscillations expressed in xOy.

The following relations (1.4.2 [13]) hold between the angles pairs in the two coordinates systems in Figure 12:

$$\begin{cases} \tan 2\psi = \tan 2\alpha \;\cos\delta \\ \sin 2\chi = \sin 2\alpha \;\sin\delta \end{cases} \tag{48}$$

About Poincaré sphere, most textbooks make clear that "parallels" and "meridians" with respect to the s_3 axis, i.e., points with the same value of χ or ψ, correspond to SOPs with the same ellipticity or tilt angle with respect to the observer's x axis, respectively.

If instead the other coordinates are considered, then points with the same polar angle 2α referred to s_1 correspond to ellipses inscribed in the same rectangle with sides a_1 and a_2. Instead, points with the same "longitude" δ are associated to polarisation states with the same phase shift value between their two components, as seen in the xOy frame.

Notice that the share of power of a SOP given by Equation (47) that is let pass by an analyser whose principal state is, for instance, H, is

$$P_{A-H} = (A \cdot H)^2 = \cos^2\alpha \tag{49}$$

that is basically Malus' law in terms of Jones vectors. Considering that the dot product between the corresponding Stokes vectors (excluding s_0) is $\vec{s}_A \cdot \vec{s}_H = \cos 2\alpha$, the same relation can be written

$$P_{A-H} = \frac{\vec{s}_A \cdot \vec{s}_H + 1}{2} \tag{50}$$

The validity of this formula is not limited to the particular SOP basis considered in the derivation, since a change of basis simply rotates the whole sphere. This fact tells us that the SOPs whose points on the sphere are apart by the same angle have the same power coupling.

2.2.2. Generalised Poincar é/Bloch Sphere

For a generic photonic circuit made up of two waveguides, one can associate [16] to the pair of complex amplitudes in the two guides the Stokes parameters as defined in Equation (41).

Therefore, one can associate a point on the Poincaré sphere (also called Bloch sphere when generalised) with a given quadruple corresponding to the given couple of complex amplitudes. As the complex amplitudes vary upon propagation along the circuit, the corresponding point undergoes a rotation (which is the product of the rotations brought about by the several circuit stages).

Note that, in this formulation, a point on the generalised sphere does not correspond to a polarisation state, in that the two fields are generally located in distinct waveguides, whereas for a plane wave (or a field pattern in free space) the two orthogonal field components are in the same place and do overlap.

However, one can reconnect to the polarisation state in this way: for any section of the circuit, to the pair of complex amplitudes in the two waveguides corresponds a certain point on the sphere. If the circuit, in the considered section, were connected to a 2D grating coupler, then the complex amplitudes of TE and TM components would equal (except for the losses) the guided ones, so the point corresponding to the polarisation state of output light would coincide with the one corresponding to the couple of guided fields.

The convenience of this approach lies in the possibility to have a visual representation of each circuit component effect, as it results in the rotation by a certain angle and around a given axis of the entire sphere.

As shown in (p. 67 [12,16]), the effect of a phase shifter is a rotation around the s_1 axis by the differential angle $\Delta\phi$ (however, the sign depends on the adopted convention) and a synchronous coupler provokes a rotation around the s_2 axis by the double of the amplitude coupling κ. For an asynchronous coupler, the rotation axis lies in the $s_1 - s_2$ plane and the rotation angle is given by the same rule as for a synchronous coupler.
In our case, the MMIs produce a $\pi/2$ rotation around s_2.

The states with $s_1 \pm 1$ correspond to all the power in the top and bottom waveguide, respectively, and are labelled as E_1 and E_2 on the figures.

2.2.3. Device Operation

The effect of our circuit is, starting from the point labelled as E_2, to rotate clockwise around s_2 by a right angle, then clockwise (because the second phase shifter in placed on the lower branch) about s_1 by an angle ϕ_2, again clockwise by a right angle around s_2, and eventually counter clockwise by ϕ_1 around s_1, respectively.

The path is travelled backwards in the compensator operation.

This is shown in Figure 14a,b, for the two solution sets.

The starting point E_2 corresponds to the Jones vector in Equation (36), a "vertical" linear polarisation. This point is brought into E_L by the first coupler, then in $P'' = (\pi/2, -\pi/2 \mp 2\alpha)$ by phase shifter ϕ_2, in $P' = (2\alpha, 0/\pi)$ (a "linear" SOP, as $s_3 = 0$, represented in the figures with a yellow dashed circle) by the other coupler and finally in $P = (2\alpha, \delta)$ by ϕ_1.

Another way of explaining the device operation is to consider it as a Mach–Zehnder Interferometer (MZI) followed by the phase shifter ϕ_1. As shown in Appendix B, a MZI behaves as a (non-endless) half-wave plate [10], thus it causes a rotation of π about an axis lying on the $s_1 - s_2$ equatorial plane, with azimuth

$$\Theta = -\phi_2/2 + \pi/2 = \pi/2 \pm \alpha - m\pi \tag{51}$$

with respect to the s_1 axis (the minus sign in front of ϕ_2 is due to the fact that it is applied to the lower waveguide).

Thus, shifter 1 has the role of carrying the SOP from the point P' on the "equator", to its destination point P, as displayed in Figure 14a,b.

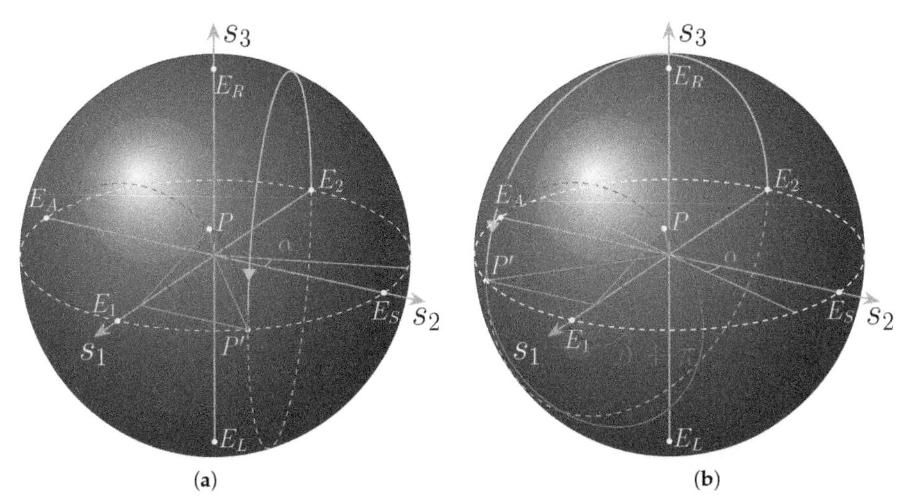

Figure 14. Action on SOP of the circuit, seen as a MZI followed by a phase shifter, for $2\alpha = 40°$, $\delta = 50°$, in the case of (**a**) the first solution set; and (**b**) the second solution set.

Even if the previous description allows a better understanding the device operation, it is of great interest to determine the overall rotation. The following derivation is based on (2.6.2 in [12]). To comply with its notation for phase shifter matrices (Table 2.1 in [12]), it is better to reformulate Equation (35) as (note that the phase shifts are opposite in sign, for the phase shifters are placed on different branches):

$$\tilde{T} = \begin{bmatrix} i\ \sin(\phi_2/2)\ e^{-i\phi_1/2} & i\ \cos(\phi_2/2)\ e^{-i\phi_1/2} \\ i\ \cos(\phi_2/2)\ e^{i\phi_1/2} & -i\ \sin(\phi_2/2)\ e^{i\phi_1/2} \end{bmatrix} =$$
$$= \begin{bmatrix} \cos\left(\frac{\phi_2}{2} - \frac{\pi}{2}\right) e^{-i(\phi_1/2-\pi/2)} & \sin\left(\frac{\phi_2}{2} - \frac{\pi}{2}\right) e^{-i(\phi_1/2+\pi/2)} \\ -\sin\left(\frac{\phi_2}{2} - \frac{\pi}{2}\right) e^{i(\phi_1/2+\pi/2)} & \cos\left(\frac{\phi_2}{2} - \frac{\pi}{2}\right) e^{i(\phi_1/2-\pi/2)} \end{bmatrix} \tag{52}$$

Now, to get the rotation matrix for Stokes' parameters, the above formula must be put in form compliant with the general one for unitary matrices given in (p. 51 in [12]):

$$U = \begin{bmatrix} e^{i\aleph}\ \cos\kappa & -e^{i\beta}\ \sin\kappa \\ e^{-i\beta}\ \sin\kappa & e^{-i\aleph}\ \cos\kappa \end{bmatrix} \tag{53}$$

Comparing the previous two equations, it is clear that:

$$\kappa = -\frac{\phi_2 - \pi}{2} \qquad \aleph = -\frac{\phi_1 - \pi}{2} \qquad \beta = -\frac{\phi_1 + \pi}{2} \tag{54}$$

In the reciprocal case where the device works as a compensator, the sign of κ and $\aleph$ changes, while β retains its sign thanks to transposition.

To determine the global rotation, we insert the parameters of Equation (54) into the general rotation matrix given in (p. 67 [12]), getting:

$$R = \begin{bmatrix} -\cos\phi_2 & \sin\phi_2 & 0 \\ \cos\phi_1\ \sin\phi_2 & \cos\phi_1\ \cos\phi_2 & \sin\phi_1 \\ \sin\phi_1\ \sin\phi_2 & \sin\phi_1\ \cos\phi_2 & -\cos\phi_1 \end{bmatrix} \tag{55}$$

For the compensator operation, the corresponding rotation matrix is the inverse of the one above, namely its transpose.

Substituting the values for the two solution sets:

$$R'^{/''}(\alpha,\delta) = \begin{bmatrix} -\cos(2\alpha) & \mp\sin(2\alpha) & 0 \\ -\cos\delta\,\sin(2\alpha) & \pm\cos\delta\,\cos(2\alpha) & \pm\sin\delta \\ -\sin\delta\,\sin(2\alpha) & \pm\sin\delta\,\cos(2\alpha) & \mp\cos\delta \end{bmatrix}$$

$$R'^{/''}(\vec{s}\,) = \begin{bmatrix} -s_1 & \mp\sqrt{s_2^2+s_3^2} & 0 \\ -s_2 & \pm\frac{s_1\,s_2}{\sqrt{s_2^2+s_3^2}} & \pm\frac{s_3}{\sqrt{s_2^2+s_3^2}} \\ -s_3 & \pm\frac{s_1\,s_3}{\sqrt{s_2^2+s_3^2}} & \mp\frac{s_2}{\sqrt{s_2^2+s_3^2}} \end{bmatrix} \tag{56}$$

One can easily check that $R'^{/''}(\alpha,\delta)$ brings E_2 into P, as it should:

$$R'^{/''}(\alpha,\delta)\begin{pmatrix}-1\\0\\0\end{pmatrix} = \begin{pmatrix}\cos(2\alpha)\\ \sin(2\alpha)\,\cos\delta \\ \sin(2\alpha)\,\sin\delta\end{pmatrix} = \begin{pmatrix}s_1\\s_2\\s_3\end{pmatrix} = P \tag{57}$$

The rotation axis is (Section 9.3.1 in [17]):

$$\vec{\Omega} = \begin{pmatrix}R_{32}-R_{23}\\R_{13}-R_{31}\\R_{21}-R_{12}\end{pmatrix} = A\begin{pmatrix}\tan(\phi_2/2)\\1\\\tan(\phi_1/2)\end{pmatrix} \tag{58}$$

the multiplicative factor A is connected to the vector norm, which, however, is irrelevant. The two solution classes have different rotation axes (Figure 15):

$$\vec{\Omega}' = A'\begin{pmatrix}-\tan(\alpha)\\1\\\tan(\delta/2)\end{pmatrix} \qquad \vec{\Omega}'' = A''\begin{pmatrix}\tan(\alpha)\\1\\-\cot(\delta/2)\end{pmatrix} \tag{59}$$

In general, they are not perpendicular to each other:

$$\vec{\Omega}'\cdot\vec{\Omega}'' \propto -\tan^2\alpha \tag{60}$$

However, each rotation axis is perpendicular to the other axis's projection on the s_2, s_3 plane:

$$\vec{\Omega}'\cdot\vec{\Omega}''_{s_2s_3} = A'\begin{pmatrix}-\tan(\alpha)\\1\\\tan(\delta/2)\end{pmatrix}\cdot A''\begin{pmatrix}0\\1\\-\cot(\delta/2)\end{pmatrix} = 0 \tag{61}$$

The rotation angle is given by (Section 9.3.1 in [17]):

$$\cos(\Gamma) = \frac{\mathrm{tr}(R)-1}{2} = 2\big(\sin(\phi_1/2)\,\sin(\phi_2/2)\big)^2 - 1 \tag{62}$$

Using a trigonometric identity, the formula reduces to:

$$\cos(\Gamma/2) = \pm\sin(\phi_1/2)\,\sin(\phi_2/2) \tag{63}$$

For the two solutions sets, it reads

$$\begin{aligned} \cos(\Gamma'/2) &= \mp \sin(\delta/2)\sin(\alpha) \\ \cos(\Gamma''/2) &= \pm \cos(\delta/2)\sin(\alpha) \end{aligned} \tag{64}$$

The overall circuit effect is shown in Figure 16.

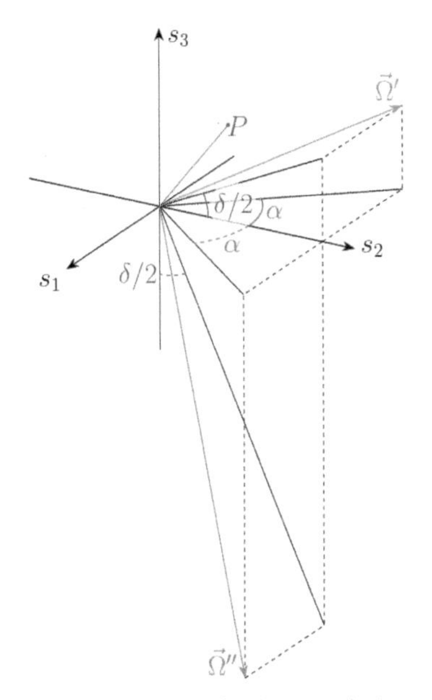

Figure 15. Rotation axes for the two solution sets.

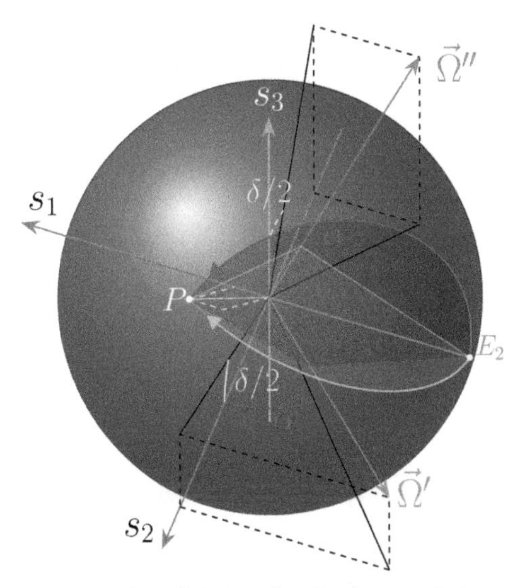

Figure 16. Overall device effect, for the two solutions.

3. Frequency Response

To deduce the light frequency dependence of the device at hand, the first order expansion for the waveguides effective index

$$n_{eff}(\lambda) = n_{eff0}\frac{\lambda}{\lambda_0} - n_{g0}\frac{\lambda - \lambda_0}{\lambda_0} \tag{65}$$

is inserted in the formula for the phase shift:

$$\phi_i = \frac{2\pi}{\lambda}\Delta_{tb}(n\,L_i) = \frac{2\pi\,\Delta_{tb}(n_{eff0})\,L_i}{\lambda_0}\left[1 - \frac{\Delta_{tb}\,n_{g0}}{\Delta_{tb}\,n_{eff0}}\left(1 - \frac{\lambda_0}{\lambda}\right)\right] \tag{66}$$

The different effective indexes of the top and bottom arms of a phase shifter result from their different temperature, since thermo optic effect is exploited in silicon. In Appendix A, it is shown that, at first order, the group index too is proportional to the temperature shift. Then,

$$\frac{\Delta_{tb}\,n_{g0}}{\Delta_{tb}\,n_{eff0}} = \frac{\alpha_g\,\Delta_{tb}(T - T_0)}{\alpha_{eff}\Delta_{tb}(T - T_0)} = \frac{\alpha_g}{\alpha_{eff}} \doteq R_{ge} \tag{67}$$

In conclusion, the phase shift has the dependence

$$\phi_i = \phi_{i_0}\left(1 - R_{ge}\frac{\Delta\lambda}{\lambda}\right) = \phi_{i_0}\left(1 + R_{ge}\frac{\Delta f}{f_0}\right) \tag{68}$$

In the (ϕ_1, ϕ_2) plane, once a given solution at f_0 has been selected, the point $(\phi_1(f), \phi_2(f))$ lies on a line passing through (ϕ_{1_0}, ϕ_{2_0}) and the origin (Figure 17):

$$\vec{\phi}(f) = \begin{pmatrix}\phi_1(f)\\ \phi_2(f)\end{pmatrix} = \begin{pmatrix}\phi_{1_0}\\ \phi_{2_0}\end{pmatrix}\left(1 + R_{ge}\frac{\Delta f}{f_0}\right) = \vec{\phi_0}\left(1 + R_{ge}\frac{\Delta f}{f_0}\right) \tag{69}$$

The "phase speed" or sensitivity to a frequency shift $\mathrm{d}f$

$$\dot{\phi_i} = \frac{\mathrm{d}\phi_i}{\mathrm{d}f} = \phi_{i_0}R_{ge}\frac{1}{f_0} \tag{70}$$

is proportional to the phase shift ϕ_{i_0} applied at the central frequency f_0, hence choosing higher order solutions, which are farther from the origin, entails a larger sensitivity to frequency, that is, a narrower bandwidth.

3.1. Compensator

The several factors appearing in the expression for the intensity surface P_L, Equation (14), are function of the difference

$$\phi_i - \phi_{i_0} = \phi_{i_0}R_{ge}\frac{\Delta f}{f_0} \tag{71}$$

To find the Free Spectral Range (FSR), i.e., the frequency spacing—if any—between two consecutive peaks of the intensity surface, the above difference should be an integer multiple of 2π, for both phase shifters. Unfortunately, since the two central frequency phase shifts are in general different, there are two different FSRs (if it makes any sense):

$$\Delta f_i = \frac{f_0}{R_{ge}}\frac{2m_i\pi}{\phi_{i_0}} \tag{72}$$

It is possible to define an FSR just in the particular case when ϕ_{1_0} and ϕ_{2_0} are commensurable

$$\frac{\phi_{2_0}}{\phi_{1_0}} = \frac{p}{q} \Rightarrow \qquad FSR = \frac{f_0}{R_{ge}} \frac{2p\pi}{\phi_{2_0}} = \frac{f_0}{R_{ge}} \frac{2q\pi}{\phi_{1_0}} \tag{73}$$

where p and q are relatively prime integers.

3.1.1. Contour Curves

In the following (Section 3.1.2), we consider contour curves, in particular the one for which $P_L(\phi_1, \phi_2) = 1/2$.

Let us consider the general case of a contour curve for the level b:

$$\cos^2(\phi_2/2 + \alpha)\, \cos^2\left(\frac{\phi_1 - \delta}{2}\right) + \cos^2(\phi_2/2 - \alpha)\, \sin^2\left(\frac{\phi_1 - \delta}{2}\right) = b \tag{74}$$

After applying some trigonometric identities, we arrive to an expression for ϕ_1 as a function of ϕ_2:

$$\cos(\phi_1 - \delta) = \frac{1 - 2b}{\sin\phi_2 \cdot \sin 2\alpha} + \cot\phi_2 \cdot \cot 2\alpha \tag{75}$$

3.1.2. Bandwidth

The -3 dB bandwidth is given by the equation

$$P_L\Big(\phi_1(f), \phi_2(f)\Big) = 1/2 \tag{76}$$

In Section 3.1.1, an expression for the contour curves of level b was found; in this particular case of $b = 1/2$, one term vanishes and the formula becomes:

$$\cos(\phi_1(f) - \delta) = \cot\phi_2(f) \cdot \cot 2\alpha \tag{77}$$

Depending on the solution set

$$\cos\left(\phi_1(f) - \delta\right) = \pm\cos\left(\phi_{1_0}^{\prime/\prime\prime} R_{ge} \frac{\Delta f}{f_0}\right) \qquad \cot\phi_{2_0}^{\prime/\prime\prime} = \mp\cot 2\alpha \tag{78}$$

The bandwidth $\Delta f_{3\,\mathrm{dB}}$ is given by the implicit equation

$$\cos\left(\phi_{1_0}^{j} R_{ge} \frac{\Delta f}{f_0}\right) = -\cot\phi_{2_0}^{j} \cdot \cot\left[\phi_{2_0}^{j}\left(1 + R_{ge}\frac{\Delta f}{f_0}\right)\right] \tag{79}$$

where j indicates the solution class.

Graphically, the equation corresponds to finding the frequency shifts for which the lines corresponding to the phase pair intersect the contour curves for the level $1/2$, which lie closest to the considered solution, as in Figure 17.

The bandwidth can be visualised as the distance between two such intersections, divided by the solution distance from the origin (because the "phase speed" is proportional to it, see Equation (70)) and multiplied by f_0.

3.1.3. Spectrum

The spectrum depends on the input SOP and is given by the expression for P_L when the frequency dependence as in Equation (68) is included

$$P_L(f) = \cos^2(\phi_2/2 + \alpha)\, \cos^2\left(\frac{\phi_1 - \delta}{2}\right) + \cos^2(\phi_2/2 - \alpha)\, \sin^2\left(\frac{\phi_1 - \delta}{2}\right) \tag{80}$$

For a better insight, the spectrum is found from the intersection with the intensity surface of the plane perpendicular to the (ϕ_1, ϕ_2) plane and passing by the line traced by $\vec{\phi}(f)$ as in Equation (69). Figure 18 displays the spectra for the same situation as in Figure 17. The relation between spectrum shapes and the corresponding intensity surface for the considered SOP would be more apparent if a linear, rather than logarithmic, vertical scale had been plotted against frequency, instead of wavelength.

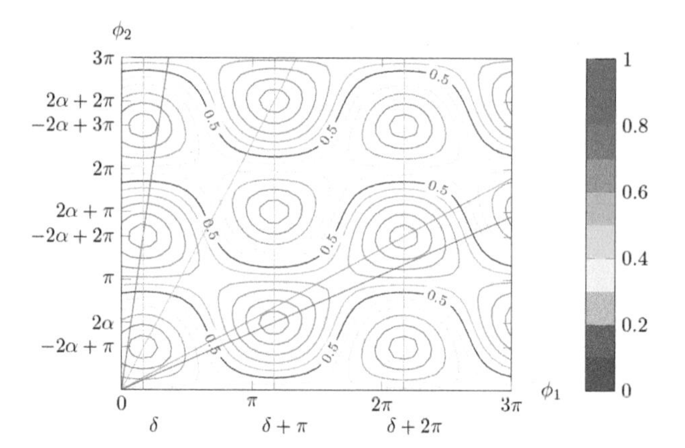

Figure 17. The pair of phase shifts as a function of frequency are lines (in different colours for each solution, refer to the legend of Figure 18) passing by the origin and the chosen peak, in the plane (ϕ_1, ϕ_2). For comparison, the contour plots for $\alpha = 55°$ and $\delta = 30°$ are included. Note that those lines in general do not pass through other maxima.

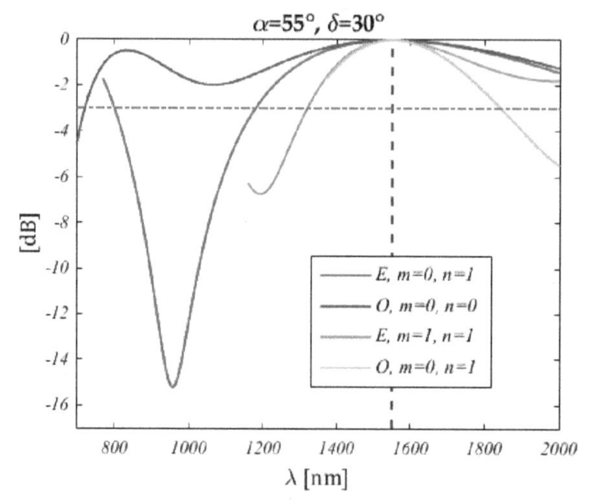

Figure 18. Spectra corresponding to the lines in Figure 17 (the colours correspond), for $f_0 = 1934$ THz (1550 nm, indicated by the black dashed vertical line). The wavelength ranges have been chosen to represent only the portion of the (ϕ_1, ϕ_2) plane shown in Figure 17. Note that the upper wavelength limit is the same, whereas the lower one increases with the solution distance from the origin. In general, the spectra are neither symmetric nor periodic. The legend lists the solution class (even or odd) and their order. The fundamental even solution is not reported, as it lies in the lower (ϕ_1, ϕ_2) half plane.

3.1.4. Effect of Unbalance

Besides the temperature difference between the two phase shifters in a given stage, differences in the applied phase shifts may arise because of differences in any of the following waveguide properties:

- length
- width
- thickness
- material composition (e.g., doping)

All of those effects are permanent, that is to say, are present even if no external control signal is applied to the circuit, thus resulting in an offset, which must be compensated, in order to apply the correct phase shift.

From the modelling perspective, just length asymmetry appears explicitly, while the other three parameters give rise to variations in quantities such as n_{eff0}, n_{g0}, α_{eff} and α_g. For the sake of simplicity, the last two contributions have been neglected.

Taking the top branch as the unbalanced one

$$\begin{aligned} L_b &= L & L_t &= L_b + \Delta l = L + \Delta l \\ n_{eff0_b} &= n_{eff0} & n_{eff0_t} &= n_{eff0} + \Delta n_{eff0} \\ n_{g0_b} &= n_{g0} & n_{g0_t} &= n_{g0} + \Delta n_{g0} \end{aligned} \tag{81}$$

the phase shift from Equation (66) becomes

$$\phi = \phi_0 + p\,\Delta f / f_0 \tag{82}$$

where

$$\begin{aligned} \phi_0 &= \frac{2\pi}{\lambda_0}\left\{ n_{eff0}\Delta l + \Delta n_{eff0} L + \alpha_{eff}\left[(T_t - T_b)\, L + (T_t - T_0)\Delta l \right] \right\} \\ p &= \frac{2\pi}{\lambda_0}\left\{ n_{g0}\Delta l + \Delta n_{g0} L + \alpha_g \left[(T_t - T_b)\, L + (T_t - T_0)\Delta l \right] \right\} \end{aligned} \tag{83}$$

while, in the balanced case, both ϕ_0 and p are proportional to ΔT (thus, one could solve for ΔT from ϕ_0 and then substitute to find the slope p), now there are two unknown variables: the two shifters' temperatures.

Nevertheless, this is no problem, in that we can consider T_b as a free parameter and express T_t as a function of it:

$$T_t = \frac{\phi_0 \frac{\lambda_0}{2\pi} - (n_{eff0}\Delta l + \Delta n_{eff0} L) + \alpha_{eff}(T_b L + T_0 \Delta l)}{\alpha_{eff}(L + \Delta l) + \Delta\alpha_{eff} L} \tag{84}$$

Thus, the slope is:

$$\begin{aligned} & p = \phi_0\, R_{ge} + \Lambda_{ge}\, \Delta l + \Gamma_{ge} L = s_0 + \Delta s \\ & \Lambda_{ge} \doteq \frac{2\pi}{\lambda_0}\left[n_{g0} - \frac{\alpha_g}{\alpha_{eff}} n_{eff0} \right] \qquad \Gamma_{ge} \doteq \frac{2\pi}{\lambda_0}\left[\Delta n_{g0} - \frac{\alpha_g}{\alpha_{eff}} \Delta n_{eff0} \right] \end{aligned} \tag{85}$$

The first factor appearing in the formula for p is the one in the ideal case or p_0, while the other two come from asymmetry, and can be gathered in a Δp term.

Considering the relation between the two phases,

$$\phi_2 = \phi_{2_0} + p_2 \frac{\Delta f}{f_0} = \phi_{2_0} + \frac{p_2}{p_1}(\phi_1 - \phi_{1_0}) \tag{86}$$

we see that the phase pair still traces a line in the (ϕ_1, ϕ_2) plane as a result of frequency shifts, but in general this line does not pass by the origin, as in general the ratio in front of $\phi_1 - \phi_{1_0}$ differs from ϕ_{2_0}/ϕ_{1_0}.

3.1.5. Bandwidth Variation

To find the bandwidth in this non-ideal case, the expression for the phase in Equation (82) is replaced in Equation (79)

$$\cos\left(p_1 \frac{\Delta f}{f_0}\right) = \mp \cot 2\alpha \cdot \cot\left(\phi_{2_0}^j + p_2 \frac{\Delta f}{f_0}\right) \tag{87}$$

As a shorthand notation, the term B_{id} is used in place of $\Delta f / f_0$ in the ideal, balanced case, whereas ΔB stands for the variation of B_{id}.

In the real case, Equation (79) can be rewritten as:

$$\begin{aligned} &\cos\left[(p_{1_0} + \Delta p_1)(B_{id} + \Delta B)\right] = \\ &= -\cot \phi_{2_0}^j \cdot \cot\left[\phi_{2_0}^j + (p_{2_0} + \Delta p_2)(B_{id} + \Delta B)\right] \end{aligned} \tag{88}$$

Neglecting higher order terms, one gets

$$\Delta B \approx -\frac{\sin(p_{1_0} B_{id}) \Delta p_1 + \cot \phi_{2_0}^j \csc^2\left(\phi_{2_0}^j + p_{2_0} B_{id}\right) \Delta p_2}{\sin(p_{1_0} B_{id}) p_{1_0} + \cot \phi_{2_0}^j \csc^2\left(\phi_{2_0}^j + p_{2_0} B_{id}\right) p_{2_0}} B_{id} \tag{89}$$

It tells us that the effect of imbalance is not necessarily detrimental, i.e., to reduce the bandwidth, provided that the ratio in front of B_{id} is negative. However, since Equation (89) is highly dependent on the input SOP, there is no easy trend that can be estimated and the only conclusion is that the bandwidth is highly dependent on the input.

3.2. Controller

Putting the frequency dependence of the phase shifts given by Equation (68) into the formula for the Stokes parameters of the SOP coming out of the circuit, we arrive to

$$\begin{aligned} s_1 &= s_0 & & \cos\left[-\phi_{2_0}\left(1 + R_{ge}\frac{\Delta f}{f_0}\right)\right] \\ s_2 &= s_0 & \cos\left[\phi_{1_0}\left(1 + R_{ge}\frac{\Delta f}{f_0}\right)\right] & \sin\left[-\phi_{2_0}\left(1 + R_{ge}\frac{\Delta f}{f_0}\right)\right] \\ s_3 &= s_0 & \sin\left[\phi_{1_0}\left(1 + R_{ge}\frac{\Delta f}{f_0}\right)\right] & \sin\left[-\phi_{2_0}\left(1 + R_{ge}\frac{\Delta f}{f_0}\right)\right] \end{aligned} \tag{90}$$

Given that the two phases ϕ_1 and ϕ_2 are proportional to each other, the motion of the SOP on Poincaré sphere can be described by a single variable Θ, for instance taken equal to ϕ_1, so that the previous equations become:

$$\begin{aligned} s_1 &= s_0 & & \cos(m\,\theta) \\ s_2 &= s_0 & \cos(\theta) & \sin(m\,\theta) \\ s_3 &= s_0 & \sin(\theta) & \sin(m\,\theta) \end{aligned} \tag{91}$$

and the ratio

$$m = -\phi_{2_0}/\phi_{1_0} \tag{92}$$

acts as a parameter. This family of curves is named Clélie and some examples are shown in Figure 19a for several values of m. Its projection on the $s_2 - s_3$ plane is a plane curve called rhodonea or rose, with polar equation

$$\rho = |\sin(m\theta)| \tag{93}$$

Polarisations whose Stokes' parameters are such that:

$$2\alpha/\delta = -m \tag{94}$$

lie on the same Clélie; the result of a frequency shift is to move the point on such curve.

Nonetheless, because of the existence of a countable infinity of phase shift pairs, which yield the same SOP, m takes on different values for the same polarisation, which means that there are many curves passing through a given point on the sphere.

Then, depending on the chosen solution, the point would follow a different path, on the corresponding curve.

If the curve is expressed as in Equation (91), then problems would arise for $\phi_{1_0} = 0$ (linear polarisations) or for values close to it, as m would diverge. This issue can be removed, however; in fact, it suffices to redefine said equation as

$$\begin{aligned} s_1 &= s_0 & & \cos(\vartheta) \\ s_2 &= s_0 & \cos(\vartheta/m) & \sin(\vartheta) \\ s_3 &= s_0 & \sin(\vartheta/m) & \sin(\vartheta) \end{aligned} \tag{95}$$

with $\vartheta = -\phi_2$. This second situation is shown in Figure 19b.

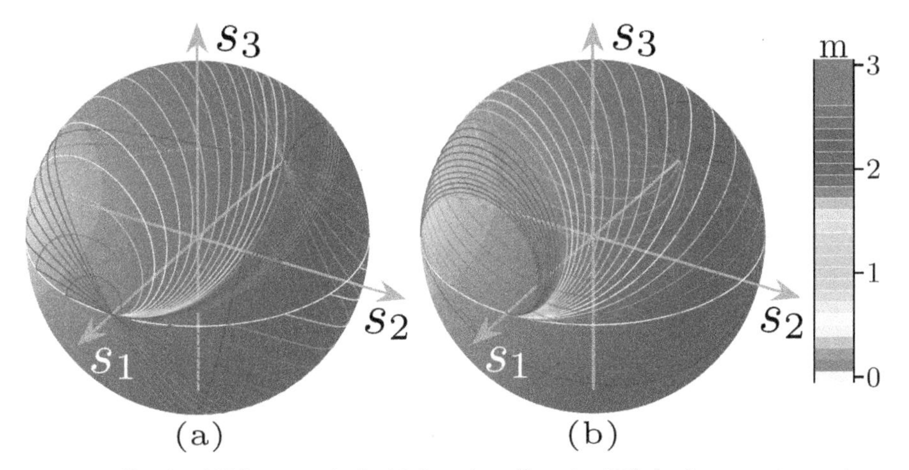

Figure 19. Family of Clélie curves, in the (**a**) direct form (Equation (91)), for the parameter ranging from $m = 0$ to 3, at steps of 1/9. Note that curves with different values of the parameter may pass by the same point, corresponding to different order solutions that have a different frequency behaviour. (**b**) The inverse form (Equation (95)), for the same parameter values of (**a**), in growing order from red to violet. The equator, in yellow, corresponds to $m = 0$, while in the previous figure it corresponds to $m \to \infty$.

3.2.1. Bandwidth

For a given central frequency SOP with Stokes parameters $(s_1, s_2, s_3)_0$, the 3 dB optical bandwidth can be visualised as the frequency excursion Δf necessary to reach a perpendicular (Section 2.2.1) Stokes vector $\vec{s}_{3\,\text{dB}}$ (from Equation (41)) lying on the same Clélie:

$$\vec{s}_{3\,\text{dB}} \cdot \vec{s}_0 = 0 \tag{96}$$

When expanded, the above relation becomes

$$\cos 2\alpha \, \cos\phi_2 - \sin 2\alpha \, \sin\phi_2 \, \cos(\phi_1 - \delta) = 0 \tag{97}$$

that is essentially Equation (79). Using Equation (50), the above procedure can be generalised to find the contour lines as in Section 3.1.1.

3.2.2. Effect of Unbalance

In the case of unbalanced shifters, as shown in Section 3.1.4, the ratio between the two phases is no longer independent from frequency, thus the above treatment is no longer valid. However, there is still a linear (or rather affine) relationship between the two phases (Equation 86)

$$\phi_2 = \phi_{2_0} + \frac{s_2}{s_1}(\phi_1 - \phi_{1_0}) = -r\phi_1 + (\phi_{2_0} + r\phi_{1_0}) = -r\phi_1 - \gamma_0 \tag{98}$$

where r is the ratio between the slopes of the two phases and in general differs from $m = -\phi_{2_0}/\phi_{1_0}$, since

$$r \doteq -\frac{s_2}{s_1} = -\frac{s_{2_0} + \Delta s_2}{s_{1_0} + \Delta s_1} \neq -\frac{s_{2_0}}{s_{1_0}} = -\frac{\phi_{2_0}}{\phi_{1_0}} = m \tag{99}$$

The evolution with frequency of the SOP can again be described with a single parameter.

$$\begin{aligned} s_1 &= s_0 & & \cos(r\,\theta + \gamma_0) \\ s_2 &= s_0 & \cos(\theta) \; & \sin(r\,\theta + \gamma_0) \\ s_3 &= s_0 & \sin(\theta) \; & \sin(r\,\theta + \gamma_0) \end{aligned} \tag{100}$$

The curve remains a Clélie, just rotated by an angle γ_0 around the s_1 axis and with a different parameter r instead of m. While previously phase pairs lying on the same line passing by the origin of the (ϕ_1, ϕ_2) plane corresponded to points on the same Clélie on Poincaré sphere, now the same is true for points situated on the line given by Equation (99).

4. Characterisation

The predictions of the above treatment were tested using the setup in Figure 20: the SOP of the laser beam was matched to the SPGC one with a fibre polarisation controller (FPC). The light from the device under test (DUT) is sent to a −10 dB fibre beam splitter. A tenth of the power went to the power meter, to capture the spectrum, while the rest was sent to the polarimeter. Another FPC was placed in front of the polarimeter for calibration purposes.

The DUT was used in the controller configuration only because it was faster to measure many SOPs, as it was not necessary to manually act on the first FPC, which would involve several trials and errors before reaching the desired SOP to feed to the DUT.

The DUT was a test structure from the second version of the miniROADM [7], substantially with the same properties as that in [5], except for a balanced arrangement, with dummy heaters on the unused arms for a broader optical bandwidth, limited just by the GCs. Moreover, the top output was also connected to a GC, while in [5] it was terminated by a monitor photodiode. Finally, −20 dB waveguide taps were coupled to each branch, for easier characterisation.

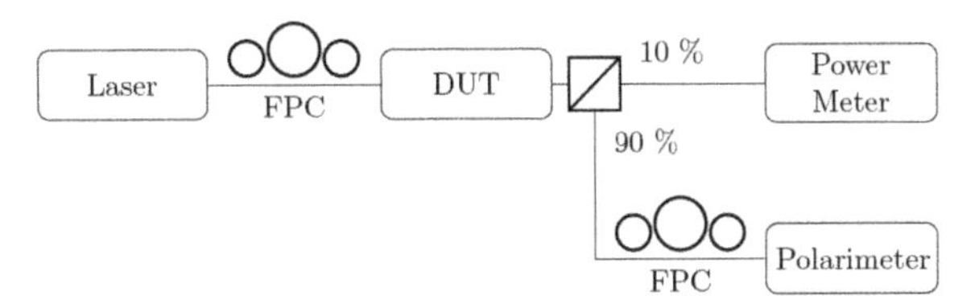

Figure 20. Setup used for the characterisation of the controller configuration. FPC, fibre polarisation controller; DUT, device under test.

Both versions were fabricated by commercial CMOS foundries, IMEC for the one of [5,7] and CMC-IME for the second one, respectively.

The hypothesis described in Section 1.1 was checked, as a preliminary stage, on a test structure consisting in a 2DGC whose outputs were both terminated by a SPGC. It was found that the SOPs that maximise the output power at either port were orthogonal to each other and that the transmission spectra corresponding to those SOPs overlapped. This confirmed that the 2DGC did not exhibit any significant Polarisation Dependent Loss (PDL).

The calibration was performed to have the polarimeter to display the actual SOP on the 2DGC, resulting in a reading consistent with Figures 12b and 19. Indeed, even with a correctly calibrated instrument, it would display a different SOP from that on the 2DGC, because of the offset introduced by the optical fibres in between and the setup in general. The calibration purpose was to remove said offset.

The effect of a birefringent element is a sphere rotation. To uniquely determine it, two pairs of points (before and after the transformation) are needed. For the first pair, we used the SOP that maximises the power coupled to a SPGC, and the second FPC was adjusted so that the polarimeter displayed a linear horizontal SOP (H, $s_1 = 1$). This step was performed on a waveguide clip, terminated with SPGCs at both ends, a test structure usually included in evaluation chips to measure GC insertion loss (IL) and waveguide propagation losses. The SOP that maximised the power from the top output of the 2DGC was then made to correspond to a linear polarisation inclined by 45° (D, $s_2 = 1$). Such a calibration should be repeated to compensate the SOP drift due to fibres, which likely caused the visible difference between measured and estimated points on the parallel with $\alpha = 30°$ in Figure 21.

In the first set of measurements, the Stokes parameters were read on the polarimeter and compared with the values obtained by putting into Equation (41) the dissipated power on each heater, as measured on the source meter. Given that the phase shift ϕ_n was produced by leveraging silicon thermo optic effect, it was assumed to be proportional to the power P_n dissipated by the heater n:

$$\phi_n = 2\pi\, P_n / P_{2\pi} \tag{101}$$

where $P_{2\pi}$ stands for the thermal efficiency, namely the power required to obtain a 2π phase shift. For data points where just one heater was active ($s_2 = 0$ and $s_3 =$ in Figure 21), this assumption was found to be valid. The above expression was used in Equation (41) to get the Stokes parameters predicted by our model.

The dissipated power and the voltage drop displayed by the power supply does not exactly correspond to that on the heater, especially when both heaters are active, because of the parasitic resistance R_G of the common ground electrode. To model this effect, we considered a star circuit, with R_G at the bottom. Changing the voltage applied to one heater while keeping the other constant results in a variation of the voltage drop on R_G, thus of the current and power on the heater, that should remain unaltered. Thus, R_G caused an electrical crosstalk between the two phase shifters. In the usual case where the voltages V_1 and V_2 applied to heater (resistor) R_1 and R_2, respectively, are both positive, increasing V_2 while keeping V_1 constant results in a reduction of P_1. Applying a current instead of a voltage bias should avoid this issue, because the power supply is in series to the heater. Moreover, it was found that the resistance increased linearly with the dissipated power:

$$R = R_0 + \gamma P \tag{102}$$

We attributed this behaviour to the fact that in a metal the resistivity is proportional to the temperature. The main consequence of this non-linearity was a smaller dissipated power for the same applied voltage.

The thermal efficiency was of $76 \pm 1\,\mathrm{mW/cycle}$. The result is shown in Figure 21.

Another useful measurement was to sweep over one of the two phase shifts while keeping the other constant.

As shown in Figure 22, the point did not exactly trace parallels and meridians on Poincaré sphere. The curves did not close on themselves after a 2π shift because of the already described electrical crosstalk due to the parasitic resistance of the common ground electrode. With reference to Figure 22a, increasing P_1 to scan in "longitude" decreased P_2, hence the polar angle, thus, the point travelled on spirals instead of parallels.

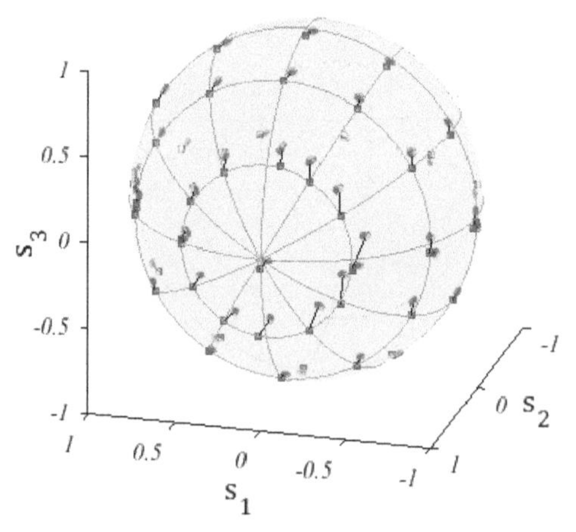

Figure 21. Comparison between experimental data points as read on the polarimeter (blue squares) and as deduced from the dissipated power on each heater (red), measured by the sourcemeter and correcting the effect of the parasitic common ground electrode resistance. The start point for P_1 and $P_2 = 0$ is that with $s_1 = -1$. Increasing P_1 produces a counter clockwise rotation around the s_2 axis, P_2 on the s_3 axis, when $P1 = 0$.

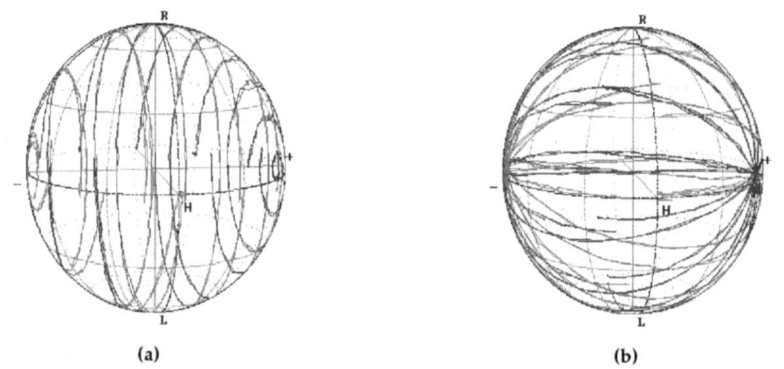

Figure 22. Curves traced on the polarimeter when scanning on (**a**) ϕ_1 and (**b**) ϕ_2, while keeping the other phase constant. The polarimeter has been calibrated so that the SOP from the SPGC is shown as a 45° linear polarisation. Red curves lie on the hemisphere on the observer's side, blue ones on the rear one. The axes are rotated by 90° around s_3, with respect to Figure 21.

When the inverse operation was performed, as in Figure 22b, a raise of the polar angle provoked a longitude diminution, which caused what would have been a meridian to precess, such as the curve of a satellite in polar orbit.

5. Conclusions

The operation of the silicon photonic circuit proposed in [6] has been analysed for both polarisation compensator and controller configurations. The behaviour at the central wavelength as well as the frequency dependence have both been considered. The whole treatment has been derived using a geometrical representation based on phasors and the Poincaré sphere. It has been shown that, thanks to this representation, the functioning of the device can be intuitively understood and analysed. This analysis has shown that any input SOP can be compensated for and that there exist two solutions for each SOP within a unit cell of the two phase shifts.

Some key results have been illustrated together with important examples. The construction of intensity surfaces as a function of the two phase shifts depends on the input SOP. It is 2π periodic in both directions, its shape depends on the Stokes parameter s_1 only, while the phase shift δ between the two components of the Jones vector only causes a shift of the surface along the ϕ_1 direction.
Conversely, it is proven that, in the controller operation, the device can generate any output SOP and that the effect of the ϕ_1 and ϕ_2 phase shifts is to trace parallels and meridians on Poincaré sphere, with respect to the s_1 axis. The rotation axis and angle are found as well.

An implicit equation for the 3 dB bandwidth, depending on the SOP and the particular choice of phase shifts pair, is derived. In general, the spectrum is not periodic and depends on both the SOP and the chosen solution. The effect on bandwidth of non-ideal factors resulting in unbalanced top and bottom phase shifter arms is studied and is proven not to be necessarily detrimental.

The curves traced with varying wavelength on Poincaré sphere by the SOP from the generator are found to be Clélie. More importantly, polarisations generated with the same ϕ_{2_0}/ϕ_{1_0} ratio have been found to lie on the same Clélie moving along it with wavelength. The effect of unbalance is just to change the parameter identifying the particular Clélie and to rotate it by a fixed angle about the s_1 axis.

This geometrical representation and the mathematical analysis of this device illustrate the power of this approach and open the route to the analysis of more complex structures as well as useful treatment of such components in the quantum domain due to the direct correspondence of the Stokes parameters and the photon density operator.

Author Contributions: Theoretical derivation, M.V.P.; experimental measurement, G.D.A. and M.V.P.; conceptualisation, V.S.; methodology, M.R.; writing—review and editing, and supervision and funding acquisition, P.V.

Funding: This research was partially funded by Tuscany Region through a POR FESR Toscana 2014–2020 grant in the context of the project SENSOR.

Acknowledgments: The authors would like to thank Fabrizio Di Pasquale for his help in the supervision of Massimo Valerio.

Conflicts of Interest: The authors declare no conflict of interest.

Abbreviations

The following abbreviations are used in this manuscript:

SOI	Silicon On Insulator
SiPho	Silicon Photonics
CMOS	Complementary Metal Oxide Semiconductor
PIC	Photonic Integrated Circuit
IL	Insertion Loss
GC	Grating Coupler
SOP	State of Polarisation
PLC	Planar Lightwave Circuit

I/O	Input/Output
TE	Transverse Electric
2DGC	Two Dimensional Grating Coupler
MMI	Multi Mode Interference
POLSK	POLarisation Shift Keying
MZI	Mach–Zehnder Interferometer
PS	Phase Shifter
SPGC	Single Polarisation GC
FPC	Fibre Polarisation Controller
DUT	Device Under Test
PDL	Polarisation Dependent Loss

Appendix A. Soi Effective and Group Indexes

Following (Equation (15) in [11]), the effective index of a straight SOI waveguide is expressed using a second-order Taylor expansion.

Comparing it with the customary formula for the group index,

$$n_g = \frac{c}{v_g} = n_{eff} - \lambda \frac{\mathrm{d}\, n_{eff}}{\mathrm{d}\lambda} \tag{A1}$$

we get:

$$n_{eff}(\lambda_0, T) = N_0(T) \qquad n_g(\lambda_0, T) = N_0(T) - \frac{\lambda_0}{\sigma_\lambda} N_1(T) \tag{A2}$$

Expanding the expression for the group index, its second-order expansion is:

$$\begin{aligned} n_g(\lambda_0, T) &= n_0 - \frac{\lambda_0}{\sigma_\lambda} n_3 + \left(n_1 - \frac{\lambda_0}{\sigma_\lambda} n_4\right)\left(\frac{T - T_0}{\sigma_T}\right) + \left(n_2 - \frac{\lambda_0}{\sigma_\lambda} n_5\right)\left(\frac{T - T_0}{\sigma_T}\right)^2 = \\ &= n_{g_0} + \alpha_g (T - T_0) + \beta_g (T - T_0)^2 \end{aligned} \tag{A3}$$

where n_i, $\sigma_{\lambda,T}$ are given in (Table III of [11]) and $\lambda_0 = 1550\,\mathrm{nm}$. Using those values, one finds:

$$\begin{cases} n_{eff_0} = 2{,}4{,}057{,}177 \\ \alpha_{eff} = 2{,}2756 \times 10^{-4} \\ \beta_{eff} = 1{,}3811 \times 10^{-7} \end{cases} \qquad \begin{cases} n_{g_0} = 4{,}3757 \\ \alpha_g = 2{,}5919 \times 10^{-4} \\ \beta_g = 1{,}006 \times 10^{-7} \end{cases} \tag{A4}$$

Whence we see that the temperature coefficient of the group index is slightly bigger than the effective index's one. In our derivation, a parameter of interest was their ratio, which happens to be:

$$R_{ge} \doteq \frac{\alpha_g}{\alpha_{eff}} = 1{,}14 \tag{A5}$$

Appendix B. Mzi as A Hwp

The Jones matrix for a Half Wave Plate is (Equation 7 [2]) (Section 4.6.1 [12])

$$HWP(\Theta) = -i \begin{bmatrix} \cos 2\Theta & \sin 2\Theta \\ \sin 2\Theta & -\cos 2\Theta \end{bmatrix} \tag{A6}$$

The one for a Mach–Zehnder interferometer:

$$H_{MZI}(\phi) = -i \begin{bmatrix} \cos(\phi/2 + \pi/2) & \sin(\phi/2 + \pi/2) \\ \sin(\phi/2 + \pi/2) & -\cos(\phi/2 + \pi/2) \end{bmatrix} \tag{A7}$$

Thus, we see that the two matrices have the same form, provided that:

$$\Theta = \phi/4 + \pi/4 \tag{A8}$$

Passing to Stokes matrices, for the HWP, we have (Equation 4.6.7 in [12])

$$R_{HWP}(\Theta) = \begin{bmatrix} \cos 4\Theta & \sin 4\Theta & 0 \\ \sin 4\Theta & -\cos 4\Theta & 0 \\ 0 & 0 & -1 \end{bmatrix} \tag{A9}$$

whose rotation axis

$$\vec{\Omega}_{HWP}(\Theta) = \begin{pmatrix} \cos 2\Theta \\ \sin 2\Theta \\ 0 \end{pmatrix} \tag{A10}$$

lies on the equatorial plane (with respect to the s_3 axis), forming an angle 2Θ with the s_1 axis.

Inserting the relation of Equation (A8) in the above formula,

$$\vec{\Omega}_{MZI} = \begin{pmatrix} \cos(\phi/2 + \pi/2) \\ \sin(\phi/2 + \pi/2) \\ 0 \end{pmatrix} \tag{A11}$$

Thus, acting on the phase shift has the result of rotating the waveplate rotation axis in the equatorial plane.

Nonetheless, a major difference with a birefringent waveplate is that, while the former is endlessly rotatable, a MZI is not, as the phase shift can be varied in a limited range only.

Given that the trace of R_{HWP} is -1, the rotation angle $\Gamma = \pi$ or an odd integer multiple thereof, because of Equation (62).

References

1. Noe, R.; Heidrich, H.; Hoffmann, D. Endless polarization control systems for coherent optics. *J. Lightware Technol.* **1988**, *6*, 1199–1208. [CrossRef]
2. Heismann, F. Analysis of a reset-free polarization controller for fast automatic polarization stabilization in fiber-optic transmission systems. *J. Lightware Technol.* **1994**, *12*, 690–699. [CrossRef]
3. Heismann, F. Integrated-optic polarization transformer for reset-free endless polarization control. *IEEE J. Quantum Electron.* **1989**, *25*, 1898–1906. [CrossRef]
4. Moller, L. WDM polarization controller in PLC technology. *IEEE Photonics Technol. Lett.* **2001**, *13*, 585–587. [CrossRef]
5. Velha, P.; Sorianello, V.; Preite, M.; De Angelis, G.; Cassese, T.; Bianchi, A.; Testa, F.; Romagnoli, M. Wide-band polarization controller for Si photonic integrated circuits. *Opt. Lett.* **2016**, *41*, 5656–5659. [CrossRef] [PubMed]
6. Caspers, J.N.; Wang, Y.; Chrostowski, L.; Mojahedi, M. Active polarization independent coupling to silicon photonics circuit. In Proceedings of the Silicon Photonics and Photonic Integrated Circuits IV, Brussels, Belgium, 13–17 April 2014, Volume 9133.
7. Sorianello, V.; Angelis, G.D.; Cassese, T.; Preite, M.V.; Velha, P.; Bianchi, A.; Romagnoli, M.; Testa, F. Polarization insensitive silicon photonic ROADM with selectable communication direction for radio access networks. *Opt. Lett.* **2016**, *41*, 5688–5691. [CrossRef] [PubMed]
8. Thaniyavarn, S. Wavelength-independent, optical-damage-immune LiNbO 3 TE–TM mode converter. *Opt. Lett.* **1986**, *11*, 39–41. [CrossRef] [PubMed]
9. Yariv, A. Coupled-mode theory for guided-wave optics. *IEEE J. Quantum Electron.* **1973**, *9*, 919–933. [CrossRef]

10. Madsen, C.K.; Oswald, P.; Cappuzzo, M.; Chen, E.; Gomez, L.; Griffin, A.; Kasper, A.; Laskowski, E.; Stulz, L.; Wong-Foy, A. Reset-free integrated polarization controller using phase shifters. *IEEE J. Sel. Top. Quantum Electron.* **2005**, *11*, 431–438. [CrossRef]
11. Rouger, N.; Chrostowski, L.; Vafaei, R. Temperature effects on silicon-on-insulator (SOI) racetrack resonators: A coupled analytic and 2-D finite difference approach. *J. Light. Technol.* **2010**, *28*, 1380–1391. [CrossRef]
12. Damask, J.N. *Polarization Optics in Telecommunications*; Springer Science & Business Media: Berlin/Heidelberg, Germany, 2004; Volume 101.
13. Born, M.; Wolf, E. *Principles of Optics: Electromagnetic Theory of Propagation, Interference and Diffraction of Light*; Pergamon Press: Oxford, UK, 1980.
14. Feynman, R.P.; Vernon, F.L.; Hellwarth, R.W. Geometrical Representation of the Schrödinger Equation for Solving Maser Problems. *J. Appl. Phys.* **1957**, *28*, 49–52. [CrossRef]
15. Fano, U. A Stokes-Parameter Technique for the Treatment of Polarization in Quantum Mechanics. *Phys. Rev.* **1954**, *93*, 121–123. [CrossRef]
16. Cherchi, M. Wavelength-flattened directional couplers: A geometrical approach. *Appl. Opt.* **2003**, *42*, 7141–7148. [CrossRef] [PubMed]
17. Vince, J. *Rotation Transforms for Computer Graphics*; Springer: London, UK, 2011.

 *micromachines*

Article

100 Gb/s Silicon Photonic WDM Transmitter with Misalignment-Tolerant Surface-Normal Optical Interfaces

Beiju Huang [1,*,†], Zanyun Zhang [2,3,*,†], Zan Zhang [4], Chuantong Cheng [1], Huang Zhang [1], Hengjie Zhang [1] and Hongda Chen [1,5]

1 State Key Laboratory on Integrated Optoelectronics, Institute of Semiconductors, Chinese Academy of Sciences, Beijing 100083, China; chengchuantong@semi.ac.cn (C.C.); zhanghuanl@semi.ac.cn (H.Z.); woniudidi@semi.ac.cn (H.Z.); hdchen@semi.ac.cn (H.C.)

2 Tianjin Key Laboratory of Optoelectronic Detection Technology and System, Tianjin Polytechnic University, Tianjin 300387, China

3 Optoelectronics Research Centre, University of Southampton, Southampton SO17 1BJ, UK

4 School of Electronic and Control Engineering, Chang'an University, Xi'an 710064, China; zhangzan@semi.ac.cn (Z.Z.)

5 College of Materials Science and Opto-Electronic Technology, University of Chinese Academy of Sciences, Beijing 100049, China

* Correspondence: bjhuang@semi.ac.cn (B.H.); zhangzanyun@tjpu.edu.cn (Z.Z.); Tel.: +86-10-8230-4531 (B.H.)

† These authors contributed equally to this paper.

Received: 29 April 2019; Accepted: 20 May 2019; Published: 22 May 2019

Abstract: A 4 × 25 Gb/s ultrawide misalignment tolerance wavelength-division-multiplex (WDM) transmitter based on novel bidirectional vertical grating coupler has been demonstrated on complementary metal-oxide-semiconductor (CMOS)-compatible silicon-on-insulator (SOI) platform. Simulations indicate the bidirectional grating coupler (BGC) is widely misalignment tolerant, with an excess coupling loss of only 0.55 dB within ±3 μm fiber misalignment range. Measurement shows the excess coupling loss of the BGC is only 0.7 dB within a ±2 μm fiber misalignment range. The bidirectional grating structure not only functions as an optical coupler, but also acts as a beam splitter. By using the bidirectional grating coupler, the silicon optical modulator shows low insertion loss and large misalignment tolerance. The eye diagrams of the modulator at 25 Gb/s don't show any obvious deterioration within the waveguide-direction fiber misalignment ranger of ±2 μm, and still open clearly when the misalignment offset is as large as ±4 μm.

Keywords: vertical grating coupler; WDM transmitter; optical interconnects; silicon photonics; silicon optical modulator

1. Introduction

The volume of the digital information captured, created, or consumed each year is expected to grow from 538 exabytes to 5.0 zettabytes between 2012 and 2020, and it will double about every two and a half years [1]. To manage this amount of information, higher data transmission method is needed, especially in the data centers and high-performance computing (HPC) systems. Unfortunately, traditional electrical interconnects based on copper have been proved to be the bottleneck for growing demand of HPC systems and high-speed data center application because electrical input/output (IO) suffers from low bandwidth, high crosstalk, high latencies and attenuation. Optical interconnects are proposed to solve the bottleneck of electrical interconnects due to its low crosstalk, low latency, high bandwidth and high energy efficiency. Optical interconnects are emerging as a better solution over electrical interconnects for rack-to rack, board-to-board, chip-to-chip, and even on-chip data communication.

Silicon photonics based optical interconnects are good approach to obtain fully integrated photonic circuits with lower cost and higher integration level because of its compatibility with silicon microelectronic process. By taking advantages of the existing CMOS-VLSI infrastructure, silicon photonics is considered to be a promising and economical platform for the monolithic integration of various active and passive optoelectronic devices used in optical interconnects, such as high speed modulators [2–6], photodetectors [7–9], grating couplers [10] and optical splitters/combiners [11,12]. Low cost and large-scale integrated photonic circuits can be achieved by silicon photonic platform to realize optical interconnects [13,14]. Now, there are several silicon platforms, including Luxtera's 130 nm Freescale platform, Mellanox's 150 mm foundry, MPW services through IMEC and IME of Singapore, 130 nm SiGe BiCMOS production line at IHP in Frankfurt, and CEA-Leti.

Impressive works in the field of optical interconnects based on silicon photonics have been reported [13–16]. Silicon transmitters are important building blocks in silicon photonic interconnects. Several WDM and pulse-amplitude-modulation (PAM) transmitters are proposed to realize high-speed optical links [17–19] by different schemes. For these high-speed silicon transmitters based on silicon photonics, coupling a light effectively from a single mode fiber (SMF, typically with core diameter of 9 μm) into the silicon waveguide (typical dimensions of 220 nm × 450 nm) in an economical way is a big challenge. Grating couplers are widely used as the optical interfaces between the SMF and the submicron silicon waveguides. However, conventional grating couplers are usually designed for tilted-fiber coupling in order to reduce the second-order back reflection. The tilted coupling scheme will lead to several problems. Firstly, the alignment is more time-consuming because of angle tuning, and the mode matching efficiency usually deviates from the design due to the unavoidable mode expansion in the fiber–chip surface cavity. Secondly, a costly angle-polishing process is required for the fiber packaging. Compared to conventional GCs, perfectly vertical grating couplers (PVGCs) enable improved alignment accuracy and relaxed packaging complexity by avoiding angle alignment errors. Because of the advantages, PVGCs are more suitable to be applied in a multichannel communication network with fan-in/fan-out array, interfacing with a multicore fiber or fiber array with more strict requirements of alignment accuracy and coupling uniformity.

In this paper, we present a 4 × 25 Gb/s ultrawide fiber misalignment-tolerant WDM transmitter based on the bidirectional grating couplers (BGCs)-based E-O modulator array and the microring multiplexer function. The BGCs exhibit the advantages of easy fabrication, perfectly vertical coupling and large misalignment tolerance, and potentially enable interfacing with a fiber array or flip-chip-integrated VCSEL laser array for low-cost photonic packaging. Benefiting from the characteristic of the BGCs, the WDM transmitter shows strong robustness to the variation of the fiber positions. Eye diagrams don't show obvious deterioration within the waveguide-direction fiber misalignment ranger of ±2 μm, and still open clearly when the misalignment offset is as large as ±4 μm.

2. Structure and Principle

As depicted in Figure 1a, a WDM optical interconnect circuit with an array of BGCs can be interfaced with a fiber array for rapid wafer-scale test and low-cost fiber packaging. We have demonstrated that a Mach–Zehnder-type E-O modulator can be built based on a BGC which functions as a 3 dB power splitter with symmetry grating design [20]. Therefore, it is experimentally feasible to realize a WDM transmitter integrated with a BGC-based modulator array and a microring (MR) array multiplexer, as schematically shown in Figure 1b. Four MZI modulators with embedded PN phase shifters in asymmetric arms are employed to generate optical modulation signals with different wavelengths. The MRs are designed with slightly different radii. Through the resonance coupling of the MR array, four optical signals can be uploaded to the bus waveguide and coupled out-of-plane with a standard GC. For wavelength alignment in measurement, each MR and MZI can be thermally tuned with integrated titanium nitride (TiN) heaters. In order to measure the optical spectra of MZI modulator and microring resonator independently, a directional coupler (DC) is inserted between the modulator and the microring to act as a −10 dB optical splitter.

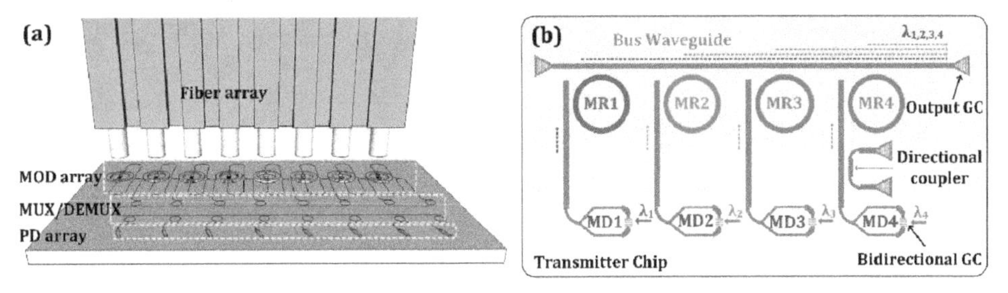

Figure 1. (**a**) Schematic diagram of the bidirectional grating coupler (BGC)-based wavelength-division-multiplex (WDM) optical interconnect tested with fiber array. (**b**) The schematic diagram of the proposed WDM transmitter.

2.1. Bidirectional Grating Coupler

The bidirectional grating coupler functions as the perfectly vertical grating coupler and power splitter in the Mach–Zehnder-type E-O modulator. According to the Bragg condition, the grating period Λ can be calculated by estimating the effective refractive index of the waveguide grating N_{effg}, which can be expressed as:

$$\Lambda = \lambda / N_{effg} \quad (1)$$

λ represents the vacuum wavelength. With the 2-D FDTD optimization, the optimal design parameters are grating period of 580 nm, etch depth of 70 nm, filling factor of 50%, period number of 22, respectively. For PVGCs, the cladding thickness have a big impact on the coupling performance as a well-designed cladding thickness can greatly reduce the upreflection loss. According to our simulations, the oxide cladding thickness of 1.3 μm offers an optimal performance with an in-plane coupling efficiency of 62% (−2.08 dB) and a 3 dB optical bandwidth of 80 nm. The upreflection power is suppressed to 16% around the center coupling wavelength, corresponding to an optical return loss of 8 dB. Although such a result is not as superior as that of a conventional GC, further improvement of the reflection loss can be achieved by utilizing an apodized grating design [21] or introducing a bilayer antireflection cladding [22].

One notable advantage of the BGC is the large misalignment tolerance along the waveguide direction. For the BGC, the grating splitting behavior is of great significance to the BGC-based modulator operation. In order to investigate the influence of fiber misalignment, the grating split ratio (defined as the single arm power coupling divided by the total in-plane power coupling) with different fiber incidence positions are calculated with 2-D FDTD method. As shown in Figure 2b, the grating functions as a perfect 3 dB splitter with the fiber placed in the grating center due to the symmetry. With certain fiber displacement, the split ratio will increase significantly, which indicates the unbalanced power coupling in opposite directions. However, it is worth noting the split ratio is quite stable near the resonant wavelength of 1546 nm and nearly immune to fiber misalignment. This characteristic can be attributed to the wavelength-dependent second order reflection and would be very useful for modulator application. It allows a BGC-based modulator robustly working at a strong coupling wavelength in case of certain fiber misalignment. Apart from the splitting behavior discussions, the coupling efficiency variation with fiber misalignment is calculated for the BGC. For comparison, a unidirectional GC with same waveguide thickness and grating etch depth is also simulated. As shown in Figure 3, the BGC has a larger optical bandwidth and misalignment tolerance in the waveguide direction. The excess coupling loss caused by fiber misalignment is 0.95 dB (−3 μm misalignment) and 1.47 dB (3 μm misalignment) for unidirectional GC, while it is only 0.55 dB (±3 μm misalignment) for the bidirectional GC.

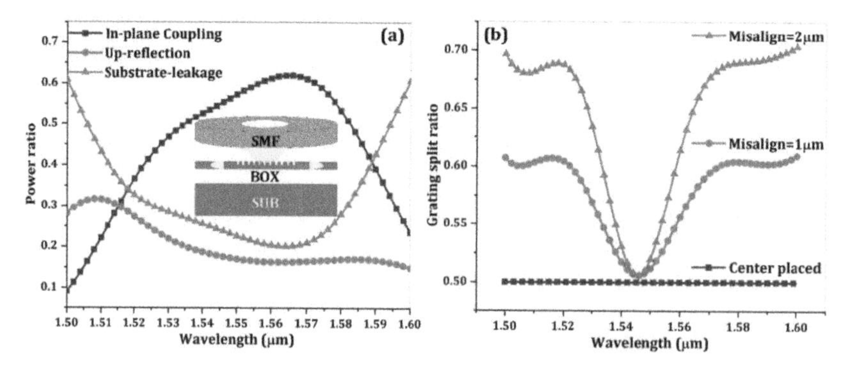

Figure 2. (**a**) The normalized optical power of different directions with perfectly vertical fiber incidence (inset picture shows the schematic of the BGC). (**b**) The wavelength-dependent grating split ratio with different fiber incident positions.

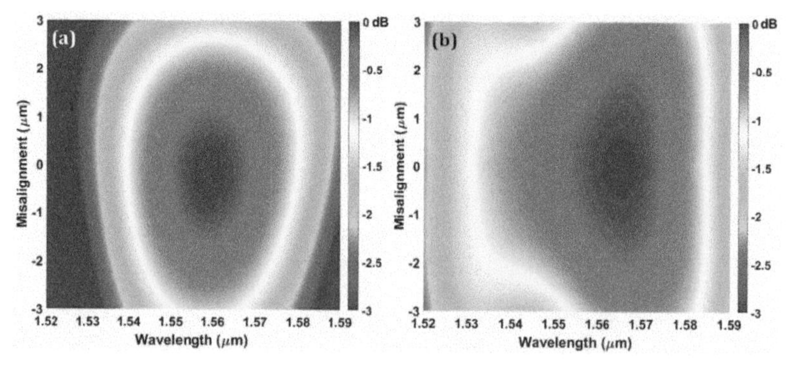

Figure 3. (**a**) The calculated coupling efficiency variation of a unidirectional grating coupler (GC) with fiber misalignment. (**b**) The calculated coupling efficiency variation of a bidirectional GC with fiber misalignment.

2.2. *BGC-Based Silicon E-O Modulator in Push-Pull Scheme*

The structural view of the bidirectional grating coupler-based optical modulator is shown in Figure 4, where the bidirectional grating coupler is used to couple light from the input fiber to the chip and split it into two arms. Comparing with traditional MZI modulators, our modulator avoids use of the conventional 3 dB beam splitter since the vertical grating coupler doubles as a fiber coupler and 3 dB optical splitter. Hence, the insertion loss of the modulator can be reasonably decreased.

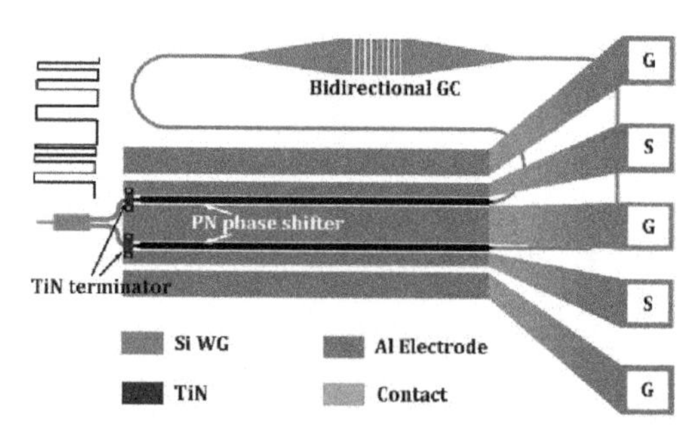

Figure 4. Schematic of demonstrated ultrawide misalignment tolerance modulator with bidirectional grating couplers.

Carrier-depletion-type optical modulators are utilized to develop our WDM transmitter chip because of their merits of high operation speed and low insertion loss. Both arms of the MZI contain a carrier-depletion-type phase shifter with length of 2 mm, cooperating with travelling wave electrode of GSGSG pattern to enable dual-port differential operation in a push-pull scheme. A low driving voltage signal with 2.5 V swing is utilized, which is CMOS-compatible and thus allows a direct integration with the electronic driver circuits. Low driving voltage can enable the use of most advanced CMOS technologies, which is beneficial for optimizing the energy efficiency. For example, the typical supply voltage for the feature size of 350 nm and 180 nm CMOS process is 3.3 V and 1.8 V. When the smaller feature size is chosen, higher speed and lower supply voltage can be obtained.

The arm length difference of the MZI is introduced to generate interference patterns in spectra similar to microring resonators. As the length difference between two arms is designed to be 248.4 μm, FSR of 2.4 nm can be obtained. The resonance behavior will enhance the modulation efficiency of MZI modulators. For asymmetrical MZI, the resonance condition is:

$$\Delta\varphi = d \cdot N_{eff} \cdot \frac{2\pi}{\lambda} = (2m+1)\pi \tag{2}$$

$\Delta\varphi$ is the phase shift, d is the length difference between the two arms, N_{eff} is the effective index of the single mode waveguide, λ is resonance wavelength, m is an integer. According to Equation (2), d is proportional to λ:

$$\Delta d/d = \Delta\lambda/\lambda \tag{3}$$

Δd and $\Delta\lambda$ are small variations of d and λ respectively. The optical phase change by Δd is:

$$\Delta\varphi = \Delta d \cdot N_{eff} \cdot \frac{2\pi}{\lambda} = 2\pi \cdot d \cdot \Delta\lambda/\lambda^2 \cdot N_{eff} \tag{4}$$

The effective refractive index of the phase shifter waveguide arms can be changed with bias voltage. The optical phase change due to bias voltage can be expressed as:

$$\Delta\varphi = \Delta N_{eff} \cdot L \cdot \frac{2\pi}{\lambda} \tag{5}$$

ΔN_{eff} is the change of the effective index. L is the length of MZI waveguide arms with embedded carrier-depletion-type phase shifter. When asymmetrical MZI resonances and destructive interference occurs, the optical phase change in Equation (4) and (5) is π. From Equation (4) and Equation (5), we can obtain:

$$2\pi \cdot d \cdot \frac{\Delta\lambda}{\lambda^2 \cdot N_{eff}} = \Delta N_{eff} \cdot L \cdot \frac{2\pi}{\lambda} \tag{6}$$

The equation can be further simplified as:

$$\Delta N_{eff} = d \cdot \Delta\lambda/\lambda \cdot N_{eff}/L \tag{7}$$

For asymmetrical MZI, the FSR can be calculated as:

$$\text{FSR} = \lambda^2/(N_g \cdot d) \tag{8}$$

N_g is the group refractive index of the single mode waveguide. The relationship between the FSR, $\Delta\lambda$ and $\Delta\varphi$ can be expressed by:

$$\Delta\varphi = 2\pi \cdot \Delta\lambda/\text{FSR} \tag{9}$$

When phase shift is π, $\Delta\lambda$ is:

$$\Delta\lambda = 0.5 \cdot \text{FSR} = 0.5 \cdot \lambda^2/(N_g \cdot d) \tag{10}$$

According to Equations (7) and (10), we obtain:

$$\Delta N_{eff} = \lambda/(2L)\cdot N_{eff}/N_g \tag{11}$$

Simulation results show that the effective index and group index for single mode rib waveguides of height 220 nm, width 500 nm and slab 60 nm is 2.52 and 4.03.

3. Experiment Results

The four-channel WDM optical transmitter is realized with the silicon photonic MPW line of IME, Singapore. Figure 5a shows the optical micrograph of the WDM transmitter implemented in a 200 mm SOI wafer with a 2 μm-thick buried oxide layer and a 220 nm-thick top silicon layer. The zoom-in picture of the functions of the MR MUX, the DC and the BGC are respectively shown in Figure 5b–d. The phase shifters are based on a ridge waveguide structure with 500 nm width and 60 nm slab thickness. Four asymmetric MZI modulators with surface-normal optical interfaces are employed to generate modulated signal at different wavelengths. According to the plasma dispersion effect, holes contribute larger index change and less absorption than electrons. The width of the p-type doping region (300 nm) is set to be larger than the n-type doping region (200 nm). The p-type and n-type doping concentration are 7×10^{17} cm^{-3} and 5×10^{17} cm^{-3}, the P^+ and N^+ regions are doped to a concentration of 10^{20} cm^{-3} to form low resistivity ohmic contact. Both arms of the MZI contain a 2 mm-long carrier-depletion type phase shifter, cooperating with GSGSG travelling wave electrode to enable dual-port low voltage differential operation. TiN resistors are integrated to form an on-chip terminator of 30 Ω to ensure impedance matching. Two thermal phase shifters are also incorporated within the interferometer to adjust the operation point at quadrature on the positive slope. Each MZI and microring resonator can be thermally tuned by an integrated heater implemented by a TiN resistor, so the working wavelengths of the modulators and the WDM function can be tuned to the same value. To make full use of the chip area, there are some other individual devices shown in the chip layout, such as three MZI modulators with different size and four Ge photodetectors.

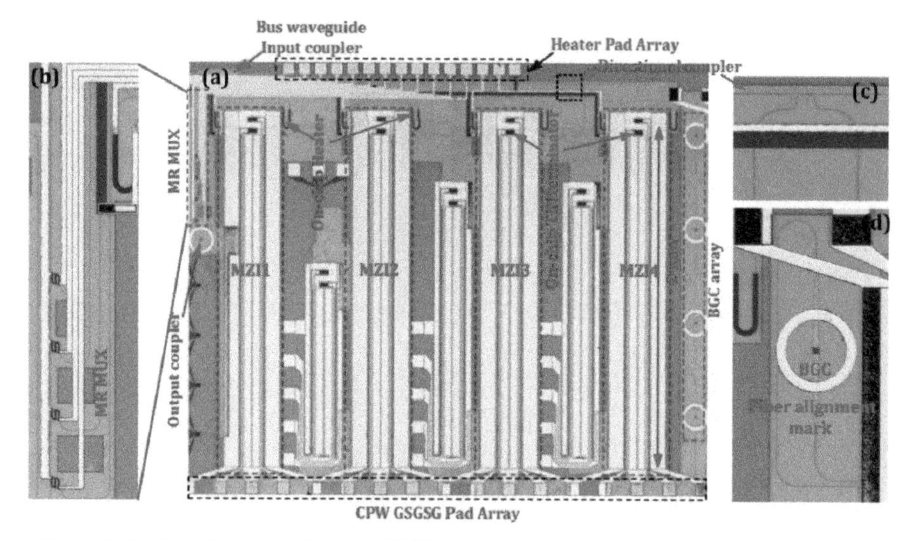

Figure 5. (**a**) Optical micrograph of the WDM transmitter with surface-normal optical interfaces. (**b**) The zoom-in picture of the MR Multiplexer. (**c**) The zoomed-in picture of the directional coupler. (**d**) The zoom-in picture of the bidirectional grating coupler-based surface-normal optical interface.

3.1. Bidirectional Grating Coupler

To investigate the fiber misalignment tolerance of the BGC, a discrete BGC with balanced arms and MMI combiner is measured with a high-precision vertical fiber alignment system with a step resolution of 20 nm to ensure a small alignment error. For comparison, the misalignment tolerance of a conventional GC (utilized for output coupling interface) is also studied. Figure 6a,b show the coupling efficiency spectra response of the fiber incidence position variations within the range of ±2 μm for the conventional GC and BGC, respectively. The measurement result of the conventional GC is obtained with a back-to-back configuration, while the result of the BGC is calculated by normalizing the output coupler loss and the MMI insertion loss. When the fiber is tuned in the optimal position, the minimum coupling loss reaches −4.8 dB and −4 dB, respectively. The measured coupling loss of the two couplers are lower than the simulation results, which is possibly due to fabrication imperfections. Interestingly, although the simulation work indicates the two kind of couplers are close in peak coupling efficiency, the measured coupling efficiency of the BGC is higher than that of a conventional GC within the same photonic platform. This may be attributed to the excess coupling loss resulting from the free-space mode expansion within a tilted fiber experimental set-up. When the fiber is slightly tuned from the optimal position, the misalignment tolerance can be analyzed. The measured results show good accordance with the simulation results shown in Figure 3. The BGC has a more symmetric and larger misalignment tolerance along the waveguide direction. The excess coupling loss of the BGC is lower than 0.7 dB within a misalignment range of ±2 μm. However, the coupling deterioration of the conventional GC is as high as 1 dB and 1.68 dB with a fiber tuning of −2 μm and +2 μm, respectively. These experimental results clearly demonstrate the improvement of fiber misalignment tolerance with a BGC design.

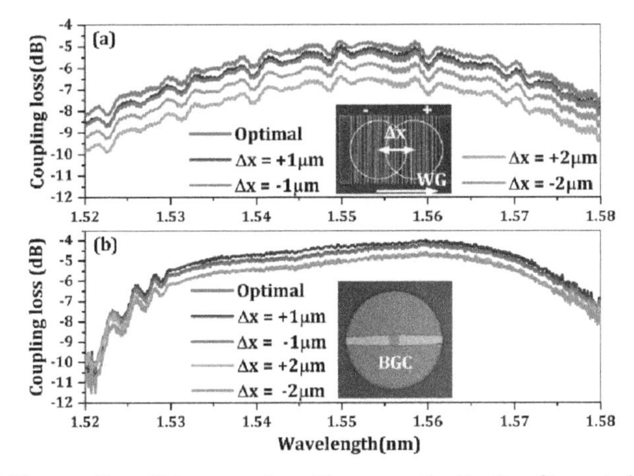

Figure 6. (**a**) The coupling efficiency spectra with waveguide-direction fiber misalignment for a conventional GC. (**b**) The coupling efficiency spectra with waveguide-direction fiber misalignment for a bidirectional GC.

3.2. BGC-Based E-O Modulator in Push-Pull Operation Mode

To investigate the optical spectrum of MZI modulator, 10% output optical power of the MZI modulator is coupled out by a directional coupler and connected with an additional output grating coupler. The measurement result is shown in Figure 7. As shown in Figure 7a, the measured FSR is about 2.5 nm, which is slightly larger than the designed value due to fabrication imperfection. As shown in Figure 7b, the transmission spectrum shifts about 0.8 nm with a bias voltage of 3 V, which corresponds to a $V\pi L$ of 0.9 V·cm. The fiber-to-fiber optical loss is 11.2 dB, including the conventional grating coupling loss of 4.8 dB and the bidirectional grating coupling loss of 4 dB, the MMI combiner insertion loss of 0.8 dB and the phase shifter insertion loss of 1.6 dB. If the new coupling method is adopted and the conventional grating coupling loss can be possibly reduced to 1.5 dB [23], the fiber-to-fiber loss of the silicon modulator can be decreased to lower than 8 dB. High on-off extinction ratio of 27 dB is obtained for the MZI optical spectra, which indicates the near perfectly 3 dB power splitting behavior of the grating.

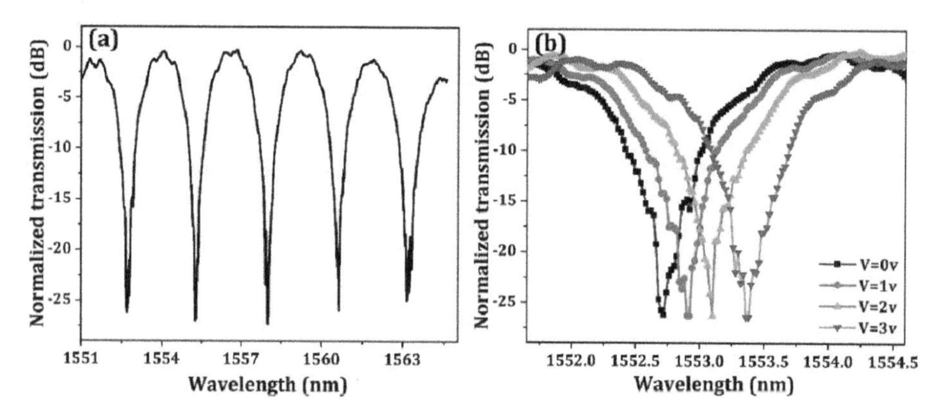

Figure 7. Static characteristics of the modulator: (**a**) Spectra response of modulator with 2.4 nm FSR, (**b**) Spectra response of modulator at different reversed voltages.

Since the designed channel space is 2.4 nm, the FSR of 4 channels WDM should be larger than 9.6 nm. Thus, the radii of the microrings should be slightly smaller than 9.88 μm, and the radii of four microrings are designed to be slightly different to obtain WDM channel spacing of 2.4 nm. Figure 8 shows the normalized optical spectra of microring multiplexer before and after thermal tuning. The MR-based WDM obtains a uniform channel space of 2.4 nm after thermal tuning, and the overall tuning power was 12.54 mW.

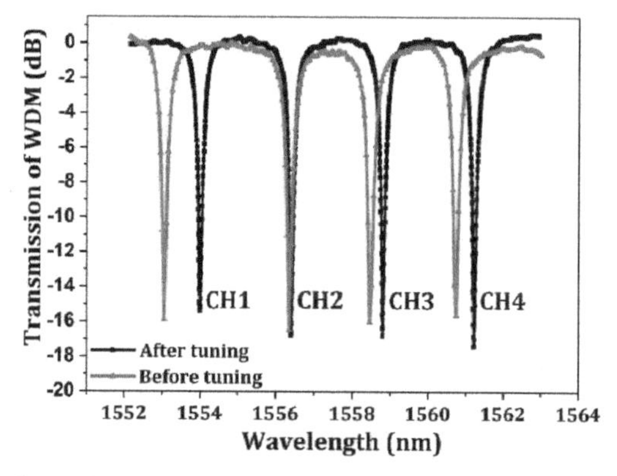

Figure 8. Normalized optical spectra of microring multiplexer before and after thermal tuning with a channel space of 2.4 nm.

The dynamic high frequency characteristics of the WDM transmitter were tested. Monochromatic light with wavelength of 1550 nm from a tunable laser is coupled to the chip by a polarization maintaining (PM) fiber. The high speed PRBS data stream generated by SHF 12104A is amplified by CENTELLAX OA4SMM4 microwave amplifier with a typical gain of 17 dB to get sufficiently high driving voltage swing. The output RF signal from microwave amplifier and DC bias are combined by Anritsu K250 to provide reversed DC bias, and then launched into the modulator using a 40 GHz microwave probe. The optical output of modulator is detected by high-speed photo-receiver 1474-A, and connected to an oscilloscope Agilent DCA 86100C for eye diagram observation. Figure 9 shows the eye diagram measurement results. With the differential RF driving signals at 25 Gb/s, we demonstrate that the optical output signals at all four channels have a clear eye opening with a healthy margin.

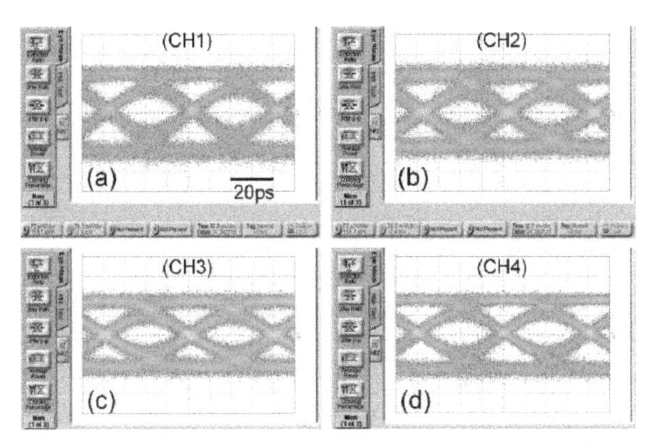

Figure 9. 25 Gb/s eye diagram of WDM transmitter with different wavelengths: (**a**) 1554 nm, (**b**) 1556.4 nm, (**c**) 1558.8 nm, (**d**) 1561.2 nm.

The fiber misalignment tolerance on the modulator dynamic performance is investigated by tuning the fiber position along the waveguide direction. Eye diagrams with fiber misalignment along the waveguide direction are shown in Figure 10. From our simulation results of the grating splitting behavior, it is indicated that the modulator performance is more robust to fiber misalignment around the grating resonant wavelength of 1546 nm. However, the coupling efficiency of the bidirectional grating coupler reaches a maximum near 1560 nm. Therefore, we choose a working wavelength of 1554 nm as a trade-off between the optical loss and the grating splitting sensitivity. The eye diagrams don't show any obvious deterioration within the waveguide direction misalignment range of ±2 μm, and still open clearly when the horizontal direction misalignment range is as large as ±4 μm. Table 1 shows the characteristic parameters of this work and the references. As comparing with references, the PVGCs-based modulator shows wider misalignment tolerance and reduced packaging difficulties by avoiding angle alignment, and has potential to provide low cost optical interconnects solutions.

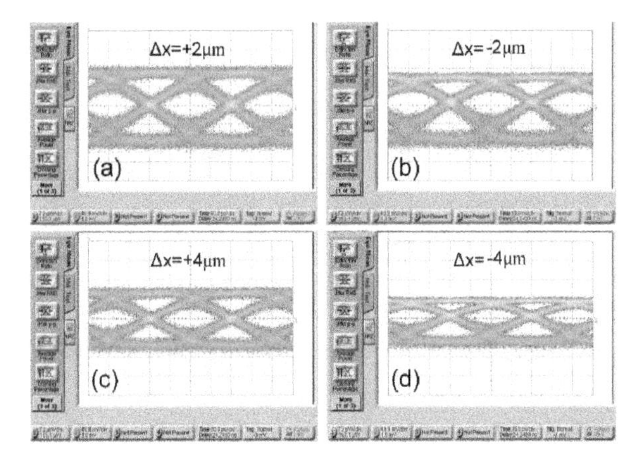

Figure 10. 25 Gb/s eye diagram of WDM transmitter at different fiber offset from the central position of the grating coupler along the horizontal direction of waveguide: (**a**) +2 μm offset, (**b**) −2 μm offset, (**c**) +4 μm offset, (**d**) −4 μm offset.

Table 1. The characteristic parameters of this work and the references.

-	Ref [14]	Ref [18]	Ref [19]	This Work
Scheme	Ring WDM	Ring WDM	Ring DWDM	MZI WDM
Speed	10 Gb/s	12.5 Gb/s	10 Gb/s	25 Gb/s
Channel Number	4	4	5	4
Channel Spacing	0.8 nm	3.8 nm	0.5 nm	2.4 nm
$V\pi$	12 V	>3 V	>4 V	4.5 V
Misalignment Tolerance	-	-	-	±4 µm

4. Conclusions

We have experimentally demonstrated a 4 × 25 Gb/s WDM transmitter based on microring multiplexer and asymmetrical MZI modulator with surface-normal optical interface. According to our derivation, the modulation efficiency of the asymmetrical MZI modulator is higher than a symmetrical MZI modulator. Benefiting from higher modulation efficiency and a differential phase shifter of two arms, low voltage operation is achieved using a 2.5 V differential driving signal. By utilizing bidirectional grating couplers, the WDM transmitter shows ultrawide misalignment tolerance. The transmitter with ±4 µm fiber misalignment tolerance provides an attractive solution to the problem of optical coupling between fiber and waveguide. The vertical grating coupler-based transmitter has potential to provide low cost optical interconnect solutions and can serve as a photonic platform for developing high bandwidth optical interconnects and optical computing systems in the next generation of optical networks.

Author Contributions: B.H. and (Zanyun Zhang) contributed equally to this paper. B.H. conceived and designed the chip. Z.Z. (Zanyun Zhang) and Z.Z. (Zan Zhang) performed the simulations and experiments. C.C., H.Z. (Huang Zhang) and H.Z. (Hengjie Zhang) analyzed the data and results. H.C. read this paper and give some useful suggestions. All authors discussed the experimental implementation and results.

Funding: This work is supported by the National Key R&D Program of China (Grant No. 2018YFA0209000), the Natural Science Foundation of China (Grant Numbers. 61675191, 61634006, 61178051), the Tianjin Research Program of application foundation and advanced technology (No. 18JCQNJC01800) and the China Scholarship Council (Award to Zanyun Zhang for 1 year's research at the University of Southampton).

Conflicts of Interest: The authors declare no conflict of interest.

References

1. Gantz, J.; Reinsel, D.; Arend, C. The digital universe in 2020: Big data, bigger digital shadows, and biggest growth in the far east-western Europe. *IDC Analyze Futrue* **2012**, *2007*, 1–16.
2. Reed, G.T.; Mashanovich, G.; Gardes, F.Y.; Thomson, D.J. Silicon optical modulators. *Nat. Photonics* **2010**, *4*, 518–526. [CrossRef]
3. Thomson, D.J.; Gardes, F.Y.; Hu, Y.; Mashanovich, G.; Fournier, M.; Grosse, P.; Fedeli, J.M.; Reed, G.T. High contrast 40Gbit/s optical modulation in silicon. *Opt. Express* **2011**, *19*, 11507–11516. [CrossRef] [PubMed]
4. Sun, J.; Kumar, R.; Sakib, M.; Driscoll, J.B.; Jayatilleka, H.; Rong, H.S. A 128 Gb/s PAM4 Silicon Microring Modulator With Integrated Thermo-Optic Resonance Tuning. *J. Lightwave Technol.* **2019**, *37*, 110–115. [CrossRef]
5. Milivojevic, B.; Wiese, S.; Anderson, S.; Brenner, T.; Webster, M.; Dama, B. Demonstration of Optical Transmission at Bit Rates of Up to 321.4 Gb/s Using Compact Silicon Based Modulator and Linear BiCMOS MZM Driver. *J. Lightwave Technol.* **2017**, *35*, 768–774. [CrossRef]
6. Tu, X.; Song, C.L.; Huang, T.Y.; Chen, Z.M.; Fu, H.Y. State of the Art and Perspectives on Silicon Photonic Switches. *Micromachines* **2019**, *10*, 51. [CrossRef] [PubMed]
7. Kang, Y.M.; Liu, H.D.; Morse, M.; Paniccia, M.J.; Zadka, M.; Litski, S.; Sarid, G.; Pauchard, A.; Kuo, Y.H.; Chen, H.W.; et al. Monolithic germanium/silicon avalanche photodiodes with 340 GHz gain-bandwidth product. *Nat. Photonics* **2009**, *3*, 59–63. [CrossRef]
8. Assefa, S.; Xia, F.N.A.; Vlasov, Y.A. Reinventing germanium avalanche photodetector for nanophotonic on-chip optical interconnects. *Nature* **2010**, *464*, 80–91. [CrossRef]

9. Simola, E.T.; De Iacovo, A.; Frigerio, J.; Ballabio, A.; Fabbri, A.; Isella, G.; Colace, L. Voltage-tunable dual-band Ge/Si photodetector operating in VIS and NIR spectral range. *Opt. Express* **2019**, *27*, 8529–8539. [CrossRef]
10. Li, C.; Zhang, H.J.; Yu, M.B.; Lo, G.Q. CMOS-compatible high efficiency double-etched apodized waveguide grating coupler. *Opt. Express* **2013**, *21*, 7868–7874. [CrossRef]
11. Piggott, A.Y.; Lu, J.; Lagoudakis, K.G.; Petykiewicz, J.; Babinec, T.M.; Vuckovic, J. Inverse design and demonstration of a compact and broadband on-chip wavelength demultiplexer. *Nat. Photonics* **2015**, *9*, 374. [CrossRef]
12. Xu, L.H.; Wang, Y.; El-Fiky, E.; Mao, D.; Kumar, A.; Xing, Z.P.; Saber, M.G.; Jacques, M.; Plant, D.V. Compact Broadband Polarization Beam Splitter Based on Multimode Interference Coupler With Internal Photonic Crystal for the SOI Platform. *J. Lightwave Technol.* **2019**, *37*, 1231–1240. [CrossRef]
13. Sun, C.; Wade, M.T.; Lee, Y.; Orcutt, J.S.; Alloatti, L.; Georgas, M.S.; Waterman, A.S.; Shainline, J.M.; Avizienis, R.R.; Lin, S.; et al. Single-chip microprocessor that communicates directly using light. *Nature* **2015**, *528*, 534. [CrossRef] [PubMed]
14. Atabaki, A.H.; Moazeni, S.; Pavanello, F.; Gevorgyan, H.; Notaros, J.; Alloatti, L.; Wade, M.T.; Sun, C.; Kruger, S.A.; Meng, H.Y.; et al. Integrating photonics with silicon nanoelectronics for the next generation of systems on a chip. *Nature* **2018**, *556*, 349–354. [CrossRef] [PubMed]
15. Kupijai, S.; Rhee, H.; Al-Saadi, A.; Henniges, M.; Bronzi, D.; Selicke, D.; Theiss, C.; Otte, S.; Eichler, H.J.; Woggon, U.; et al. 25 Gb/s Silicon Photonics Interconnect Using a Transmitter Based on a Node-Matched-Diode Modulator. *J. Lightwave Technol.* **2016**, *34*, 2920–2923. [CrossRef]
16. Streshinsky, M.; Novack, A.; Ding, R.; Liu, Y.; Lim, A.E.J.; Lo, P.G.Q.; Baehr-Jones, T.; Hochberg, M. Silicon Parallel Single Mode 48 × 50 Gb/s Modulator and Photodetector Array. *J. Lightwave Technol.* **2014**, *32*, 3768–3775. [CrossRef]
17. Li, C.; Bai, R.; Shafik, A.; Tabasy, E.Z.; Wang, B.H.; Tang, G.; Ma, C.; Chen, C.H.; Peng, Z.; Fiorentino, M.; et al. Silicon Photonic Transceiver Circuits With Microring Resonator Bias-Based Wavelength Stabilization in 65 nm CMOS. *IEEE J. Solid-St. Circ.* **2014**, *49*, 1419–1436. [CrossRef]
18. Manipatruni, S.; Chen, L.; Lipson, M. Ultra high bandwidth WDM using silicon microring modulators. *Opt. Express* **2010**, *18*, 16858–16867. [CrossRef] [PubMed]
19. Chen, C.H.; Seyedi, M.A.; Fiorentino, M.; Livshits, D.; Gubenko, A.; Mikhrin, S.; Mikhrin, V.; Beausoleil, R.G. A comb laser-driven DWDM silicon photonic transmitter based on microring modulators. *Opt. Express* **2015**, *23*, 21541–21548. [CrossRef]
20. Zhang, Z.Y.; Huang, B.J.; Zhang, Z.; Cheng, C.T.; Chen, H.D. Bidirectional grating coupler based optical modulator for low-loss Integration and low-cost fiber packaging. *Opt. Express* **2013**, *21*, 14202–14214. [CrossRef]
21. Chen, X.; Li, C.; Fung, C.K.Y.; Lo, S.M.G.; Tsang, H.K. Apodized Waveguide Grating Couplers for Efficient Coupling to Optical Fibers. *IEEE Photonics Technol. Lett.* **2010**, *22*, 1156–1158. [CrossRef]
22. Zhang, Z.Y.; Huang, B.J.; Zhang, Z.; Cheng, C.T.; Liu, H.W.; Li, H.Q.; Chen, H.D. Highly efficient vertical fiber interfacing grating coupler with bilayer anti-reflection cladding and backside metal mirror. *Opt. Laser Technol.* **2017**, *90*, 136–143. [CrossRef]
23. Fang, Q.; Liow, T.Y.; Song, J.F.; Tan, C.W.; Bin Yu, M.; Lo, G.Q.; Kwong, D.L. Suspended optical fiber-to-waveguide mode size converter for Silicon photonics. *Opt. Express* **2010**, *18*, 7763–7769. [CrossRef] [PubMed]

Article

Silicon Quantum Dot Light Emitting Diode at 620 nm

Hiroyuki Yamada [1,2] and Naoto Shirahata [1,2,3,*]

1 International Center for Materials Nanoarchitectonics (MANA), National Institute for Materials Science (NIMS), 1-1 Namiki, Tsukuba 305-0044, Japan; yamada.hiroyuki2@nims.go.jp
2 Department of Physics, Chuo University, 1-13-27 Kasuga, Bunkyo, Tokyo 112-8551, Japan
3 Graduate School of Chemical Sciences and Engineering, Hokkaido University, Sapporo 060-0814, Japan
* Correspondence: shirahata.naoto@nims.go.jp; Tel.: +81-29-859-2743

Received: 29 April 2019; Accepted: 10 May 2019; Published: 11 May 2019

Abstract: Here we report a quantum dot light emitting diode (QLED), in which a layer of colloidal silicon quantum dots (SiQDs) works as the optically active component, exhibiting a strong electroluminescence (EL) spectrum peaking at 620 nm. We could not see any fluctuation of the EL spectral peak, even in air, when the operation voltage varied in the range from 4 to 5 V because of the possible advantage of the inverted device structure. The pale-orange EL spectrum was as narrow as 95 nm. Interestingly, the EL spectrum was narrower than the corresponding photoluminescence (PL) spectrum. The EL emission was strong enough to be seen by the naked eye. The currently obtained brightness (~4200 cd/m^2), the 0.033% external quantum efficiency (EQE), and a turn-on voltage as low as 2.8 V show a sufficiently high performance when compared to other orange-light-emitting Si-QLEDs in the literature. We also observed a parasitic emission from the neighboring compositional layer (i.e., the zinc oxide layer), and its intensity increased with the driving voltage of the device.

Keywords: quantum dot; silicon nanocrystals; light emitting diode

---

1. Introduction

Solid-state lighting in the form of light emitting diodes (LEDs) is expected to reduce global energy consumption in the lighting industry [1,2]. Unlike phosphor-coated chips that control the current commercialized LEDs, electric-driven LEDs offer advantageous properties including structurally admissible heavy carrier injection compared to phosphor-coated devices [3–5].

Devices with active layers of colloidal quantum dots (QDs) of semiconductors offer an advantageous electroluminescence (EL) performance, including color purity, high luminance (~200,000 cd/m^2), narrower spectra for emission (full-width at half maximum, fwhm < 40 nm), spectral tunability of the emissions over a broad wavelength range from ultraviolet to near-infrared through to full-color of visible, an operation voltage as low as 3 V, a stable emission under long-term operation even at high current-density conditions, and a solution-based processability [6–8]. The best values of external quantum efficiencies (EQEs) of red-emitting quantum dot light emitting diodes (QLEDs) with conventional and inverted structures are currently 20.5% and 18.0%, respectively [9,10]. These magnitudes are close to an energy conversion efficiency of a mercury lamp which works as a benchmark for the industry. The high values of EQE require optically active (or emission) layers of cadmium-based QDs such as CdSe covered with a shell of crystalline ZnS. However, Hazardous Substances (RoHS) strongly restrict the use of toxic elements, including Cd, for electronic products. Due to the complete ban of these elements in the future, recent efforts have shifted toward fabricating heavy-metal-free QLEDs. More recently, Yang and co-workers reported the InP/ZnSe/ZnS-based red QLED with a 6.6% EQE [11]. Xu and co-workers reported a pale-orange-emitting (λ_{em} = 625 nm) QLED in which colloidal QDs consisting of $CuInS_2$-ZnS-alloyed (ZCIS) cores and ZnSe/ZnS double shells work as an active layer [12]. The 6.6% EQE is the current record of EQE for solution-processed QLEDs exhibiting visible emission.

Silicon (Si), which is abundant, is poised to become a safe alternative to Cd-based QDs [13]. Many studies have concluded that Si is nontoxic to the environment and the human body [14]. Bulk crystalline Si exhibits poor optical performance due to the indirect bandgap nature, but the confined carriers in the nanocrystal with a diameter smaller than the bulk exciton Bohr radius (~5 nm) induce a change in the energy structure. This situation allows for the overlapped wave functions of spatially confined carriers, leading to zero-phonon optical interband transitions for recombination, as a result of the relaxation of the k-selection rule due to the Heisenberg uncertainty relation [15]. Passivation of freestanding silicon quantum dots (SiQDs) with hydrogen atoms is the simplest way to form a surface that negligibly influences the optical properties. Hydrogen-capping gives nonpolarity to a surface, allowing the highest coverage while avoiding a surface that can remain unpassivated. Therefore, most studies on theoretical modeling use the hydrogenated surface to investigate the effect of quantum confinement (QC) generated in a "pure" SiQD. Analogous to the QDs of other semiconductors, the space-confinement-induced changes of the energy structure is expected to enhance the photoluminescence quantum yield (PLQY), but the values remain low (~5%) for ncSi:H [16].

Simply substituting the SiQDs' surface hydrogen atoms with alkyl chains, which yields a covalent carbon–silicon linkage, their PLQYs increase to ~65% at maximum [17–20]. Such an enhancement has been arguably observed for alkyl-terminated SiQDs with size-dependent PL bands peaking in the 590–1130 nm range [16,21,22]. Currently, the enhancement is postulated to arise from an increase in the radiative recombination rate [19], a dramatic reduction of the nonradiative channels [20,23], or a bandgap modulation from indirect to direct transitions [24]. Such high values of the PLQY are suitable as active layers for the QLED. To date, the EL spectra over a wavelength range from 625 to 850 nm have been reported from Si-QLED devices. The record values of EQE are as high as 8.6% for near-infrared EL [25], 6.2% for red EL [26], 0.03% for orange EL [27], and 0.03% for white EL emissions [28]. The shortest emission wavelength is currently around 625 nm for a Si-QLED, but its EQE is as low as 0.0006% [27]. In order to form an image of superior color rendering, the enhanced EQE of a pale-orange-light emitter (i.e., 600–630 nm range) is a challenging task.

In this study, we synthesized the colloidal ink of a pale-orange fluorescent SiQD with an 8% PLQY. The QD was used for the preparation of a Si-QLED with an inverted device structure. The Si-QLED exhibits the EL spectrum peaking at 620 nm, which is included in the pale-orange emission wavelength range.

2. Materials and Methods

2.1. Reagents and Materials

Triethoxysilane (TES) was purchased from TCI chemicals (Tokyo, Japan). 1-Decene was purchased from Sigma-Aldrich (Saint Louis, MO, USA) and was used as received. Electronic grade hydrofluoric acid (49% aqueous solution, Kanto Chemical, Tokyo, Japan), Toluene (High Performance Liquid Chromatography, HPLC, grade), dicholrobenzene, ethanol, and methanol were purchased from Wako chemical (Tokyo, Japan). Colloidal ink of zinc oxide (ZnO, Sigma-Aldrich), 4,4′-bis(carbazole-9-yl)biphenyl (CBP, 99.9% trace metals basis, Sigma-Aldrich), and molybdenum (VI) oxide (MoO_3, 99.97% trace metals basis, Sigma-Aldrich) were used as received. Water was purified and deionized using a Sartorius (Arium 611 UV, Sartorius AG, Göttingen, Germany) water purification system.

2.2. Preparation of Silicon Quantum Dots (SiQDs)

The synthesis of SiQDs was performed in a two-step process, according to our previous papers [28,29]. Typically, TES was employed as a starting precursor. The hydrolysis product, i.e., $(HSiO_{1.5})_n$, of the TES was thermally disproportionated at 1050 °C for 2 h in 5%/95% H_2/Ar atmosphere, yielding SiQDs dispersed in a SiO_2 matrix. After cooling to room temperature, 300 mg of the brown solid (i.e., Si/SiO_2 composite) as powder was mechanically ground in an agate mortar with a pestle. The fine powder obtained was stirred for ~1 h in a mixture of ethanol and 48% HF (aq) to liberate the QDs from the oxide. Then, the acidic solution was centrifuged at 15,000 rpm and washed with ethanol and acetonitrile in that order. According to the analysis with Fourier transform infrared (FTIR) spectroscopy, the precipitated product was a SiQD terminated with hydrogen atoms. Thermal hydrosilylation was carried out in 1-decene at 200 °C. We obtained a transparent brown-colored solution at the same time as the solution temperature reached 200 °C. The unreacted 1-decenes were removed by a vacuum evaporator. Finally, the chloroform solution of the product was purified and separated by gel permeation chromatography (GPC). The substitution of hydrogen atoms by a 1-decane monolayer, yielding a carbon–silicon covalent linkage (i.e., decane-terminated SiQD, SiQD-De), was experimentally confirmed by FTIR. The GPC-treated samples were dried under vacuum conditions and stored in Ar atmosphere prior to use for device fabrication.

2.3. Device Fabrication

Devices were fabricated on a glass substrate. A 150 nm thin film of indium tin oxide (ITO) uniformly sputtered on the glass gives a resistivity of 10–14 Ω/sq, which is good value for EL device fabrication. ITO-coated substrates were prepared in a manner similar to conventional device fabrication. Next, the colloidal ink of ZnO was spin-coated with a rotation speed of 2000 rpm. After baking the film at 120 °C in air, the emission layer of the SiQD-De was spin-coated with a concentration of 10 mg/mL in toluene with a speed of 1500 rpm. Then an organic layer of CBP with a thickness of 40 nm was thermally evaporated. A 30 nm MoO_3 layer was deposited with a vacuum level of 10^{-5} Pa by thermal evaporation. Then the Al top electrode with a thickness of 150 nm was deposited to mask over the film.

2.4. Optical Properties

Optical absorbance spectra were recorded using an ultraviolet-to-visible (UV–VIS) spectrophotometer (JASCO V-650, Tokyo, Japan) with an integrated sphere by diffuse reflection setup. A PL measurement at room temperature was carried out with a spectrofluorometer (NanoLog, Horiba Jovin Yvon, Tokyo, Japan). For measuring the PL and PL excitation (PLE) spectra, we prepared a chloroform solution of SiQD-De. Absolute PLQYs were measured by the standardized integrating sphere method (C9920-02, Hamamatsu Photonics, Hamamatsu, Japan). To avoid the solvent effect on the PLQY, the values of the PLQYs were measured using the SiQD-De films deposited onto quartz glass substrates. The peak value of the PLQY was estimated to be ~8%. This value is lower than those of red-emitting SiQDs and near-infrared (NIR)-emitting SiQDs but is close to the values for the reported SiQDs which exhibit PL spectra peaking in the pale-orange color range.

2.5. Calculation of External Quantum Efficiency (EQE)

EQE was calculated as the ratio, per unit time, of the number of forward-emitted photons to the number of injected electrons, $I_d/|e|$, where I_d is the current passing through the QLED device at an applied bias, V. We can express this as

$$\mathrm{EQE}(\%) = N_{\mathrm{phot}} \times |e| I_{\mathrm{d}} \times g \times 100 \quad (1)$$

where N_{phot} is the number of forward-emitted photons actually collected by the photodiode and the geometric factor, g, accounts for the solid angle of the EL profile (assumed to be Lambertian) subtended by the photodiode, $\Omega = \pi/g$:

$$g = (a^2 + L^2)/a^2 \quad (2)$$

where a is the diameter of the active area of the photodiode and L is the distance between the emitting QLED pixel and the photodiode. N_{phot} was calculated from the photocurrent output of the photodiode in response to the detected EL. The photodiode current, divided by the responsivity value of the photodiode at the peak wavelength of the EL curve, gives the light output power from the LED device. Then the number of photons is calculated by just dividing with hc/λ. Therefore, the simplified formula we used for EQE is

$$\text{EQE}(\%) = \frac{I(\text{Photodiode}) \times \lambda(\text{peak}) \times \text{electron charge} \times g \times 100}{R(\lambda) \times hc \times I(\text{device})} \tag{3}$$

where I(Potodiode) is the photocurrent, in which the dark current is subtracted, detected by the photodiode (Hamamatsu S1336-8BQ) placed just below the EL device; I(device) is the device current; R is the responsivity of the photodiode; and g is the configuration factor of our measurement setup, which we estimated was 4.

Brightness was also calculated from the EQE value and the EL spectrum. EL × CIE (Commission Internationale de I'Eclairage) gives the typical human response. Total luminance intensity was calculated by integrating the EL spectra over the whole wavelength range. Brightness was then calculated by dividing by the active device area and 2π, as shown in the equation below:

$$\text{Brightness or Luminance}\left(\text{Cd/m}^2\right) = \frac{683.002 \times \text{EL area under the curve normlaized} \times \text{CIE} \times I\ (\text{device}) \times EQE}{2\pi \times \text{Device Area}} \tag{4}$$

3. Results and Discussion

The combined UV–VIS absorbance and scattering were studied via the Kubelka–Munk analysis, as shown in Figure 1. We prepared a SiQD-De on a film covering a 0.5 mm thick quartz glass substrate. The very broad spectrum with a peak at 340 nm might be due to some polydispersion of SiQD-De. The PL spectrum peak at 617 nm exhibits a narrow emission line but has a tail in a longer wavelength, possibly due to polydispersion. The SiQD-De has a diameter of 1.8 nm according to the Scherrer broadening analysis, corresponding to the PL band peak at 617 nm and 8% of an absolute PLQY. The decay time of the PL was approximately ~15 s due to the energy structure retaining the indirect bandgap character, corresponding to the literature [23].

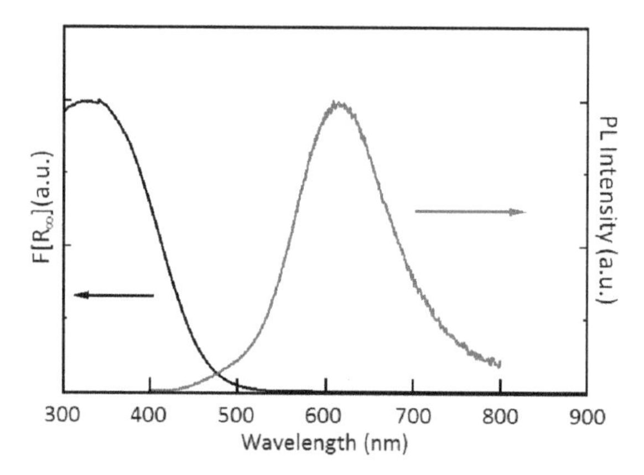

Figure 1. Photoluminescence (PL) and plots of $F[R_\infty]$ vs. wavelength for the powder form of the decane-terminated silicon quantum dot (SiQD-De) specimen. $F[R_\infty]$ of the powder form is the Kubelka–Munk function, with $F[R_\infty] = (1 - R_\infty)^2/2R_\infty$.

Figure 2 schematically illustrates a device architecture and its flat energy band diagram for our Si-QLED that exhibits the EL spectrum peaking at 620 nm. The device was constructed on a 150 nm thick layer of the ITO cathode (resistivity = 10–14 Ω/sq). The anode was a 150 nm thick Al film deposited under a vacuum. As shown in the drawing, the QLED consisted of an inverted device architecture with a multilayer structure, as reported in the literature [30,31]. In contrast to the conventional device structure, the electrons and holes were injected from the ITO and Al electrodes in the inverted device structure, respectively. The multilayered structure used here has constituent layers in the following order: ITO/ZnO/SiQD-De/CBP/MoO_3/Al. The local surface work function of the MoO_3 film was measured by photoelectron yield spectroscopy as 6.2 eV. According to the previous paper [30], the values of the work function of MoO_3 films are influenced by the chemical composition of the films. In this work, a 30 nm thick MoO_3 film deposited in 10^{-5} Pa vacuum conditions provided a local surface work function as low as ~6.2 eV. The electrons are transported to the SiQD layer through the electron injection and transportation layers (EIL/ETL) of the ZnO nanocrystal layer. On the other hand, the holes are transported to the SiQD layer through the hole injection and transportation layers (HIL/HTL) of the MoO_3 and CBP layers. As illustrated in the energy diagram, the layer of ZnO exhibits a low electron affinity (−4.3 eV), being close to the value of the ITO, to facilitate the injection of electrons, leading to a low turn-on voltage. Furthermore, we expect that a low value in the work function of the ZnO layer (−7.7 eV) possibly suppresses a hole leakage current.

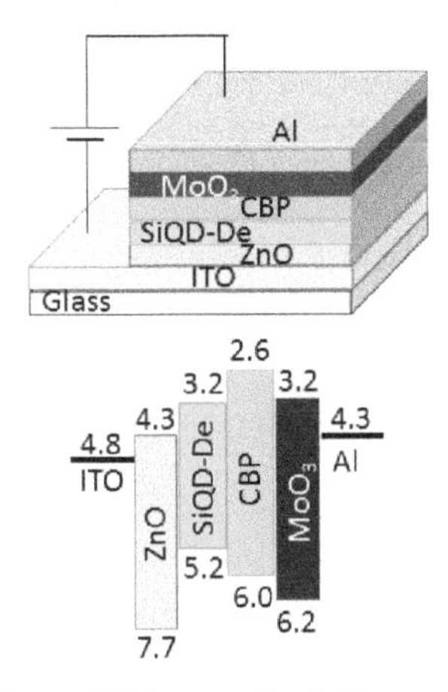

Figure 2. Schematic representation and flat energy band diagram of the pale-orange-light-emitting silicon quantum dot light emitting diode (Si-QLED) with an inverted device structure.

Figure 3 illustrates a typical performance of the Si-QLED in terms of the device current−voltage, the photocurrent−voltage, the luminance−voltage, and the EL spectrum compared to the PL spectrum. Figure 3a shows the current−voltage (I−V) characteristics along with the photodiode J−V characteristics. A calibrated Si photodetector (Hamamatsu S1336 8BQ, Hamamatsu Photonics) coupled with a Keithley 2423 was used for this measurement. The number of photons, collected directly with the photodetector, emitted from the ITO side, increased with the current. The turn-on voltage, which is defined as the minimum applied bias where the QLED starts to emit light, was estimated from the photodiode J−V characteristics. The estimated turn-on voltage was 2.8 V, which is as small as the value of the normal Si-QLED emitting the light peaking at 625 nm [27]. It is reported that the turn-on voltage is influenced by the device composition rather than the QD size [27]. Therefore, the observation of the low turn-on voltage for our device confirms the presence of a small barrier height for the charge injection into the photoactive layer from the electrodes. Figure 3b plots the typical luminance curves as a function of the applied voltage. The luminance reaches a value as high as 4200 cd/m^2 at 5 V. The EL spectrum has a peak at 620 nm which shifts 3 nm to red when compared to the corresponding PL spectrum, as shown in Figure 3c. The EL spectrum is as narrow as 95 nm (~283 meV) fwhm, as evidenced in Figure 3c, and is slightly narrower than the PL spectrum. The observation of the negligible spectral shift (~3 nm) and the value of fwhm smaller than the PL linewidth indicates the effective suppression of the quantum-confined Stark effect, although further study is needed for clarification of the mechanism. The inset of Figure 3c shows a photograph of the QLED operating at 5 V, indicating that the light emitted from the QLED is sufficiently bright and vivid to be visible to the naked eye even in an illuminated room. However, the color purity of the pale orange was lower than our expectation in spite of the narrow linewidth of the spectrum without an emission tail. The low purity of color might be due to the appearance of another EL spectrum peaking at around 420 nm, while such a blue emission is not observed in the PL spectrum. The EL spectra exhibit the voltage dependence, as shown in Figure 4. The spectral profile at 4 V bias shows the EL peak at 620 nm, and we see a small luminance contribution (λ_{em} = ~420 nm) from a layer among the multilayers, leading to the low purity of the pale-orange color. This parasitic EL emission at 420 nm might originate from the neighboring compositional layer of ZnO. As the voltage increases, the parasitic EL intensity from the layer of ZnO increases. On the other hand, the EL intensity originating from the ncSi-De layers also grows under increasing applied bias voltage. There are two possible reasons to explain the appearance of the parasitic emission. First, according to the flat energy band diagram, there is an energy gap of 0.9 eV between the ZnO and the ncSi-De layers. Due to this energy barrier that blocks the carrier transportation, some of the electrons injected from the ITO electrode remained in the conduction minimum of ZnO, yielding the recombination for blue EL. The other possible reason is that this unwanted emission in the blue range appears due to an insufficient thickness of ncSi-De [28]. A further extension of voltage is needed to discuss the underlying physics in which the parasitic emission happens. Figure 4 demonstrates a couple of advantages of the inverted device structure. First, there is no shift of the EL peak with increasing operation voltage, indicating that the EL emission at 620 nm originates solely from the QD layer even in the high applied voltage range. In contrast, it is well known that the EL peak shift to blue for conventional structures of the QLED of semiconductors includes Si [25,26,32]. Second, the EL spectral shape is independent of driving voltage. It is important to note that emission spectral characteristics, such as EL peak position and shape, are not influenced by the presence of the parasitic emission spectrum. Therefore, the observed stability of the EL characteristics could be due to a good conductivity in the band alignment of HTL which leads to the difficulty in the buildup of the band-filling. In terms of the EL stability under increasing operation voltage, the use of metal oxide layers (i.e., ZnO and MoO_3) for the EIL/ETL and the HTL/HIL takes advantage of the inherent robustness and protection of the interlayers from oxidation. We measured the photocurrent as the EL output with a photodetector to estimate the EQE and plotted the values in Figure 5 as a function of the injected current density. The peak value of EQE was estimated to be 0.033%, which is currently a record value for a Si-QLED operating in the pale-orange emission range. An enhancement of the EQE might be obtained by a good band alignment of our inverted device structure. Decreasing the degree of the charged QDs, which leads to carrier loss due to Auger recombination, would contribute to the enhanced EQE [33].

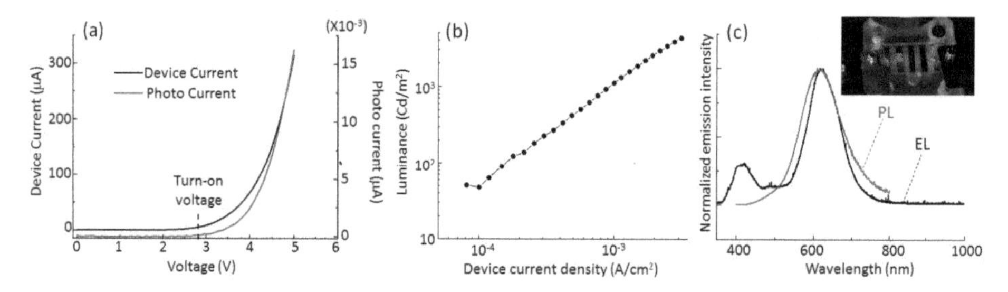

Figure 3. (**a**) Device I–V characteristics (black line) and photodiode I–V characteristics (red line), (**b**) luminance–current density characteristics, and (**c**) a typical electroluminescence (EL) spectrum at the operation voltage of 5 V (PL spectrum of the corresponding SiQD-De dispersed in chloroform). A photograph demonstrates a representative pale-orange-light-emitting quantum dot light emitting diode (QLED).

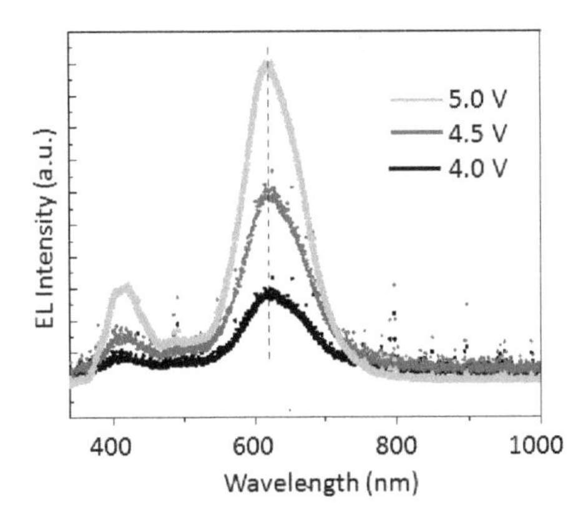

Figure 4. EL spectra at three different bias voltages.

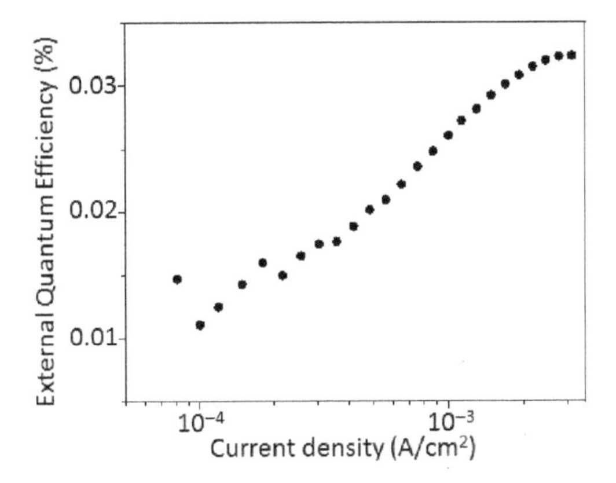

Figure 5. External quantum efficiency (EQE) versus device current density.

4. Conclusions

A SiQD-based QLED, exhibiting a narrow EL spectrum peaking at 620 nm, was produced via the solution-processed method. In this work, we synthesized a pale-orange luminescent sample of a SiQD with 8% PLQY and used it for an active layer of QLED. The QLED, consisting of multilayers of an inverted device structure, emits a pale-orange emission with 0.03% EQE, which is bright enough for confirming the emission by the naked eye even in an illuminated room. This work expands the tunable emission range that is currently limited to between 640 nm and 850 nm. The coefficient of optical absorption for the SiQD is significantly lower than those of QDs of II–VI and III–V semiconductors, because SiQDs retain the indirect bandgap structure, consistent with the experimental observation of a long PL decay time on a microsecond scale. Therefore, it is reasonable to consider that a SiQD is not a good phosphor adaption for phosphor-coated chips in the development of a liquid crystal display. However, a Si-QLED has the potential to become a good light emitter, because its poor optical absorption character does not influence the EQE. Further improvement of the optical performance of a Si-QLED toward the first commercialization of a heavy-metal-free QLED requires a dramatic enhancement of PLQYs even for visible-light-emitting SiQDs.

Author Contributions: N.S. designed the research, H.Y. performed research, and H.Y. and N.S. completed the manuscript preparation.

Funding: A-step (JPMJTS1619) from Japan Science and Technology Agency and the Izumi Science and Technology Foundation (2018-J-053)

Acknowledgments: The authors would like to thank the World Premier International Research Center Initiative Program.

Conflicts of Interest: The authors declare no conflict of interest.

References

1. Crawford, M.H. LEDs for Solid-State Lighting: Performance Challenges and Recent Advances. *IEEE J. Sel. Top. Quantum Electron.* **2009**, *15*, 1028–1040. [CrossRef]
2. Pimputkar, S.; Speck, J.S.; DenBaars, S.P.; Nakamura, S. Prospects for LED lighting. *Nat. Photon.* **2009**, *3*, 180–182. [CrossRef]
3. Babu, S.S.; Aimi, J.; Ozawa, H.; Shirahata, N.; Saeki, A.; Seki, S.; Ajayaghosh, A.; Möhwald, H.; Nakanishi, T. Solvent-free Luminescent Organic Liquids. *Angew. Chem. Int. Ed.* **2012**, *51*, 3391–3395. [CrossRef] [PubMed]
4. Song, W.S.; Yang, H. Efficient White-Light-Emitting Diodes Fabricated from Highly Fluorescent Copper Indium Sulfide Core/Shell Quantum Dots. *Chem. Mater.* **2012**, *24*, 1961–1967. [CrossRef]
5. Ghosh, B.; Ogawara, M.; Sakka, Y.; Shirahata, N. White-light Emitting Liquefiable Si Nanocrystals. *Chem. Lett.* **2012**, *41*, 1157–1159. [CrossRef]
6. Caruge, J.M.; Halpert, J.E.; Bulovic, V.; Bawendi, M.G. Colloidal Quantum-Dot Light-Emitting Diodes with Metal-Oxide Charge Transport Layers. *Nat. Photon.* **2008**, *2*, 247–250. [CrossRef]
7. Yang, J.; Choi, M.K.; Kim, D.H.; Hyeon, T. Designed Assembly and Integration of Colloidal Nanocrystals for Device Applications. *Adv. Mater.* **2016**, *28*, 1176–1207. [CrossRef]
8. Kwak, J.; Bae, W.K.; Lee, D.; Park, I.; Lim, J.; Park, M.; Cho, H.; Woo, H.; Yoon, D.Y.; Char, K.; Lee, S.; Lee, C. Bright and Efficient Full-Color Colloidal Quantum Dot Light-Emitting Diodes Using an Inverted Device Structure. *Nano Lett.* **2012**, *12*, 2362–2366. [CrossRef]
9. Dai, X.; Zhang, Z.; Jin, Y.; Niu, Y.; Cao, H.; Liang, X.; Chen, L.; Wang, J.; Peng, X. Solution-Processed, High Performance Light Emitting Diodes Based on Quantum Dots. *Nature* **2014**, *515*, 96–99. [CrossRef] [PubMed]
10. Mashford, B.S.; Stevenson, M.; Popovic, Z.; Hamilton, C.; Zhou, Z.; Breen, C.; Steckel, J.; Bulovic, V.; Bawendi, M.; Coe-Sullivan, S.; Kazlas, P.T. High-Efficiency Quantum-Dot Light-Emitting Devices with Enhanced Charge Injection. *Nat. Photon.* **2013**, *7*, 407–412. [CrossRef]
11. Cao, F.; Wang, S.; Wang, F.; Wu, Q.; Zhao, D.; Yang, X. A Layer-by-Layer Growth Strategy for Large-Size InP/ZnSe/ZnS Core–Shell Quantum Dots Enabling High-Efficiency Light-Emitting Diodes. *Chem. Mater.* **2018**, *30*, 8002–8007. [CrossRef]
12. Tan, Z.; Zhang, Y.; Xie, C.; Su, H.; Liu, J.; Zhang, C.; Dellas, N.; Mohney, S.E.; Wang, Y.; Wang, J.; Xu, J. Near-Band-Edge Electroluminescence from Heavy-Metal-Free Colloidal Quantum Dots. *Adv. Mater.* **2011**, *23*, 3553–3558. [CrossRef]
13. Ghosh, B.; Shirahata, N. Colloidal Silicon Quantum Dots: Synthesis and Tuning the Emission in the Ranging from near-UV through Visible to near-IR. *Sci. Technol. Adv. Mater.* **2014**, *15*, 014207. [CrossRef]
14. Chinnathambi, S.; Shirahata, N. Recent Advances on Fluorescent Biomarkers of Near-Infrared Quantum Dots for *In Vitro* and *In Vivo* Imaging. *Sci. Technol. Adv. Mater.* **2019**, *20*, 337–355. [CrossRef]
15. Dohnalová, K.; Gregorkiewicz, T.; Kůsová, K. Silicon Quantum Dots: Surface Matters. *J. Phys.: Condens. Matter* **2014**, *26*, 173201. [CrossRef]
16. Hessel, C.M.; Reid, D.; Panthani, M.G.; Rasch, M.R.; Goodfellow, B.; Wei, J.; Fujii, H.; Akhavan, V.; Korgel, B.A. Synthesis of Ligand-Stabilized Silicon Nanocrystals with Size-Dependent Photoluminescence Spanning Visible to Near-Infrared Wavelengths. *Chem. Mater.* **2012**, *24*, 393–401. [CrossRef]
17. Miller, J.B.; Van Sickle, A.R.; Anthony, R.J.; Kroll, D.M.; Kortshagen, U.R.; Hobbie, E.K. Ensemble Brightening and Enhanced Quantum Yield in Size-Purified Silicon Nanocrystals. *ACS Nano* **2012**, *6*, 7389–7396. [CrossRef]
18. Dasog, M.; Kehrle, J.; Rieger, B.; Veinot, J.G.C. Silicon Nanocrystals and Silicon-Polymer Hybrids: Synthesis, Surface Engineering, and Applications. *Angew. Chem. Int. Ed.* **2015**, *54*, 2–20. [CrossRef] [PubMed]
19. Mastronardi, M.L.; Maier-Flaig, F.; Faulkner, D.; Henderson, E.J.; Kübel, C.; Lemmer, U.; Ozin, G.A. Size-Dependent Absolute Quantum Yields for Size-Separated Colloidally-Stable Silicon Nanocrystals. *Nano Lett.* **2012**, *12*, 337–342. [CrossRef] [PubMed]

20. Ghosh, B.; Hamaoka, T.; Nemoto, Y.; Takeguchi, M.; Shirahata, N. Impact of Anchoring Monolayers on the Enhancement of Radiative Recombination in Light-Emitting Diodes Based on Silicon Nanocrystals. *J. Phys. Chem. C* **2018**, *122*, 6422–6430. [CrossRef]
21. Chandra, S.; Ghosh, B.; Beaune, G.; Nagarajan, U.; Yasui, T.; Nakamura, J.; Tsuruoka, T.; Baba, Y.; Shirahata, N.; Winnik, F.M. Functional double-shelled silicon nanocrystals for two-photon fluorescence cell imaging: spectral evolution and tuning. *Nanoscale* **2016**, *8*, 9009–9019. [CrossRef]
22. Jurbergs, D.; Mangolini, E.R.L.; Kortshagen, U.R. Silicon nanocrystals with ensemble quantum yields exceeding 60%. *Appl. Phys. Lett.* **2006**, *88*, 233116. [CrossRef]
23. Ghosh, B.; Takeguchi, M.; Nakamura, J.; Nemoto, Y.; Hamaoka, T.; Chandra, S.; Shirahata, N. Origin of the Photoluminescence Quantum Yields Enhanced by Alkane-Termination of Freestanding Silicon Nanocrystals: Temperature-Dependence of Optical Properties. *Sci. Rep.* **2016**, *6*, 36951. [CrossRef]
24. Dohnalová, K.; Poddubny, A.N.; Prokofiev, A.A.; DAM de Boer, W.; Umesh, C.P.; Paulusse, J.M.J.; Zuilhof, H.; Gregorkiewicz, T. Surface Brightens Up Si Quantum Dots: Direct Bandgap-like Size-Tunable Emission. *Light: Sci. Appl.* **2013**, *2*, e47. [CrossRef]
25. Cheng, K.Y.; Anthony, R.; Kortshagen, U.R.; Holmes, R.J. High-Efficiency Silicon Nanocrystal Light-Emitting Devices. *Nano Lett.* **2011**, *11*, 1952–1956. [CrossRef] [PubMed]
26. Liu, X.; Zhao, S.; Gu, W.; Zhang, Y.; Qiao, X.; Ni, Z.; Pi, X.; Yang, D. Light-Emitting Diodes Based on Colloidal Silicon Quantum Dots with Octyl and Phenylpropyl Ligands. *ACS Appl. Mater. Interf.* **2018**, *10*, 5959–5966. [CrossRef] [PubMed]
27. Maier-Flaig, F.; Rinck, J.; Stephan, M.; Bocksrocker, T.; Bruns, M.; Kübel, C.; Powell, A.K.; Ozin, G.A.; Lemmer, U. Multicolor Silicon Light-Emitting Diodes (SiLEDs). *Nano Lett.* **2013**, *13*, 475–480. [CrossRef] [PubMed]
28. Ghosh, B.; Masuda, Y.; Wakayama, Y.; Imanaka, Y.; Inoue, J.; Hashi, K.; Deguchi, K.; Yamada, H.; Sakka, Y.; Ohki, S.; et al. Hybrid White Light Emitting Diode Based on Silicon Nanocrystals. *Adv. Funct. Mater.* **2014**, *24*, 7151–7160. [CrossRef]
29. Chandra, S.; Masuda, Y.; Shirahata, M.; Winnik, F.M. Transition Metal Doped NIR Emitting Silicon Nanocrystals. *Angew. Chem. Int. Ed.* **2017**, *56*, 6157–6160. [CrossRef]
30. Ghosh, B.; Yamada, H.; Chinnathambi, S.; Özbilgin, I.N.G.; Shirahata, N. Inverted Device Architecture for Enhanced Performance of Flexible Silicon Quantum Dot Light-Emitting Diode. *J. Phys. Chem. Lett.* **2018**, *9*, 5400–5407. [CrossRef] [PubMed]
31. Yao, L.; Yu, T.; Ba, L.; Meng, H.; Fang, X.; Wang, Y.; Li, L.; Rong, X.; Wang, S.; Wang, X.; Ran, G.; Pi, X.; Qin, G. Efficient Silicon Quantum Dots Light Emitting Diodes with an Inverted Device Structure. *J. Mater. Chem. C* **2016**, *4*, 673–677. [CrossRef]
32. Gu, W.; Liu, X.; Pi, X.; Dai, X.; Zhao, S.; Yao, L.; Li, D.; Jin, Y.; Xu, M.; Yang, D. Silicon-Quantum-Dot Light-Emitting Diodes with Interlayer-Enhanced Hole Transport. *IEEE Photo. J.* **2017**, *9*, 4500610. [CrossRef]
33. Bae, W.K.; Park, Y.S.; Lim, J.; Lee, D.; Padilha, L.A.; McDaniel, H.; Robel, I.; Lee, C.; Pietryga, J.M.; Klimov, V.I. Controlling the Influence of Auger Recombination on the Performance of Quantum-Dot Light-Emitting Diodes. *Nat. Commun.* **2013**, *4*, 2661. [CrossRef] [PubMed]

micromachines

Letter

Silicon Optical Modulator Using a Low-Loss Phase Shifter Based on a Multimode Interference Waveguide

Daisuke Inoue *, Tadashi Ichikawa, Akari Kawasaki and Tatsuya Yamashita

Toyota Central R&D Labs., Inc., 41-1, Yokomichi, Nagakute, Aichi 480-1192, Japan
* Correspondence: daisuke-i@mosk.tytlabs.co.jp; Tel.: +81-5-6163-4300

Received: 25 June 2019; Accepted: 15 July 2019; Published: 18 July 2019

Abstract: We have developed a novel phase modulator, based on fin-type electrodes placed at self-imaging positions of a silicon multimode interference (MMI) waveguide, which allows reduced scattering losses and relaxes the fabrication tolerance. The measured propagation losses and spectral bandwidth are 0.7 dB and 33 nm, respectively, on a 987 μm-long phase shifter. Owing to the self-imaging effect in the MMI waveguide, the wave-front expansion to the electrode was counteracted, and therefore, the scattering loss caused by electrode fins was successfully mitigated. As a proof-of-concept for the MMI-based phase modulator applications, we performed optical modulation based on Mach–Zehnder interferometers (MZIs). The π shift current of the modulator was 1.5 mA.

Keywords: silicon photonics; modulator; multimode interferometer; photonics integrated circuit; carrier plasma; Mach–Zehnder interferometers

1. Introduction

Silicon (Si) photonics densely integrate optical devices, incorporating P–N junctions that enable the development of advanced devices, such as high-speed modulators, matrix switches, and variable attenuators [1–7].

One of the most common problems in Si photonics is the combination of channel waveguides and rib waveguides. Rib structures need large bending radii, which can cause large footprints. Combining a channel waveguide and rib waveguide is ideal for realizing more compact Si photonic circuits. However, a major issue in silicon photonics is the cost. Additional masks and processes are required to fabricate a rib structure [8].

Therefore, in this paper, we propose a novel optical phase modulator with a fin-type electrode based on a multimode interferometer (MMI) waveguide. It is noteworthy that our phase modulator has a wide working wavelength range. Furthermore, it can be fabricated at a low cost, because the minimum line width for processing the modulator is 0.4 μm.

2. Principle and Design

2.1. Princple

We used the carrier-injection method for fabricating a phase modulator composed of an MMI, as shown in Figure 1. The phase modulator is comprised of a multimode waveguide and is connected to a single-mode waveguide. The MMI causes periodic self-imaging of an input field profile [9].

As shown in Figure 1b, the electrode fins are placed at a distance where the self-imaging of an input field occurs. As a result, the scattering loss in the phase modulator caused by the electrode fins is reduced.

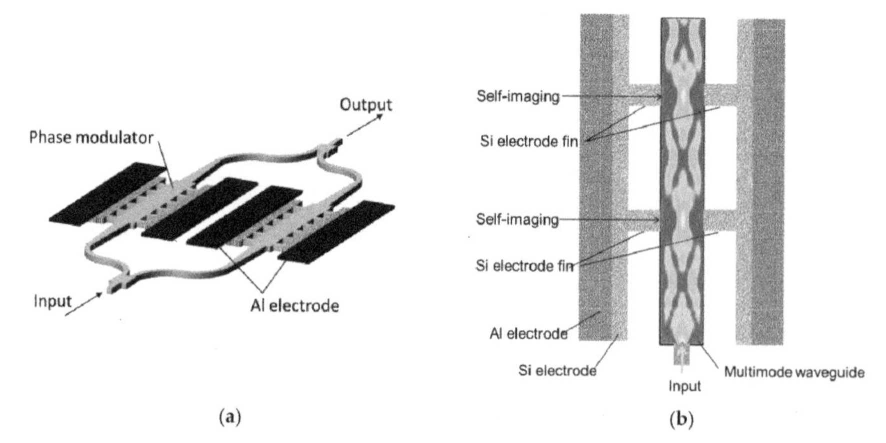

Figure 1. Schematic of (**a**) a Mach–Zehnder interferometer modulator with a multimode interferometer (MMI) phase modulator, and (**b**) a phase modulator based on MMI.

2.2. Design of the Phase Modulator

First, we determine the MMI length of the phase modulator via finite-difference time domain calculations under transverse electric-like polarization. The width of the MMI waveguide is ~1.2 μm, and the Si layer thickness is 0.2 μm, which provides a period length of 2.35 μm for self-imaging. A taper was inserted between the single-mode and multimode waveguides for reducing the connection loss. The length of the taper was 9 μm, and the width increased from 0.4 μm to 0.85 μm.

Next, we simulated propagation loss in the waveguide with electrode fins. Figure 2a shows the calculation model of the MMI phase modulator; the electrode width is defined in Figure 2b. Figure 2c shows the propagation loss in the multimode waveguide as a function of the electrode width. Figure 2c indicates that increasing the electrode width increases the propagation loss. However, the propagation loss in the waveguide with an electrode width of 0.4 μm is sufficiently thin for practical application. In addition to the propagation loss, the fabrication process was optimized for a line width of 0.4 μm. We fixed the width of the electrode fin to 0.4 μm for the above reasons.

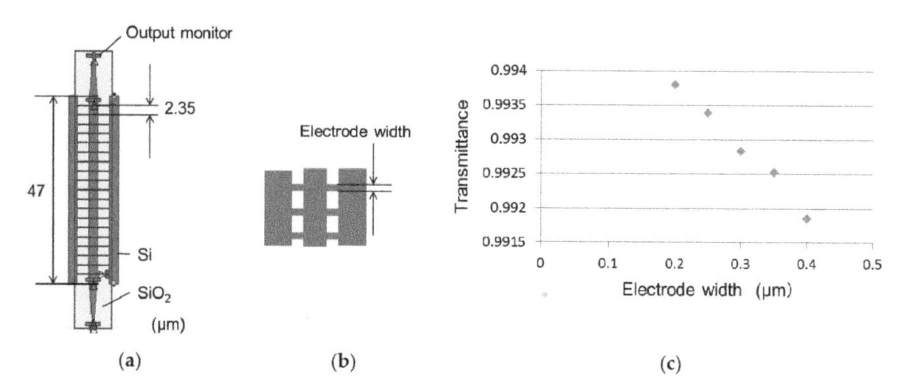

Figure 2. (**a**) Numerical calculation model of MMI phase modulator, (**b**) magnified drawing of electrode fin, and (**c**) transmittance (based on calculation results).

2.3. Device Structure of the Optical Modulator

Our optical modulator is based on Mach–Zehnder interferometers (MZIs). We fabricated our MZI-based modulator using the MMI phase modulator with fin-type electrodes, as shown in Figure 1a.

We fabricated two types of MZI modulators. One modulator had a different optical path, referred so as an asymmetric MZI. The other has the same optical path, called a symmetric MZI.

Figure 3 shows the layout of the phase modulator based on an MMI. The length of the electrode fin is 1.4 μm. The P–N region at the side of the MMI phase modulator comprises highly doped P (4×10^{15} cm^{-2}) and N (1×10^{15} cm^{-2}) regions. This modulator only needs P and N regions. The gap between the P-type and N-type areas is 2 μm. A 0.5-μm-thick aluminum (Al) electrode is deposited on silicon (Si). The Si waveguide and Al electrode are covered with a 3-μm-thick oxide silicon layer and a 4-μm-thick SiO_2 layer as the upper cladding layer. A 0.3-μm-thick tantalum layer was deposited on SiO_2 as the heater.

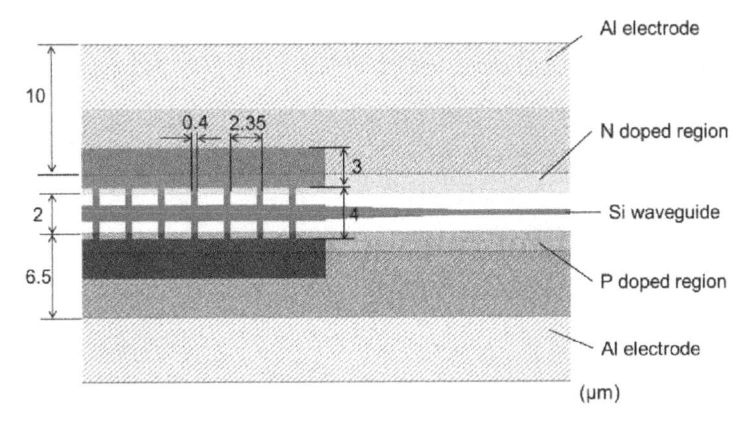

Figure 3. Layout of a phase modulator based on MMI.

Figure 4a shows the scanning electron microscope (SEM) images of the MMI phase modulator with fins before the deposition of the oxide silicon layer. Figure 4b shows the optical microscope image of the MZI-based modulator formed using this MMI phase modulator with electrode fins.

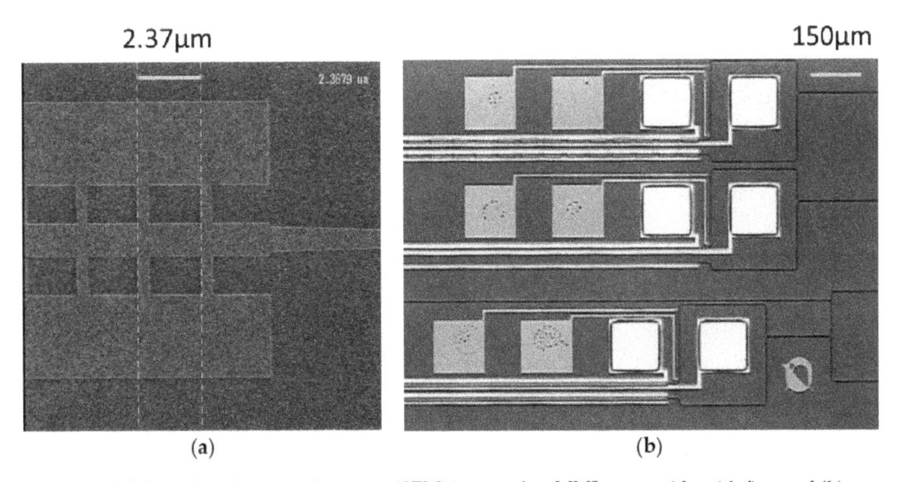

Figure 4. (**a**) Scanning electron microscope (SEM) image of an MMI waveguide with fins, and (**b**) an optical microscope image of a Mach–Zehnder interferometer (MZI) modulator formed using an MMI phase modulator. Orange lines indicate the scale.

3. Experimental Results

3.1. Propagation Loss and Transmission Spectrum of a Multimode Interference Phase Shifter

We outsourced the fabrication of the photonics-integrated circuit. The foundry used electron beam lithography for the fabrication. The pitch here is about 0.4 µm, and we can achieve this with deep ultraviolet (UV) lithography, which has low cost for mass production. We fabricated four types of MMI waveguides with fins for measuring the propagation loss, as shown in Figure 5; in this figure, #1 represents 47 µm of the MMI waveguide length with fins, which is the same shape as was obtained via numerical calculation. Furthermore, #2 represents the 20 stages of cascade connections for 47 µm of the MMI waveguide length with fins. It is difficult to estimate small propagation losses in these samples. Therefore, we prepared #2 as a cascade connection of #1. In addition, #3 represents 987 µm of the MMI waveguide length with fins. Finally, #4 represents 1927 µm of the MMI waveguide length with fins.

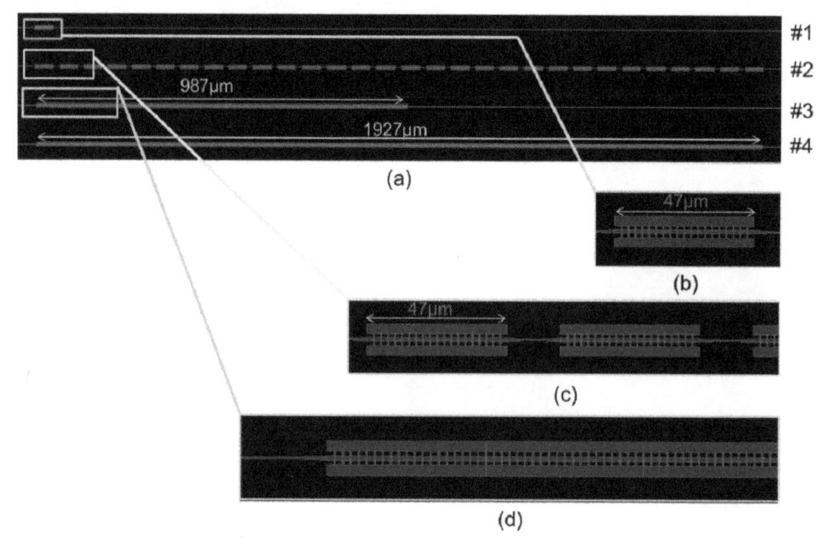

Figure 5. (**a**) Four types of MMI waveguides with fins for measuring propagation loss, and magnified drawing of (**b**) waveguide #1, (**c**) waveguide #2, and (**d**) waveguide #3.

The light of the laser source is injected in the input facet of the device by butt coupling through a focused single mode fiber, and it is collected with a multimode fiber. The wavelength of the laser light was 1550 nm. For comparison, we measured the insertion loss in a single-mode waveguide with different lengths on the same wafer. To extract the propagation losses of the MMI waveguide, we drew a linear regression of the output power versus the length of devices 1, 3, and 4, as shown in Figure 6. We evaluated the insertion loss of an MMI waveguide with fins as 1.1 dB/mm from the coefficient of the result of the linear regression. The offset value of the linear regression of 9.2 dB, including the connection loss of a pair of tapers and the coupling loss at both the input and output ends between the spot size converter (SSC) and fiber. Assuming that the insertion losses and the propagation losses of the MMI waveguide are the same, we can predict the insertion loss of an MMI waveguide with 0.94 µm of length as 10.28 dB. However, the insertion loss of #2 was 11.74 dB, as shown in Figure 6. We consider that the difference is caused by the cascade connection of tapers. The differential number of tapers between waveguide #2 and #3 is 19 pairs. We estimated that the connection loss of a pair of tapers is 0.08 dB. The propagation loss in the straight waveguide on the same wafer was 0.32 dB/mm; there are 422 fins per 1 mm of MMI waveguide. We evaluated the scattering loss of a fin as 0.0018 dB.

We considered that the insertion loss of the MMI waveguide with fins is considered sufficient for practical use.

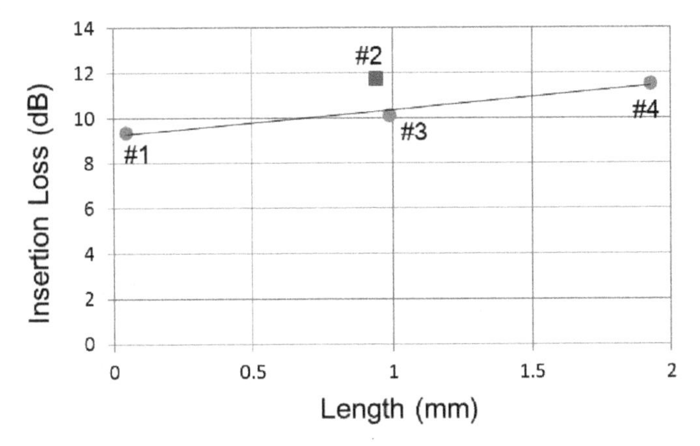

Figure 6. Length dependence of the insertion loss of an MMI waveguide with fins.

We measured the transmission spectrum for 987 μm of the MMI waveguide length with fins. We input light waves emitted from a super-luminescent laser diode (SLD) into the MMI waveguide with fins with a tapered fiber, and measured the transmission spectrum using an optical spectrum analyzer (OSA). As a reference, we measured the spectrum of the SLD. Figure 7 shows the transmission spectrum of the MMI waveguide with fins. The full width at half maximum (FWHM) in the transmission spectrum of the MMI waveguide with fins was 33 nm. We decided that 987 μm of length would be appropriate for phase modulators, owing to the aforementioned reason.

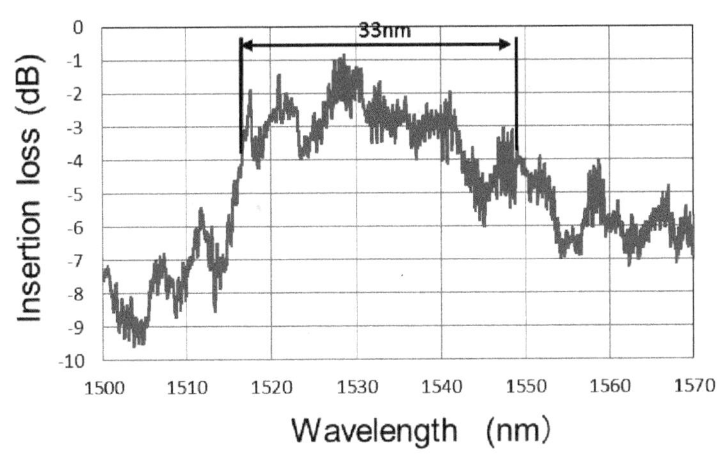

Figure 7. Transmission spectrum of an MMI waveguide with fins.

3.2. Shift in Transmission Spectrum and Modulation by Current Injection

First, we measured the shift in the transmission spectrum of the asymmetric MZI modulator. We contacted the P and N electrodes with needle probes and injected the forward bias current. We input light waves delivered from the SLD into the MZI modulator with a tapered fiber, and measured the transmission spectrum with the OSA. Figure 8 shows the output spectrum of the MZI modulator under injection currents of 0 mA and 10 mA. On increasing the injection current, the peaks shifted toward shorter wavelengths. When the carrier plasma effect occurs, increasing the carrier density decreases the refractive index. On the other hand, increasing the temperature will increase the refractive index. We confirmed that this wavelength shift was caused by the carrier plasma effect.

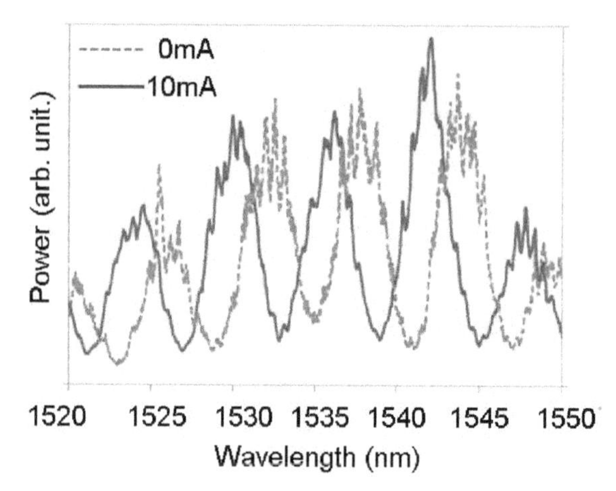

Figure 8. Shift in transmission spectrum of an MZI-based modulator with an MMI phase modulator caused by current injection.

Next, we attempted to modulate light waves. We input laser light into the symmetric MZI modulator with a tapered fiber and coupled the output light with a multimode fiber. The wavelength of the light was 1550 nm. We applied alternating current (AC) at 10 MHz to the MZI modulator. The optical output was detected using an optical/electrical converter (1444-50, Newport, Irvine, CA, USA). Electrical signals were captured by using a sampling oscilloscope (MSO4104, Agilent, Santa Clara, CA, USA). Figures 9 and 10 show the measurement results of the modulated light waves. We measured output power, as a function of the injection current. It appears that the injection current reached the half-wave current, because both the ceiling and floor of the waveforms were distorted.

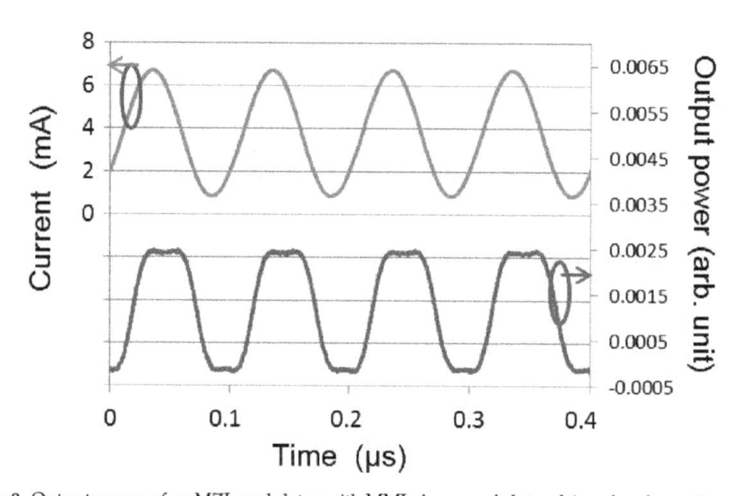

Figure 9. Output power of an MZI modulator with MMI phase modulator driven by alternating current (AC) power.

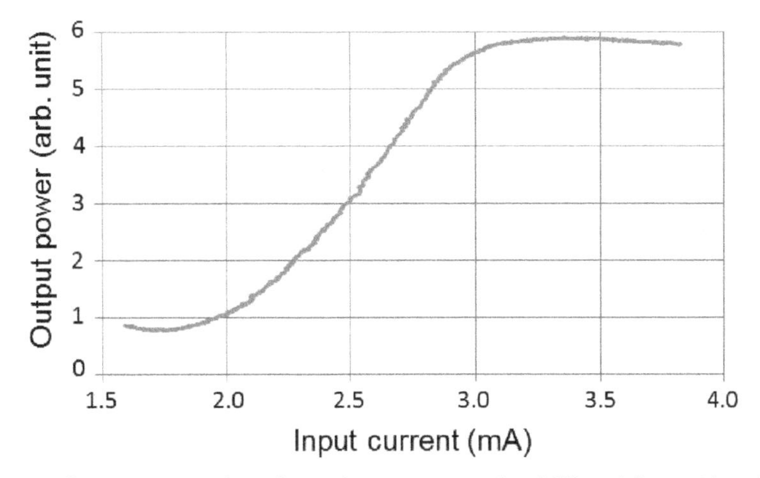

Figure 10. Injection current dependence of output power of an MZI modulator with an MMI phase modulator.

4. Discussion

Here, we discuss the issues pertaining to MMI phase shifters. When light passes through an MMI structure, internal reflection occurs. In our experiment, we observed some ripples, which are indicated in the transmission spectrum shown in Figure 8.

The free spectrum range (FSR) of a Fabry–Perot resonator is expressed as

$$\mathrm{FSR} = \frac{\lambda^2}{2n_r l},$$

where λ is the wavelength, and n_r and l represent the refractive index and length of the cavity, respectively. The period of the ripples in the transmission spectrum of the MMI waveguide shown in Figure 8 is 0.55 nm; the measurement wavelength is 1.55 µm. The length of the phase shifter is 987 µm. The refractive index of the silicon waveguide was calculated to be 2.1. Since not one mode but a few modes propagate through an MMI waveguide, we calculated the FSR of the MMI waveguide with an average refractive index; the FSR was predicted to be 0.57 nm. The calculated FSR agrees with the period of the ripples in the transmission spectrum.

However, the transmission spectrum of the MMI waveguide shown in Figure 7 has two types of ripples. The period for one type of ripple is 5 nm, and that for the other type is 0.55 nm. The ripple with a 0.55 nm period was attributed to the internal reflection in the MMI waveguide. The ripple with a 5 nm period corresponds to 11 µm of cavity length. The length of one of the tapers is 9 µm, while the other is close to 11 µm. The ripple with a 5 nm period is considered to be caused by a taper.

The period of the fins in the MMI phase shifter was 2.35 µm. The FSR for a period of 2.35 µm is expected to be 24 nm. However, 24 nm ripple periods are not observed in either Figure 7 or Figure 8. Hence, we assume the design of the fins to be satisfactory.

5. Conclusions

We fabricated and demonstrated a phase-shifting structure suitable for silicon photonics circuits. A phase-shifter based on an MMI was constructed from multimode waveguides and electrode fins placed at a distance where the self-imaging of an input field occurs. Self-imaging at the center of the MMI-based phase shifter counteracted the wave-front expansion, thus mitigating the scattering loss caused by the electrode fins.

The proposed phase modulator does not need the half-etching process. The minimum linewidth was 0.4 μm, and a low-cost lithography process was sufficient for this modulator. In addition to the low minimum line width, this phase modulator required only P+ and N+ regions. Therefore, the fabrication process was simplified, and the costs were reduced.

We fabricated MMI-based multimode waveguides with fins and evaluated the insertion loss into them. The propagation loss for 987 μm of the MMI waveguide length with fins was 0.7 dB higher than that for the same length of a single-mode waveguide. The FWHM in the transmission spectrum of the phase shifter with electrode fins was 33 nm. The propagation loss was small enough, and the FWHM in the transmission spectrum of the phase-shifter with electrode fins was wide enough for practical applications.

We injected current into the fabricated modulator and measured the change in the transmission spectrum. The shift in the spectrum indicates that the working of this modulator is governed by the carrier plasma effect. We demonstrated the modulation of light waves by applying AC to the modulator.

Author Contributions: Conceptualization, D.I., T.I., A.K. and T.Y.; Methodology, D.I. and T.Y.; Validation, D.I., T.I. and A.K.; Formal Analysis, D.I. and T.Y.; Data Curation, D.I.; Writing-Original Draft Preparation, D.I.; Writing-Review & Editing, T.Y.; Supervision, T.Y.

Funding: This research received no external funding.

Conflicts of Interest: The authors declare no conflict of interest.

References

1. Kondo, K.; Baba, T. Slow-light-induced Doppler shift in photonic-crystal waveguides. *Phys. Rev. A* **2016**, *93*, 011802. [CrossRef]
2. Terada, Y.; Miyasaka, K.; Ito, H.; Baba, T. Slow-light effect in a silicon photonic crystal waveguide as a sub-bandgap photodiode. *Opt. Lett.* **2016**, *41*, 289–292. [CrossRef] [PubMed]
3. Kinugasa, S.; Ishikura, N.; Ito, H.; Yazawa, N.; Baba, T. One-chip integration of optical correlator based on slow-light devices. *Opt. Express* **2015**, *23*, 20767–20773. [CrossRef] [PubMed]
4. Xu, F.; Poon, A.W. Silicon cross-connect filters using microring resonator coupled multimode-interference-based waveguide crossings. *Opt. Express* **2008**, *16*, 8649–8657. [CrossRef] [PubMed]
5. Suzuki, K.; Nguyen, H.C.; Tamanuki, T.; Shinobu, F.; Saito, Y.; Yuya, S.; Baba, T. Slow-light-based variable symbol-rate silicon photonics DQPSK receiver. *Opt. Express* **2012**, *20*, 4797–4804. [CrossRef] [PubMed]
6. Nguyen, H.C.; Sakai, Y.; Shinkawa, M.; Ishikura, N.; Baba, T. Photonic crystal silicon optical modulators: Carrier-injection and depletion at 10 Gb/s. *IEEE J. Quantum Electron.* **2012**, *48*, 210–220. [CrossRef]
7. Akiyama, S.; Baba, T.; Imai, M.; Akagawa, T.; Noguchi, M.; Saito, E.; Noguchi, Y.; Hirayama, N.; Horikawa, T.; Usuki, T. 50-Gb/s silicon modulator using 250-μm-long phase shifter based-on forward-biased pin diodes. In Proceedings of the 9th IEEE International Conference on Group IV Photonics (IEEE, 2012), San Diego, CA, USA, 29–31 August 2012; pp. 192–194.
8. Jensen, A.S. Fabrication and Characterization of Silicon Waveguide for High-Speed Optical Signal Processing. Ph.D. Thesis, Technical University of Denmark, Lyngby, Denmark, 2015.
9. Soldano, L.B.; Pennings, E.C.M. Optical multi-mode interference devices based on self-imaging: Principles and applications. *J. Lightwave Technol.* **1995**, *13*, 615–627. [CrossRef]

MDPI
St. Alban-Anlage 66
4052 Basel
Switzerland
Tel. +41 61 683 77 34
Fax +41 61 302 89 18
www.mdpi.com

Micromachines Editorial Office
E-mail: micromachines@mdpi.com
www.mdpi.com/journal/micromachines

CPSIA information can be obtained
at www.ICGtesting.com
Printed in the USA
LVHW070945170920
665299LV00059B/699

9 783039 369089